Dynamical Systems and Microphysics

GEOMETRY AND MECHANICS

ACADEMIC PRESS RAPID MANUSCRIPT REPRODUCTION

The Proceedings of the 2nd International Seminar
on Mathematical Theory of Dynamical Systems and Microphysics
held at the International Center for Mechanical Sciences at Udine, Italy
from September 1 to 11, 1981

Dynamical Systems and Microphysics

GEOMETRY AND MECHANICS

EDITED BY

André Avez
Université Pierre et Marie Curie
Paris, France

Austin Blaquière
Université de Paris 7
Paris, France

Angelo Marzollo
Université de Paris 7, Paris, France
and *University of Udine, Udine, Italy*

1982

ACADEMIC PRESS
A Subsidiary of Harcourt Brace Jovanovich, Publishers
New York London
Paris San Diego San Francisco São Paulo Sydney Tokyo Toronto

ACADEMIC PRESS, INC.
111 Fifth Avenue, New York, New York 10003

United Kingdom Edition published by
ACADEMIC PRESS, INC. (LONDON) LTD.
24/28 Oval Road, London NW1 7DX

LIBRARY OF CONGRESS CATALOG CARD NUMBER: 82-11650

ISBN: 0-12-068720-8

PRINTED IN THE UNITED STATES OF AMERICA

82 83 84 85 9 8 7 6 5 4 3 2 1

Contents

III. Lagrangian and Hamiltonian Formulations of Mechanics

IV. Perturbations

V. Some Problems in Quantum Mechanics

VI. Contributed Papers

Contributors

Numbers in parentheses indicate the pages on which the authors's contributions begin.

André Avez (353), Université Pierre et Marie Curie, Paris, France

I. M. Benn (403), Department of Physics, University of Lancaster, Lancaster, United Kingdom

M. S. Berger (109), Center for Applied Mathematics and Mathematical Science, University of Massachusetts, Amherst, Massachusetts 01003

A. Blaquière (143, 209), Laboratoire d'Automatique Théorique, Université de Paris 7, 75005 Paris, France

S. Ciulli (409), Dublin Institute for Advanced Studies, Dublin, Ireland

M. Courbage (415), Université Libre de Bruxelles, B-1050 Bruxelles, Belgique

Giacomo Della Riccia (281), Istituto de Matematica, Informatica e Sistemistica dell' Universita, 33100 Udine, Italy, and Department of Mathematics, the Ben-Gurion University of the Negev, Beer-Sheva, Israel

T. Dereli (403), Department of Physics, University of Lancaster, Lancaster, United Kingdom

Francis Fer (303), Ecole Nationale Supérieure des Mines, 75006 Paris, France

V. A. Franke (389), Department of Theoretical Physics, University of Leningrad, Leningrad, U.S.S.R.

Gian Carlo Ghirardi (367), Istituto di Fisica Teorica dell' Università, Trieste, Italy

Riccardo Goldoni (425), University of Pisa, Istituto Matematico L. Tonelli, Pisa, Italy

André Heslot (431), Faculté des Sciences et Techniques de Monastir, Tunisie

G. Leitmann (119), Department of Mechanical Engineering, University of California, Berkeley, California 94720

André Lichnerowicz (27), Chaire de Physique Mathématique, Collège de France, Paris, France

Georges Lochak (321, 329), Foundation Louis de Broglie, 75003 Paris, France

C. M. Marle (61), Université Pierre et Marie Curie, Paris, France

A. Marzollo (209), Faculté des Sciences, Université de Udine, Udine, Italie, and Laboratoire d'Automatique Théorique, Université de Paris 7, 75005 Paris, France

Yu. V. Novozhilov (389), Department of Theoretical Physics, University of Leningrad, Leningrad, U.S.S.R., and Science Sector, UNESCO, 75700 Paris, France

Z. Oziewicz (437), Institute of Theoretical Physics, Wroclaw University, Poland

E. V. Prokhvatilov (389), Department of Theoretical Physics, University of Leningrad, Leningrad, U.S.S.R.

Pham Mau Quan (91), Département de Mathématiques, Université de Paris-Nord, 93430 Villetaneuse, France

Antonio F. Rañada (443), Department of Theoretical Physics, University of Leningrad, Leningrad, U.S.S.R.

Alberto Rimini (367), Istituto di Fisica dell-Università, Salerno, Italy

A. J. van der Schaft (233), Mathematics Institute, The Netherlands

Paolo Serafini (449), University of Udine, Institute of Mathematics, Computer and System Science, Udine, Italy

Wiktor Szczyrba (455), Institute of Mathematics, Polish Academy of Sciences, Warsaw, Poland

W. Thirring (75), Institut für Theoretische Physik der Universität Wien, Wien, Austria

R. Thom (267), Institut des Hautes Etudes Scientifiques, Bures-sur-Yvette, France

R. W. Tucker (403), Department of Physics, University of Lancaster, Lancaster, United Kingdom

W. M. Tulczyjew (3), Department of Mathematics and Statistics, The University of Calgary, Calgary, Alberta, Canada

José Vassalo-Pereira (321, 343), Facultate de Ciêncas de Lisboa, Portugal

B. Vujanović (293), Faculty of Technical Sciences, University of Novi Sad, 21000 Novi Sad, Yugoslavia

J. C. Willems (233), Mathematics Institute, The Netherlands

Kunio Yasue (461), Département de Physique Théorique, Université de Genève, Genève, Suisse

Preface

An increasing number of physicists are becoming aware of the importance of some not yet fully solved problems concerning the conceptual and mathematical foundations of quantum mechanics, relativity, thermodynamics, and other areas of physics. At the same time, it has become more and more apparent that some of these problems have significant links with corresponding questions faced by other disciplines, in particular by mathematical system theory. For example, the interconnections between the Hamilton-Jacobi theory of mechanics and optimal control, and between quantum mechanics and stochastic control have recently inspired a number of interesting studies. These studies have helped to clarify some basic issues that turn out to be common to physics and system theory, despite the different terminologies used in the two fields.

Moreover, both physics and contemporary mechanics are stimulating sources of mathematical research—for example, in the qualitative theory of differential equations, symplectic structures, and differential geometry.

The above considerations induced the International Center for Mechanical Sciences (CISM) to organize in September 1979 in Udine, Italy, an international Seminar on Dynamical Systems and Microphysics. This seminar, held under the honorary chairmanship of Louis de Broglie, enjoyed the sponsorship and support of the United Nations Educational, Scientific and Cultural Organization (UNESCO), of the International Federation for Automatic Control (IFAC) and of the Italian National Research Council (CNR).

The success of that interdisciplinary seminar, as well as the renewed encouragement of Louis de Broglie and of the above mentioned institutions, motivated CISM to organize a second international Seminar on Dynamical Systems and Microphysics for September 1981. This time the seminar focused on geometry and mechanics.

This volume contains the texts of all invited papers and of a selected number of contributed papers, in the second seminar. It is primarily directed to researchers and graduate students in theoretical physics, mechanics, control and system theory, and mathematics. It may also be profitably read by philosophers of science and, to some extent, by persons who have a keen interest in basic questions of contemporary mechanics and physics and some background in the physical and mathematical sciences. We are convinced that adherence to mathematical rigor and to logical coherence are essential for approaching modern physics (and system theory), and hope to have succeeded in following this guideline; however, the level of the mathematical techniques used in most contributions should make this volume accessible to persons who do not have a knowledge of advanced mathematics.

For the convenience of the reader, we have divided the material into six parts. Invited papers appear in Parts I-V and contributed papers are collected in Part VI.

Part I deals with geometric structures in mechanics and with general relativity. Part II is devoted to system-theory approaches to mechanics. Part III deals with Lagrangian and Hamiltonian formulations of mechanics. Part IV is concerned with perturbations. Part V deals with some problems in quantum mechanics. As we shall see below, the distribution of contributed papers of Part VI follows the same pattern. We turn now to a brief summary of the papers.

The papers by Tulczyjew, Lichnerowicz, Heslot, Marle, Thirring, Pham Mau Quan, Berger, Goldoni, Szczyrba and by Benn, Dereli, and Tucker deal with geometric structures, mechanics, and general relativity. Tulczyjew's paper provides a good background to the others and introduces new results. Lichnerowicz's approach to quantum mechanics, via deformation of the symplectic structure, gives an entirely new and striking insight into the correspondence principle Heslot's paper builds another bridge on the same road from symplectic to quantum-mechanics. Mechanical systems with time-dependent constraints require a generalization of symplectic manifolds, a subject which is examined by Marle; as a bonus, the results of Kirilov and Souriau are recovered.

Some papers are more directly related to group invariance. The one by Thirring exposes the gauge invariance of the Einstein field. Pham Mau Quan gives a group treatment of the space of the orbits in the Kepler problem. Using gauge invariance, Berger studies the stable configurations in nonlinear problems arising from physics.

Finally, three papers specifically deal with general relativity. The one by Goldoni tries to reconcile the global character of Mach's principle with the local one of Einstein's equations. Szczyrba puts these equations in symplectic form, and Benn, Dereli, and Tucker generalize them to include the fundamental interactions.

The papers by Leitmann, Blaquière and Marzollo, and by Willems and van der Schaft deal with system theory approaches to mechanics. Leitmann's paper provides some background in the theories of (Closed-loop) optimal control and of controllability (reachability), in the framework of a geometric approach. It is a good introduction to the paper by Blaquière and Marzolli to examine the relationship between optimal control theory and some aspects of the calculus of variations with respect to their applications to classical mechanics. Special attention is devoted to the possibility of expressing Hamilton's principle in the case of nonholonomic constraints in the framework of optimization theory. In the paper by Willems and van der Schaft a system-theory approach to mechanics is based on the use of external and internal variables for modeling dynamical systems, with applications to Hamiltonian systems.

The papers by Thom, Della Riccia, Oziewicz, Vujanović, Yasue, Fer, Courbage, and Serafini are concerned with Lagrangian and Hamiltonian formulations of mechanics, the limits of such formulations, and, beyond these limits, the relationship between reversible and irreversible processes.

Thom's paper contains some considerations about the Hamiltonian formalism, in an attempt to answer the questions "What makes Hamiltonian systems so important?" The obverse problem of determining a Lagrangian and its properties for a differential system is analyzed by Della Riccia and by Oziewicz. Vujanovic studies the possibility of finding the conservation laws of classical nonconservative dynamical systems and presents a method for solving the canonical (Hamilton's) differential equations of motion of nonconservative systems, similar to the method of Hamilton-Jacobi. Yasue reports an approach to quantum mechanics based on the stochastic Lagrangian formalism. Fer discusses the limits of Hamiltonian mechanics for dealing with irreversible phenomena; the relationship between reversible and irreversible processes are also dealt with by Courbage in the framework of thermodynamics and by Serafini in the framework of system theory.

The papers by Lochak and Vassalo-Pereira, Lochak, Vassalo-Pereira, and Avez are concerned with perturbations. The joint paper by Lochak and Vassalo-Pereira and the one by Lochak study the adiabatic invariants of classical systems submitted to time-periodic perturbations. A nonlinear version of the Floquet theorem and some applications are given. The paper by Vassalo-Pereira gives a theory of Huygen's effect of two interacting clocks. Finally, Avez examines the existence of aperiodic orbit of a perturbed biharmonic oscillator.

A sample of studies dealing with crucial questions in quantum mechanics is presented in the papers by Ghirardi and Rimini, by Franke, Novozhilov, and Prokhvatikov, and by Rañada. The first of these discusses key problems in the quantum theory of measurement, in particular schemes for ideal measurement processes; the second shows how to solve difficulties linked with quantization on a plane tangential to the light cone; and the last proposes a model in which an extended particle with structure is represented by a solitary wave.

Finally, a contribution by Cuilli deserves a special place, as it is related to almost all of the topics already mentioned, and particularly epistemology.

We take this opportunity to express our gratitude to CISM, which played a major role in the organization of the seminar that inspired this volume. We also wish to acknowledge the financial assistance of UNESCO and CNR. In particular, the former enabled several researchers from developing countries to attend and to participate actively in the seminar.

PART I

GEOMETRIC STRUCTURES, MECHANICS, AND GENERAL RELATIVITY

LAGRANGIAN SUBMANIFOLDS, STATICS AND DYNAMICS OF MECHANICAL SYSTEMS

W.M. Tulczyjew
Department of Mathematics and Statistics
The University of Calgary
Calgary, Alberta, Canada

Istituto di Fisica Matematica "J.-L. Lagrange"
Torino, Italy

Introduction.

The behaviour of a static mechanical system is described by a configuration-force relation. For simple systems this relation is represented by sections of the force bundle. For more general systems it is necessary to represent the configuration-force relations by Lagrangian submanifolds of the force bundle. Such systems are studied by catastrophe theory. A similar generalization is necessary in the theory of dynamical systems. Usually considered dynamical systems are represented by Hamiltonian vector fields. Representation of more general systems requires the introduction of Lagrangian submanifolds of suitably constructed symplectic manifolds.

These lectures provide the fundamentals of the geometry of Lagrangian submanifolds with examples of applications to mechanics. A list of references is given at the end of these notes.

ISBN 0-12-068720-8

1. Lagrangian submanifolds of cotangent bundles.

Let Q be a manifold of dimension m. We consider the tangent bundle

$$\tau_Q : TQ \to Q$$

and the cotangent bundle

$$\pi_Q : T^*Q \to Q.$$

We have the canonical 1-form θ_Q on T^*Q defined by

$$\langle u, \theta_Q \rangle = \langle T\pi_Q(u), \tau_{T^*Q}(u) \rangle,$$

where u is an element of the tangent bundle TT^*Q, $\tau_{T^*Q} : TT^*Q \to T^*Q$ is the tangent bundle projection and $T\pi_Q : TT^*Q \to TQ$ is the tangent mapping of π_Q.

PROPOSITION 1.1. *For each section (1-form) $\sigma : Q \to T^*Q$ we have*

$$\sigma^*\theta_Q = \sigma,$$

where $\sigma^*\theta_Q$ is the pull-back of θ_Q from T^*Q to Q.

Proof. If v is an element of TQ then

$$\begin{aligned}\langle v, \sigma^*\theta_Q \rangle &= \langle T\sigma(v), \theta_Q \rangle \\ &= \langle T\pi_Q(T\sigma(v)), \tau_{T^*Q}(T\sigma(v)) \rangle \\ &= \langle T(\pi_Q \circ \sigma)(v), \sigma(\tau_Q(v)) \rangle \\ &= \langle v, \sigma \rangle.\end{aligned}$$

Hence $\sigma^*\theta_Q = \sigma$. ■

Local coordinates (q^i); $i = 1,\dots,m$ in Q induce coordinates (q^i, f_j); $i,j = 1,\dots,m$ in T^*Q such that $\theta_Q = f_i dq^i$. In TQ we have induced coordinates $(q^i, \delta q^j)$; $i,j = 1,\dots,m$ such that $\delta q^j(v) = \langle v, dq^j \rangle$.

DEFINITION 1.1. A submanifold $S \subset T^*Q$ is said to be

Lagrangian if $dim(S) = m$ and $\omega_Q|S = 0$. Here $\omega_X = d\theta_X$.

If coordinates (s^α); $\alpha = 1,\ldots,m$ are introduced in a submanifold $S \subset T^*Q$ then in terms of coordinates (q^i, f_j) the submanifold S is described by equations

$$q^i = \xi^i(s^\alpha),\ f_j = \eta_j(s^\alpha).$$

The condition $\omega_Q|S = 0$ is expressed by relations

$$\frac{\partial\xi^i}{\partial s^\alpha}\frac{\partial\eta i}{\partial s^\beta}\, ds^\alpha \wedge ds^\beta = 0$$

equivalent to

$$\frac{\partial\xi^i}{\partial s^\alpha}\frac{\partial\eta i}{\partial s^\beta} - \frac{\partial\xi^i}{\partial s^\alpha}\frac{\partial\eta i}{\partial s^\beta} = 0.$$

Expressions

$$[s^\alpha, s^\beta] = \frac{\partial\xi^i}{\partial s^\alpha}\frac{\partial\eta i}{\partial s^\beta} - \frac{\partial\xi^i}{\partial s^\alpha}\frac{\partial\eta i}{\partial s^\beta}$$

are known as Lagrange brackets.

PROPOSITION 1.2. *The image of a section* $\sigma:Q \to T^*Q$ *is a Lagrangian submanifold if and only if* σ *is a closed* 1-*form.*

Proof. $dim(im(\sigma)) = m$ and the condition $\omega_Q|im(\sigma) = 0$ is equivalent to $\sigma^*\omega_Q = 0$. From

$$d\sigma = d\sigma^*\theta_Q = \sigma^* d\theta_Q = \sigma^*\omega_Q$$

it follows that $im(\sigma)$ is Lagrangian if and only if σ is closed. ■

DEFINITION 1.2. A Lagrangian submanifold $S \subset T^*Q$ is said to be *regular* if S is the image $im(\sigma)$ of a section $\sigma:Q \to T^*Q$.

If $S \subset T^*Q$ is the image of section $\sigma:Q \to T^*Q$ then it is described in terms of coordinates (q^i, f_j) by equations

$$f_j = \sigma_j(q^i).$$

The condition $\omega_Q|S = 0$ is expressed by relations

$$\frac{\partial \sigma_j}{\partial q^i} dq^i \wedge dq^j = 0$$

equivalent to

$$\frac{\partial \sigma_j}{\partial q^i} - \frac{\partial \sigma_i}{\partial q^j} = 0.$$

COROLLARY 1.1. *If $S \subset T^*Q$ is a regular Lagrangian submanifold then there exist functions $U:Q \to \mathbf{R}$ such that locally $S = im(-dU)$.*

DEFINITION 1.3. A function $U:Q \to \mathbf{R}$ such that $S = im(-dU)$ is called a *generating function* of S. The Lagrangian submanifold S is said to be *generated* by U.

A regular Lagrangian submanifold $S \subset T^*Q$ generated by a function $U:Q \to \mathbf{R}$ is described in terms of coordinates (q^i, f_j) by equations

$$- f_j = \frac{\partial U}{\partial q^j} .$$

Let $\overline{Q}$ be a submanifold of Q. We denote by $TQ|\overline{Q}$ and $T^*Q|\overline{Q}$ the restrictions

$$TQ|\overline{Q} = \{v \in TQ;\ \tau_Q(v) \in \overline{Q}\},$$

$$T^*Q|\overline{Q} = \{f \in T^*Q;\ \pi_Q(f) \in \overline{Q}\}$$

of TQ and T^*Q to $\overline{Q}$. We identify the tangent bundle $T\overline{Q}$ with a subbundle of $TQ|\overline{Q}$ and denote by $T^\circ\overline{Q}$ the polar

$$T^\circ\overline{Q} = \{f \in T^*Q|\overline{Q};\ \langle v,f\rangle = 0 \text{ for each } v \in T_q\overline{Q},\ q = \pi_Q(f)\}$$

of $T\overline{Q}$ in $T^*Q|\overline{Q}$. The cotangent bundle $T^*\overline{Q}$ is canonically isomorphic with the quotient bundle $T^*Q|\overline{Q}/_{T^\circ\overline{Q}}$. Let

$$\gamma : T^*Q|\overline{Q} \to T^*\overline{Q}$$

denote the canonical projection defined by

$$\langle v, \gamma(f)\rangle = \langle v, f\rangle$$

for each $v \in T_q\overline{Q} \subset T_qQ$, $q = \pi_Q(f)$. Let $\pi_Q|\overline{Q}$ be the projection of $T^*Q|\overline{Q}$ onto $\overline{Q}$ induced by π_Q. Then

$$\pi_{\overline{Q}}\circ\gamma = \pi_Q|\overline{Q}.$$

For each $u \in T(T^*Q|\overline{Q}) \subset TT^*Q$ we have

$$\begin{aligned}\langle u, \gamma^*\theta_{\overline{Q}}\rangle &= \langle T\gamma(u), \theta_{\overline{Q}}\rangle \\ &= \langle T\pi_{\overline{Q}}(T\gamma(u)), \tau_{T^*\overline{Q}}(T\gamma(u))\rangle \\ &= \langle T(\pi_{\overline{Q}}\circ\gamma)(u), \gamma(\tau_{T^*Q}(u))\rangle \\ &= \langle T\pi_Q(u), \tau_{T^*Q}(u)\rangle \\ &= \langle u, \theta_Q\rangle.\end{aligned}$$

Hence,

$$\gamma^*\theta_{\overline{Q}} = \theta_Q|(T^*Q|\overline{Q}).$$

PROPOSITION 1.3. *Let $S \subset T^*Q$ be a regular Lagrangian submanifold generated by a function $U : Q \to \mathbf{R}$. Then*

$$\overline{S} = \gamma(S \cap T^*Q|\overline{Q}) \subset T^*\overline{Q}$$

is a regular Lagrangian submanifold generated by the function $\overline{U} = U|\overline{Q}$.

Proof. Since S is the image of a section $\sigma : Q \to T^*Q$ it follows that $S \cap T^*Q|\overline{Q}$ is the image of the section $\sigma' : \overline{Q} \to T^*Q|\overline{Q}$: $q \mapsto \sigma(q)$ and $\overline{S}$ is the image of the section $\overline{\sigma} = \gamma\circ\sigma'$. For each $q \in \overline{Q}$ and each $v \in T_q\overline{Q} \subset T_qQ$ we have $\langle v, \overline{\sigma}\rangle = \langle v, \overline{\sigma}(q)\rangle = \langle v, \gamma(\sigma'(q))\rangle = \langle v, \sigma'(q)\rangle = \langle v, \sigma(q)\rangle = \langle v, \sigma\rangle = -\langle v, dU\rangle = -\langle v, d\overline{U}\rangle$. Hence $\overline{\sigma} = -d\overline{U}$. ∎

PROPOSITION 1.4. *Let $\overline{S} \subset T^*\overline{Q}$ be a regular Lagrangian submanifold generated by a function $\overline{U}:\overline{Q} \to \mathbf{R}$. Then*

$$S = \gamma^{-1}(\overline{S}) \subset T^*Q$$

is a Lagrangian submanifold.

Proof. It is clear that S is a submanifold of dimension m contained in $T^*Q|\overline{Q}$. The 1-form $\theta_{\overline{Q}}$ restricted to $\overline{S}$ is exact:

$$\theta_{\overline{Q}}|\overline{S} = - dF,$$

where $F:\overline{S} \to \mathbf{R}$ is defined by $F\circ\sigma = \overline{U}$. If $u \in TS \subset T(T^*Q|\overline{Q})$ then $T\gamma(u) \in T\overline{S} \subset TT^*\overline{Q}$ and

$$\begin{aligned} \langle u,\theta_Q\rangle &= \langle u,\gamma^*\theta_{\overline{Q}}\rangle \\ &= \langle T\gamma(u),\theta_{\overline{Q}}\rangle \\ &= -\langle T\gamma(u),dF\rangle \\ &= -\langle u,\gamma^*(dF)\rangle \\ &= -\langle u,d(F\circ\gamma)\rangle. \end{aligned}$$

Hence $\theta_Q|S = - d(F\circ\gamma)$ and $\omega_Q|S = 0$. ∎

Let $(\overline{q}^a)$; $a = 1,\ldots,k$ be coordinates in $\overline{Q}$ and let $\overline{Q}$ be described by equations

$$q^i = \phi^i(\overline{q}^a).$$

Then the coordinate expression of γ in terms of coordinates (q^i,f_j) in T^*Q and coordinates $(\overline{q}^a,\overline{f}_b)$ in $T^*\overline{Q}$ is

$$\overline{f}_b = f_i \frac{\partial\phi^i}{\partial\overline{q}^b}.$$

The Lagrangian submanifold $\overline{S} \subset T^*\overline{Q}$ of Proposition 1.3 is described by

$$\overline{f}_b = - \frac{\partial\overline{U}}{\partial\overline{q}^b} = - \frac{\partial U}{\partial q^i}\frac{\partial\phi^i}{\partial\overline{q}^b},$$

and the Lagrangian submanifold $S \subset T^*Q$ of Proposition 1.4 is described by

$$q^i = \phi^i(\bar{q}^a),$$

$$f_j \frac{\partial \phi^j}{\partial \bar{q}^b} = - \frac{\partial \bar{U}}{\partial \bar{q}^b} .$$

Let f be an element of T^*Q and let M be a subspace of the tangent space $T_f T^*Q$. We denote by $M^{¶}$ the subspace of $T_f T^*Q$ defined by

$$M^{¶} = \{u \in T_f T^*Q;\ \forall_{v \in M} \langle u \wedge v, \omega_Q \rangle = 0\}.$$

We have the following easy to verify relations

$$M^{¶¶} = M,$$

$$(M + N)^{¶} = M^{¶} \cap N^{¶},$$

$$(M \cap N)^{¶} = M^{¶} + N^{¶},$$

$$O^{¶} = T_f T^*Q,$$

$$(T_f T^*Q)^{¶} = O,$$

where M and N are subspaces of $T_f T^*Q$ and O is the subspace of $T_f T^*Q$ containing only the 0 vector.

PROPOSITION 1.5. *A submanifold $S \subset T^*Q$ is Lagrangian if and only if $(T_f S)^{¶} = T_f S$ for each f in S.*

Proof. The condition $\omega_Q | S = 0$ is equivalent to $T_f S \subset \subset (T_f S)^{¶}$ for each f in S, and $dim(T_f S) + dim((T_f S)^{¶}) = 2m$. ■

Let $\zeta : \hat{Q} \to Q$ be a differential fibration. We denote by $V\hat{Q} \subset T\hat{Q}$ the subbundle of vertical vectors and by $V^\circ\hat{Q} \subset T^*\hat{Q}$ the polar of $V\hat{Q}$. There is a canonical projection

$$\varepsilon : V^\circ\hat{Q} \to T^*Q$$

defined by

$$\langle v, \varepsilon(\hat{f})\rangle = \langle w, \hat{f}\rangle,$$

where $\hat{f} \in V^\circ\hat{Q}$, $v \in TQ$ and $w \in T\hat{Q}$ satisfy $\tau_{\hat{Q}}(w) = \pi_{\hat{Q}}(f)$ and $T\zeta(w) = v$. Relations

$$\varepsilon^*\theta = \theta_{\hat{Q}}|V^\circ\hat{Q}$$

and

$$\varepsilon^*\omega = \omega_{\hat{Q}}|V^\circ\hat{Q}$$

follow directly from the definition of ε. The second of these relations implies $ker(T_{\hat{f}}\varepsilon) \subset (T_{\hat{f}}V^\circ\hat{Q})^¶$ for each $\hat{f} \in T^*\hat{Q}$. If $dim(Q) = m$ and $dim(\hat{Q}) = \hat{m}$ then $dim((T_{\hat{f}}V^\circ\hat{Q})^¶) = \hat{m} - m$ and $dim(ker(T_{\hat{f}}\varepsilon)) = m - \hat{m}$. Hence

$$ker(T_{\hat{f}}\varepsilon) = (T_{\hat{f}}V^\circ\hat{Q})^¶.$$

PROPOSITION 1.6. *Let $\hat{S} \subset T^*\hat{Q}$ be a regular Lagrangian submanifold generated by a function $\hat{U}:\hat{Q} \to \mathbf{R}$. If $\hat{S}$ is transverse to $V^\circ\hat{Q}$ then*

$$S = \varepsilon(\hat{S} \cap V^\circ\hat{Q}) \subset T^*Q$$

is an immersed Lagrangian submanifold.

Proof. The transversality condition means $T_{\hat{f}}\hat{S} + T_{\hat{f}}V^\circ\hat{Q} = T_{\hat{f}}T^*\hat{Q}$ for each $\hat{f} \in \hat{S} \cap V^\circ\hat{Q}$ and implies that $\hat{S} \cap V^\circ\hat{Q}$ is a submanifold of $V^\circ\hat{Q}$ of dimension m. Since $(T_{\hat{f}}\hat{S})^¶ = T_{\hat{f}}\hat{S}$ and $(T_{\hat{f}}V^\circ\hat{Q})^¶ = ker(T_{\hat{f}}\varepsilon)$ we have $T_{\hat{f}}\hat{S} \cap ker(T_{\hat{f}}\varepsilon) = 0$. It follows that $S = \varepsilon(\hat{S} \cap V^\circ\hat{Q})$ is an immersed submanifold of T^*Q of dimension m. The condition $\omega_Q|S = 0$ is a consequence of $\omega_{\hat{Q}}|\hat{S} \cap V^\circ\hat{Q} = 0$ and $\omega_{\hat{Q}}|V^\circ\hat{Q} = \varepsilon^*\omega_Q$. ■

DEFINITION 1.4. If assumptions of Proposition 1.6 are satisfied then the function $U:Q \to \mathbf{R}$ is called a *Morse family* and

S is said to be generated by $\hat{U}$.

It is known that each Lagrangian submanifold is locally generated by a Morse family.

Let (q^i); $i = 1,\dots,m$ be coordinates in Q and (q^i, q^A); $i = 1,\dots,m$; $A = 1,\dots,\hat{m} - m$ adapted coordinates in $\hat{Q}$ such that ζ maps a point with coordinates (q^i, q^A) onto the point with coordinates (q^i). We use in $T^*\hat{Q}$ coordinates (q^i, q^A, f_j, f_B) such that $\theta_{\hat{Q}} = f_i dq^i + f_A dq^A$. The submanifold $V^\circ\hat{Q}$ is characterized by $f_A = 0$. Coordinates (q^i, q^A, f_j) will be used in $V^\circ\hat{Q}$. The projection ε maps a point with coordinates (q^i, q^A, f_j) onto the element of T^*Q with coordinates (q^i, f_j). A function $\hat{U}(q^i, q^A)$ is a Morse family if and only if the matrix

$$\left\| \frac{\partial^2 \hat{U}}{\partial q^i \partial q^B}, \frac{\partial^2 \hat{U}}{\partial q^A \partial q^B} \right\|$$

is of rank $\hat{m} - m$. The Lagrangian submanifold S generated by $\hat{U}$ is the set of elements of T^*Q with coordinates (q^i, f_j) satisfying

$$f_i = -\frac{\partial \hat{U}}{\partial q^i}, \quad \frac{\partial \hat{U}}{\partial q^A} = 0$$

for some values of q^A.

If $\hat{S} = im(-d\hat{U})$ then $K = \pi_{\hat{Q}}(\hat{S} \cap V^\circ\hat{Q})$ is called the *critical set* of the function $\hat{U}: \hat{Q} \to \mathrm{R}$ along the fibres of ζ. If $\hat{U}$ is a Morse family then the critical set is a submanifold of $\hat{Q}$ of dimension m.

PROPOSITION 1.7. *If $S \subset T^*Q$ is a Lagrangian submanifold generated by a Morse family $\hat{U}: \hat{Q} \to \mathrm{R}$ and the critical set K is the image of a section $\kappa: Q \to \hat{Q}$ of the fibration $\zeta: \hat{Q} \to Q$ then*

$$S = im(-\, dU),$$

where $U{:}Q \to \mathbf{R}$ *is defined by* $U = \hat{U}\circ\kappa$.

Proof. Relation $\hat{S} \cap V^\circ\hat{Q} = \varepsilon(-\, d\hat{U}(K))$ follows from the definition of K. If $K = im(\kappa)$ then $S = im(-\, \varepsilon\circ d\hat{U}\circ\kappa)$. If $v \in TQ$ and $w = T\kappa(v)$ then

$$\langle v, \varepsilon\circ d\hat{U}\circ\kappa\rangle = \langle w, d\hat{U}\rangle$$

and

$$\langle v, dU\rangle = \langle v, d(\hat{U}\circ\kappa)\rangle = \langle v, \kappa^* d\hat{U}\rangle = \langle w, d\hat{U}\rangle.$$

Hence $\varepsilon\circ d\hat{U}\circ\kappa = dU$ and $S = im(-\, dU)$. ■

In terms of coordinates (q^i, q^A) in Q the critical set K is characterized by equations

$$\frac{\partial\hat{U}}{\partial q^A} = 0.$$

If K is the image of a section of ζ then these equations can be solved for q^A in terms of q^i:

$$q^A = \kappa^A(q^i).$$

The function $U(q^i)$ generating S is obtained from

$$U(q^i) = \hat{U}(q^i, \kappa^A(q^j)).$$

The Lagrangian submanifold $\overline{S}$ in Proposition 1.3 is said to be obtained from S by a reduction with respect to the submanifold $\overline{Q} \subset Q$ and the Lagrangian submanifold S in Proposition 1.6 is said to be obtained by a reduction with respect to the fibration $\zeta{:}\hat{Q} \to Q$.

In applications to physics the manifold Q appears as the configuration space of a static system and T^*Q is the generalized force bundle. The system is characterized by its configuration-force relation represented geometrically by a submanifold S of T^*Q. If the system is reciprocal S is a Lagrangian submanifold and is usually generated by a potential energy function $U: Q \to \mathrm{R}$. If S is not regular then it is related to a regular Lagrangian submanifold by a suitable reduction.

EXAMPLE 1.1. Let Q be the Euclidean plane with coordinates (x,y). Coordinates (x,y,f,g) will be used in T^*Q. Equations

$$f = -k(x-a),$$

$$g = -ky$$

describe a regular Lagrangian submanifold $S \subset T^*Q$ which represents the position-force relation for a point subject to a siple restoring force with the centre of attraction at $(a,0)$. The submanifold S is generated by the potential energy function

$$U(x,y) = \frac{k}{2}[(x-a)^2 + y^2].$$

EXAMPLE 1.2. Let Q be the Euclidean plane. Equations

$$x = a\cos\theta,$$

$$y = a\sin\theta,$$

$$f = \lambda\cos\theta,$$

$$g = \lambda\sin\theta$$

describe a Lagrangian submanifold S of T^*Q with coordinates (θ,λ); $0 \leq \theta < 2\pi$, $-\infty < \lambda < \infty$. This submanifold represents the position-force relation for a point constrained to remain on the circle $\overline{Q} \subset Q$ described by $x^2 + y^2 = a^2$ or by $x = a\cos\theta$, $y = a\sin\theta$, where θ is a coordinate in $\overline{Q}$. The potential energy function $\overline{U}(\theta) = 0$ generates a regular Lagrangian submanifold $\overline{S}$ of $T^*\overline{Q}$ described in terms of coordinates (θ,τ) in $T^*\overline{Q}$ by the equation $\tau = 0$. The submanifold S is related to $\overline{S}$ as in Proposition 1.4.

EXAMPLE 1.3. A modified version of Example 1.2 is provided by $S \subset T^*Q$ described by equations

$$x = a\cos\theta,$$

$$y = a\sin\theta,$$

$$f = -\,ka(\cos\theta - 1) + \lambda\cos\theta,$$

$$g = -\,ka\sin\theta + \lambda\sin\theta.$$

This Lagrangian submanifold represents a point constrained to remain on the circle $\overline{Q}$ and is also subject to a simple restoring force with its centre of attraction at $(a,0)$. The potential energy function $\overline{U}(\theta) = ka^2(1 - \cos\theta)$ generates a regular Lagrangian submanifold $\overline{S}$ of $T^*\overline{Q}$ described by $\tau = -\,ka^2\sin\theta$. The submanifold S is obtained from $\overline{S}$ as in Proposition 1.4.

EXAMPLE 1.4. Let Q be again the Euclidean plane. Equations

$$x = r\cos\theta,$$

$$y = r\sin\theta,$$

$$f = -\,k(r - a)\cos\theta,$$

$$g = -\,k(r - a)\sin\theta$$

describe a Lagrangian submanifold S of T^*Q with coordinates (θ,r); $0 \le \theta < 2\pi$, $-\infty < r < \infty$. The submanifold S is not regular. It represents the position-force relation for a point subject to a simple restoring force whose centre of attraction is allowed to move freely on the circle $x^2 + y^2 = a^2$. A Morse family for S defined on the manifold $\hat{Q} = Q \times S^1$ with coordinates (x,y,θ) is given by

$$\hat{U}(x,y,\theta) = \frac{k}{2}\,[(x - a\cos\ )^2 + (y - a\sin\ )^2].$$

All examples given above can be derived by suitable reductions from the following single example.

EXAMPLE 1.5. Let Q be the product of two Euclidean planes with coordinates (x_1,y_1,x_2,y_2). We use coordinates $(x_1,y_1,x_2,y_2,f_1,g_1,f_2,g_2)$ in T^*Q. Equations

$$x_2 = a\cos\theta,$$
$$y_2 = \ \sin\ ,$$
$$f_1 = -\,k(x_1 - a\cos\theta),$$
$$g_1 = -\,k(y_1 - a\sin\theta),$$
$$f_2 = k(x_1 - a\cos\theta) + \lambda\cos\theta,$$
$$g_2 = k(y_1 - a\sin\theta) + \lambda\sin\theta$$

describe a Lagrangian submanifold S of T^*Q with coordinates (x_1,y_1,θ,λ). This submanifold represents the position-force relation for two points with coordinates (x_1,y_1) and (x_2,y_2) respectively interacting by a simple restoring force. The second of these points is constrained to remain on the circle $x_2{}^2 + y_2{}^2 = a^2$. Let $\overline{Q}$ be the submanifold of Q with coordinates (x_1,y_1,θ) described by equations

$$x_2 = a\cos\theta,$$
$$y_2 = a\sin\theta.$$

The potential energy function defined on $\overline{Q}$ is

$$\overline{U}(x_1,y_1,\theta) = \frac{k}{2}\,[(x_1 - a\cos\theta)^2 + (y_1 - a\sin\theta)^2].$$

This function generates a regular Lagrangian submanifold $\overline{S}$ of $T^*\overline{Q}$ described by equations

$$f_1 = -k(x_1 - a\cos\theta),$$
$$g_1 = -k(y_1 - a\sin\theta),$$
$$\tau = ka(x_1\sin\theta - y_1\cos\theta)$$

in terms of coordinates $(x_1,y_1,\theta,f_1,g_1,\tau)$ in $T^*\overline{Q}$. The submanifold S is related to $\overline{S}$ as in Proposition 1.4.

It is left to the reader to verify that the following reductions of Lagrangian submanifolds S and $\overline{S}$ of Example 1.5 reproduce Examples 1.1 through 1.4 respectively.

1. Reduction of $\overline{S}$ with respect to the submanifold Q_1 of $\overline{Q}$ defined by $\theta = 0$.

2. Reduction of S with respect to the fibration

$$\zeta_2 : Q \to Q_2 : (x_1,y_1,x_2,y_2) \mapsto (x_2,y_2).$$

3. Reduction of S with respect to the submanifold Q_3 of Q defined by $x_1 = a$, $y_1 = 0$.

4. Reduction of $\overline{S}$ with respect to the fibration

$$\zeta_4 : \overline{Q} \to Q_4 : (x_1,y_1,\theta) \mapsto (x_1,y_1).$$

Note that the generating function of $\overline{S}$ is exactly the Morse family of Example 1.4. Reduction 2 will produce the potential energy function of Example 1.2 if Proposition 1.7 is used.

2. Hamiltonian systems.

DEFINITION 2.1. A *symplectic manifold* (P,ω) is a manifold P of even dimension $2m$ and a non-degenerate closed 2-form ω on P.

DEFINITION 2.2. A submanifold $N \subset P$ is said to be *Lagrangian* if $dim(N) = m$ and $\omega|N = 0$.

A theorem of Darboux guarantees the existence of local coordinates (q^i, p_j); $i,j = 1,\ldots,m$ such that $\omega = dp_i \wedge dq^i$. A Lagrangian submanifold $N \subset P$ described by equations

$$q^i = \xi^i(t^\alpha), \quad p_j = \eta_j(t^\alpha)$$

satisfies the Lagrange bracket conditions

$$[t^\alpha, t^\beta] = \frac{\partial \xi^i}{\partial t^\alpha}\frac{\partial \eta_i}{\partial t^\beta} - \frac{\partial \xi^i}{\partial t^\beta}\frac{\partial \eta_i}{\partial t^\alpha} = 0,$$

where (t^α); $\alpha = 1,\ldots,m$ are coordinates in N.

The mapping

$$\beta_P : TP \to T^*P : u \mapsto i_u\omega$$

is a vector bundle isomorphism. If χ_P denotes the pull-back $\beta_P{}^*\theta_P$ of the canonical 1-form θ_P from T^*P to TP and $\rho_P = d\chi_P$ then (TP, ρ_P) is a symplectic manifold.

Coordinates (q^i, p_j) in P induce in TP coordinates $(q^i, p_j, \dot{q}^k, \dot{p}_l)$; $i,j,k,l = 1,\ldots,m$ such that $\dot{q}^k(u) = \langle u, dq^k\rangle$ and $\dot{p}_l(u) = \langle u, dp_l\rangle$. In terms of these coordinates we have

$$\chi_P = \dot{p}_i dq^i - \dot{q}^i dp_i$$

and

$$\rho_P = d\dot{p}_i \wedge dq^i + dp_i \wedge d\dot{q}^i.$$

DEFINITION 2.3. A vector field $X:P \to TP$ is said to be *Hamiltonian* if the 1-form $i_X\omega$ is exact. A function $H:P \to \mathbf{R}$ such that $i_X\omega = - dH$ is called a *Hamiltonian* of X. The field X is said to be *locally Hamiltonian* if $i_X\omega$ is closed.

Since $di_X\omega = 0$ implies $\pounds_X\omega = 0$ a locally Hamiltonian vector field is also called an *infinitesimal symplectic transformation.*

PROPOSITION 2.1. *A vector field $X:P \to TP$ is locally Hamiltonian if and only if $D = im(X)$ is a Lagrangian submanifold of the symplectic manifold (TP, ρ_P).*

Proof. $dim(D) = 2m = \frac{1}{2}dim(TP)$ and

$$\begin{aligned} X^*\rho_P &= d(X^*\chi_P) \\ &= d(X^*\beta_P{}^*\theta_P) \\ &= d(\beta_P \circ X)^*\theta_P \\ &= d(\beta_P \circ X) \\ &= di_X\omega. \end{aligned}$$

Hence $\rho_P | D = 0$ if and only if $di_X\omega = 0$. ∎

If $X:P \to TP$ is Hamiltonian and $X \quad \omega = - dH$ then $D = im(X$ is the image by β_P of the regular Lagrangian submanifold $N \subset T^*P$ generated by $H:P \to \mathbf{R}$.

DEFINITION 2.4. If $X \lrcorner\, \omega = - dH$ then X is said to be *generated* by H and H is called a *Hamiltonian* of X.

If X is a Hamiltonian vector field generated by a Hamiltonian $H:P \to \mathbf{R}$ then $D = im(X)$ is described in terms of coordinates $(q^i, p_j, \dot{q}^k, \dot{p}_l)$ of TP by equations

$$\dot{q}^i = \frac{\partial H}{\partial p_i},$$

$$\dot{p}_j = -\frac{\partial H}{\partial q^j}.$$

DEFINITION 2.6. A submanifold $K \subset P$ is called a *coisotropic submanifold* of (P,ω) if at each $p \in K$

$$(T_pK)^{\P} \subset T_pK,$$

where

$$(T_pK)^{\P} = \{u \in T_pP;\ \langle u \wedge w,\omega\rangle = 0 \text{ for each } w \in T_pP\}.$$

PROPOSITION 2.2. *If $K \subset P$ is coisotropic then*

$$E = \cup_{p\in K}(T_pK)^{\P} \subset TK$$

is an involutive distribution.

Proof. The set E is obviously a distribution of dimension equal to the codimension of K in P. The image of a vector field $X{:}K \to TK$ is contained in E if and only if $\langle X \wedge Z,\omega\rangle = 0$ for each vector field $Z{:}K \to TK$. If $X{:}K \to TK$ and $Y{:}K \to TK$ both satisfy this condition then $\langle[X,Y] \wedge Z,\omega\rangle = 0$ follows from the identity $\langle X \wedge Y \wedge Z,d\omega\rangle = \langle X,d\langle Y \wedge Z,\omega\rangle\rangle + \langle Y,d\langle Z \wedge X,\omega\rangle\rangle + \langle Z,d\langle X \wedge Y,\omega\rangle\rangle - \langle[X,Y] \wedge Z,\omega\rangle - \langle[Y,Z] \wedge X,\omega\rangle - \langle[Z,X] \wedge Y,\omega\rangle$. Hence E is involutive. ■

Let $K \subset P$ be a coisotropic submanifold such that the set $\overline{P}$ of maximal integral manifolds of the distribution E is a manifold, there is a symplectic form $\overline{\omega}$ on $\overline{P}$ and $\kappa^*\overline{\omega} = \omega|K$, where $\kappa{:}K \to \overline{P}$ is the canonical projection.

DEFINITION 2.7. The symplectic manifold $(\overline{P},\overline{\omega})$ is called the *reduced symplectic manifold* corresponding to the coisotropic submanifold K of (P,ω).

PROPOSITION 2.3. *If K is a coisotropic submanifold of (P,ω) and $(\overline{P},\overline{\omega})$ is the corresponding reduced symplectic manifold then TK is a coisotropic submanifold of (TP,ρ_P) and*

$$(T\kappa)^*\rho_{\overline{P}} = \rho_P|TK.$$

Proof. Let $w \in TTK \subset TTP$ then

$$\begin{aligned}
\langle w,(T\kappa)^*\chi_{\overline{P}}\rangle &= \langle TT\kappa(w),\chi_{\overline{P}}\rangle \\
&= \langle TT\kappa(w),\beta_{\overline{P}}{}^*\theta_{\overline{P}}\rangle \\
&= \langle T\beta_{\overline{P}}(TT\kappa(w)),\theta_{\overline{P}}\rangle \\
&= \langle T\pi_{\overline{P}}(T\beta_{\overline{P}}(TT\kappa(w))),\tau_{T^*\overline{P}}(T\beta_{\overline{P}}(TT\kappa(w)))\rangle \\
&= \langle T\kappa(T\tau_K(w)),\beta_{\overline{P}}(T\kappa(\tau_{TK}(w)))\rangle \\
&= \langle T\kappa(\tau_{TK}(w)) \wedge T\kappa(T\tau_K(w)),\overline{\omega}\rangle \\
&= \langle \tau_{TK}(w) \wedge T\tau_K(w),\kappa^*\overline{\omega}\rangle \\
&= \langle \tau_{TK}(w) \wedge T\tau_K(w),\omega\rangle \\
&= \langle \tau_{TP}(w) \wedge T\tau_P(w),\omega\rangle
\end{aligned}$$

and

$$\begin{aligned}
\langle w,\chi_P|TK\rangle &= \langle w,\chi_P\rangle \\
&= \langle w,\beta_P{}^*\theta_P\rangle \\
&= \langle T\beta_P(w),\theta_P\rangle \\
&= \langle T\pi_P(T\beta_P(w)),\tau_{T^*P}(T\beta_P(w))\rangle \\
&= \langle T\tau_P(w),\beta_P(\tau_{TP}(w))\rangle \\
&= \langle \tau_{TP}(w) \wedge T\tau_P(w),\omega\rangle.
\end{aligned}$$

Hence $(T\kappa)^*\chi_{\overline{P}} = \chi_P|TK$. Consequently $(T\kappa)^*\rho_{\overline{P}} = \rho_P|TK$. This last equation implies $ker(T_uT\kappa) \subset (T_uTK)^¶$ for each $u \in TK$. If $dim(K) = 2m - k$ then $dim((T_uTK)^¶) = 2k$ and $dim(ker(T_uT\kappa)) = 2k$. Hence $ker(T_uT\kappa) = (T_uTK)^¶$. Since $ker(T_uT\kappa) \subset T_uTK$ it follows that TK

is coisotropic. ■

PROPOSITION 2.4. *If* $\overline{X}:\overline{P} \to T\overline{P}$ *is a Hamiltonian vector field generated by a Hamiltonian* $\overline{H}:\overline{P} \to \mathbf{R}$ *then* $D = (T\kappa)^{-1}(im(\overline{X}))$ *is a Hamiltonian system in* (P,ω).

The proof of this proposition is analogous to that of Proposition 1.4.

DEFINITION 2.8. The Hamiltonian system D constructed in Proposition 2.4 is said to be *generated* by the function $H = \overline{H}\circ\kappa$ and H is called a *Hamiltonian* of D.

A coisotropic submanifold $K \subset P$ can be described by equations $K^A(q^i,p_j) = 0$; $A = 1,\ldots,k$ such that the Poisson brackets

$$\{K^A,K^B\} = \frac{\partial K^A}{\partial q^i}\frac{\partial K^B}{\partial p_j} - \frac{\partial K^B}{\partial q^i}\frac{\partial K^A}{\partial p_j}$$

vanish on K. The distribution E is described by

$$\dot{q}^i = \lambda_A \frac{\partial K^A}{\partial p_i},$$

$$\dot{p}_j = -\lambda_A \frac{\partial K^A}{\partial q^j},$$

$$K^A(q^i,p_j) = 0$$

for some values of λ_A. If $\overline{H}$ is a function on $\overline{P}$ and $\tilde{H}$ a function on P such that $\overline{H}\circ\kappa = \tilde{H}|K$ then the Hamiltonian system $D \subset TP$ generated by $H = \overline{H}\circ\kappa$ is described by

$$\dot{q}^i = \frac{\partial \tilde{H}}{\partial p_i} + \lambda_A \frac{\partial K^A}{\partial p_i},$$

$$\dot{p}_j = -\frac{\partial\tilde{H}}{\partial q^j} - \lambda_A \frac{\partial K^A}{\partial q^j} ,$$

$$K^A(q^i,p_j) = 0$$

for some values of λ_A.

Let $\hat{K} \subset \hat{P}$ be a coisotropic submanifold of a symplectic manifold $(\hat{P},\hat{\omega})$ such that there exists the reduced symplectic manifold (P,ω) associated with $\hat{K}$ and let $\hat{\kappa}:\hat{K} \to P$ be the canonical projection.

PROPOSITION 2.5. *Let $\hat{X}:\hat{P} \to T\hat{P}$ be a Hamiltonian vector field generated by a Hamiltonian $\hat{H}:\hat{P} \to \mathbf{R}$. If $im(\hat{X})$ is transverse to $T\hat{K} \subset T\hat{P}$ then $D = im(T\hat{\kappa}\circ\hat{X}) \subset TP$ is a Hamiltonian system in (P,ω).*

The proof of this proposition parallels that of Proposition 1.6.

DEFINITION 2.9. The Hamiltonian system constructed in Proposition 2.5 is said to be *generated* by the function $\hat{H}$ and $\hat{H}$ is called a *generalized Hamiltonian* of D.

Let $\zeta:Z \to P$ be a fibration, let $G:Z \to \mathbf{R}$ be a Morse family and let $S \subset T^*P$ be the Lagrangian submanifold generated by G.

DEFINITION 2.10. The Hamiltonian system $D = \beta_P^{-1}(S)$ is said to be *generated* by G and G is called a *Morse family* of D.

It can be shown that if D is generated by a generalized Hamiltonian $\hat{H}$ then $H = \hat{H}|K$ is a Morse family of D.

Let $\hat{K} \subset \hat{P}$ be a coisotropic submanifold and let (q^i,p_j) be coordinates in the associated reduced symplectic manifold P. Adapted coordinates (q^i,q^A,p_j,p_B) can be introduced in $\hat{P}$ such that

$\hat{K}$ is described by $p_A = 0$ and (q^i, q^A, p_j) are coordinates in $\hat{K}$, the canonical projection $\hat{\kappa}:\hat{K} \to P$ is described by $\hat{\kappa}:(q^i, q^A, p_j) \mapsto (q^i, p_j)$ and $\hat{\omega} = dp_i \wedge dq^i + dp_A \wedge dq^A$ on $\hat{K}$. In terms of coordinates $(q^i, q^A, p_j, p_B, \dot{q}^k, \dot{q}^C, \dot{p}_l, \dot{p}_D)$ in $T\hat{P}$ the submanifold $T\hat{K}$ is described by $p_A = 0$, $\dot{p}_B = 0$ and $T\hat{\kappa}$ is given by $T\hat{\kappa}:(q^i, q^A, p_j, \dot{q}^k, \dot{q}^B, \dot{p}_l) \mapsto$ $\mapsto (q^i, p_j, \dot{q}^k, \dot{p}_l)$. If $\hat{X}$ is a Hamiltonian vector field generated by a function $\hat{H}(q^i, q^A, p_j, p_B)$ and $im(\hat{X})$ is transverse to $T\hat{K}$ then the matrix

$$\left\| \frac{\partial^2 H}{\partial q^i \partial q^B} \, , \, \frac{\partial^2 H}{\partial p_j \partial q^B} \, , \, \frac{\partial^2 H}{\partial q^A \partial q^B} \right\|$$

is of rank $\hat{k}$. Here $H(q^i, q^A, p_j) = \hat{H}(q^i, q^A, p_j, 0)$ and $\hat{k} = codim(\hat{K})$. The Hamiltonian system $D \subset TP$ generated by the generalized Hamiltonian $\hat{H}$ is described by

$$\dot{q}^i = \frac{\partial H}{\partial p_i} \, ,$$

$$\dot{p}_j = - \frac{\partial H}{\partial q^j} \, ,$$

$$\frac{\partial H}{\partial q^A} = 0$$

for some values of q^A. The function $H(q^i, q^A, p_j)$ is a Morse family of D.

If $\gamma:\mathbf{R} \to P$ is a differentiable curve we denote by $\dot{\gamma}(t)$ the vector tangent to γ at $\gamma(t)$ and introduce the prolongation $\dot{\gamma}$ of γ:

$$\dot{\gamma}:\mathbf{R} \to TP:t \mapsto \dot{\gamma}(t).$$

DEFINITION 2.11. A curve $\gamma:\mathbf{R} \to P$ is called an *integral curve* of a Hamiltonian system $D \subset TP$ if $im(\dot{\gamma}) \subset D$.

DEFINITION 2.12. A Hamiltonian system D is said to be *integrable* if for each $u \in D$ there is an integral curve γ of D such that $\dot{\gamma}(0) = u$.

In applications to physics Hamiltonian systems represent dynamics of mechanical systems. Integrability is an important property of Hamiltonian systems from the point of view of such applications. It is known that Hamiltonian vector fields and Hamiltonian systems constructed as in Proposition 2.4 are integrable. Little is known about criteria of integrability of more general Hamiltonian systems.

EXAMPLE 2.1. Let Q be the four-dimensional space-time manifold of general relativity with coordinates (x^κ). Let $g_{\kappa\lambda}$ and $g^{\mu\nu}$ denote the components of the covariant and the contravariant metric tensors respectively and let $\Gamma^{\kappa}_{\lambda\mu}$ be the Christoffel symbols. The cotangent bundle $P = T^*Q$ with coordinates (x^κ, p_λ) represents the space-time-momentum-energy manifold of a relativistic particle. The dynamics of a particle of mass m is represented by a Hamiltonian system $D \subset TP$ described by equations

$$p_\kappa = m(g_{\mu\nu}\dot{x}^\mu\dot{x}^\nu)^{-\frac{1}{2}} g_{\kappa\lambda}\dot{x}^\lambda,$$
$$\dot{p}_\kappa + \Gamma^{\mu}_{\kappa\lambda} p_\mu \dot{x}^\lambda = 0,$$
$$g_{\mu\nu}\dot{x}^\mu\dot{x}^\nu > 0$$

in terms of coordinates $(x^\kappa, p_\lambda, \dot{x}^\mu, \dot{p}_\nu)$ in TP. This Hamiltonian system is generated by the Morse family

$$H(x^\kappa, y\ , p_\lambda) = y[(g^{\mu\nu}p_\mu p_\nu)^{\frac{1}{2}} - m].$$

To prove integrability of D we observe that D is an open submanifold of a Hamiltonian system generated as in Proposition 2.4. Let a coisotropic submanifold K of P be defined by $(g^{\mu\nu}p_\mu p_\nu)^{\frac{1}{2}} = m$. The Hamiltonian $H = 0$ defined on P generates a Hamiltonian system described by

$$\dot{x}^\kappa = \frac{y}{m} g^{\kappa\lambda} p_\lambda \ ,$$
$$\dot{p}_\kappa + \Gamma_{\kappa\lambda}{}^{\mu} p_\mu \dot{x}^\lambda = 0,$$
$$(g^{\mu\nu}p_\mu p_\nu)^{\frac{1}{2}} = m$$

for some y. The Hamiltonian system D is obtained by restricting y to positive values.

EXAMPLE 2.2. Let Q be the Euclidean plane with coordinates (x,y). We consider in Q a particle of mass m subject to a simple restoring force whose centre of attraction is allowed to move freely on the circle $x^2 + y^2 = a^2$. The phase space of the particle is the cotangent bundle $P = T^*Q$ with coordinates (x,y,p,r) and the dynamics is a Hamiltonian system D described by

$$x = \lambda\cos\theta,$$
$$y = \lambda\sin\theta,$$
$$p = m\dot{x},$$
$$r = m\dot{y},$$
$$\dot{p} = -\ k(\lambda - a)\cos\theta,$$
$$\dot{r} = -\ k(\lambda - a)\sin\theta$$

in terms of coordinates $(x,y,p,r,\dot{x},\dot{y},\dot{p},\dot{r})$ in TP and coordinates $(\lambda,\theta,\dot{x},\dot{y})$ in D with $0 \leq \theta < 2\pi$, $-\infty < \lambda < \infty$. The function

$$H(x,y,\theta,p,r) = \frac{1}{2m}(p^2 + r^2) + \frac{k}{2}[(x - a\cos\theta)^2 + (y - a\sin\theta)^2]$$

is a Morse family of D.

References.

[1] R. Abraham and J.E. Marsden, *Foundations of Mechanics*, Benjamin-Cummings (1978).

[2] S. Benenti and W.M. Tulczyjew, *The geometrical meaning and globalization of the Hamilton-Jacobi method*, Lecture Notes in Mathematics, 836 (1980), pp. 9-21.

[3] J. Kijowski and W.M. Tulczyjew, *A symplectic framework for field theories*, Lecture Notes in Physics, 107, Springer-Verlag (1979).

[4] A. Lichnerowicz, C. R. Acad. Sci. Paris, 280 (1975), pp. 37-40.

[5] M.R. Menzio and W.M. Tulczyjew, *Infinitesimal symplectic relations and generalized Hamiltonian dynamics*, Ann. Inst. H. Poincare, 28 (1978), pp. 349-367.

[6] W.M. Tulczyjew, *Hamiltonian systems, Lagrangian systems and the Legendre transformation*, Symposia Mathematica, 14 (1974), pp. 247-258.

[7] W.M. Tulczyjew, *The Legendre transformation*, Ann. Inst. H. Poincare, 27 (1977), pp. 101-114.

[8] A. Weinstein, *Lectures on Symplectic Manifolds*, CBMS Regional Conferences, 29 (1976), A.M.S.

DEFORMATIONS AND QUANTIZATION

André Lichnerowicz

Chaire de Physique Mathématique

Collège de France

Paris, France

It is well-known that it is possible to give a complete description of Classical Mechanics in terms of symplectic geometry and Poisson bracket. It is the essence of the Hamiltonian formalism. In a common program with Flato, Fronsdal, J. Vey and others, we have studied properties and applications of the deformations of the associative algebra defined by the usual product of functions. Such deformations give a new invariant approach of Quantum Mechanics which corresponds to an autonomous generalization of the Weyl-Wigner quantization. The quantization of a system described by a symplectic manifold (phase space) is given by the choice of a so-called *star-product*. I consider here only dynamical systems with a finite number of degrees of freedom, but the

DYNAMICAL SYSTEMS
AND MICROPHYSICS

ISBN 0-12-068720-8

approach and a significative part of the results may be extended to physical fields.

1 - CLASSICAL DYNAMICS AND SYMPLECTIC GEOMETRY

a) Let (W,F) be *a smooth symplectic manifold* of dimension $2n$. We denote by $b_k(W)$ the $k^{\underline{th}}$ Betti number of W. For simplicity, we put $N = C^\infty(W, \mathbb{R})$. The symplectic structure is defined on W by the closed 2-form of rank $2n$ (F^n is $\neq 0$ everywhere). Consider the isomorphism of vector bundles $\mu = TW \to T^*W$ defined by $\mu(X) = -i(X)F$ (where $i(\ .\ )$ is the inner product); this isomorphism may be extended to tensors in a natural way. We denote by Λ the antisymmetric contravariant 2-tensor $\mu^{-1}(F)$.

A *symplectic vector field* is a vector field such that $\mathcal{L}(X)F = 0$ (where $\mathcal{L}(\ .\ )$ is the Lie derivative). It defines an infinitesimal automorphism of the structure. We denote by L the (infinite dimensional) Lie algebra of the symplectic vector fields. If X belongs to L we have :

$$\mathcal{L}(X)F = (d\ i(X) + i(X)\ d)F = d\ i(X)F = -\ d\mu(X) = 0$$

and the 1-form $\mu(X)$ is closed. If $X, Y \in L$, we have :

$$\mu([X,Y]) = d\ i(\Lambda)\ (\mu(X) \wedge \mu(Y)) \qquad (1\text{-}1)$$

Let $L^{\star}$ be the subspace of L defined by the converse image of the exact 1-forms ($X_u = \mu^{-1}(du)$; $u \in N$). An element of $L^{\star}$ is a *hamiltonian vector field*. Consider the commutator ideal $[L,L]$ of L : each element of $[L,L]$ is, by definition, a finite sum of brackets of elements of L. It follow trivially from (1-1) that $[L,L] \subset L^{\star}$. It has been proved (Arnold, Calabi, myself) that we have exactly $[L,L] = L^{\star}$ and it is clear that $\dim L/L^{\star} = b_1(W)$. We are led to introduce the Poisson bracket

$$(1\text{-}2) \qquad \{u,v\} = i(\Lambda)(du \wedge dv) = \mathcal{L}(X_u)v = P(u,v) \qquad (u,v \in N)$$

where the Poisson operator P is a bidifferential operator of order 1 for each argument, null on the constants; P defines on N a structure of Lie algebra and (N,P) is the Poisson Lie algebra of the manifold and we have a natural homomorphism of (N,P) onto $L^{\star}$ since $X_{\{u,v\}} = [X_u, X_v]$.

It is known from a long time that a symplectic manifold admits atlases of so-called *canonical charts* $\{x^i\} = \{x^{\alpha}, x^{\bar{\alpha}}\}$ ($i,j,\ldots = 1,\ldots,2n$; $\alpha = 1,\ldots,n$; $\bar{\alpha} = \alpha + n$). On the domain U of a canonical chart, we have :

$$F|_U = \sum_{\alpha} dx^{\alpha} \wedge dx^{\bar{\alpha}}$$

and

$$\{u,v\}|_U = \sum_{\alpha} (\partial_{\alpha} u\, \partial_{\bar{\alpha}} v - \partial_{\bar{\alpha}} u\, \partial_{\alpha} v) \qquad (\partial_i = \partial/\partial x^i)$$

b) Consider a dynamical system with time independent constraints and n degrees of freedom. The corresponding configuration space is an arbitrary differentiable manifold M of dimension n. It is well-known that the cotangent bundle T^*M admits a natural symplectic structure defined by the 2-form $F = d\omega$, where ω is the Liouville 1-form which may be written locally in terms of classical variables $\omega = \sum_\alpha p_\alpha \, dq^\alpha$, so that $F = \sum_\alpha dp_\alpha \wedge dq^\alpha$ $(\alpha = 1,..,n)$. For the Hamiltonian formalism, a dynamical state of the system is nothing other as a point of $W = T^*M$ which is the usual phase space of the mechanicians and physicists. The analysis of the equations of Mechanics has showed, from a long time, that it is essential to may introduce changes of the variables (q^α, p_α) which do not respect the cotangent structure of W. We are thus led to introduce as *phase space* a symplectic manifold (W,F) of dimension 2n. We can put in a canonical chart $q^\alpha = x^{\bar{\alpha}}, p_\alpha = x^\alpha$ and so obtain the interpretations which are usual in Classical Dynamics.

On the manifold (W,F), Dynamics is determined by a function $H \in N$, *the classical Hamiltonian* of the system, which defines a hamiltonian vector field X_H. *A motion of the dynamical system is given by an integral curve c(t) of* X_H*, the parameter t being the time.* Such is the geometrical meaning of the classical equations of Hamilton.

c) We can adopt another viewpoint. The space N admits the following two algebraic structures :

1) a structure of *associative algebra* given by the usual product of functions (which is commutative)

2) a structure of *Lie algebra* given by the Poisson bracket.

According to (1-2), the Poisson bracket defines derivations of the product. Consider a family u_t of elements of N satisfying the differential equation :

(1-3) $$du_t/dt = \{H, u_t\}$$

and taking the initial value u_o at t = 0. We see that the evolution in the time of u_t arises from the flow of X_H which appears in the first viewpoint; (1-3) may be considered as the intrinsic equation of Classical Dynamics.

We have completely described Classical Mechanics in terms of the two laws of composition defined on N. It is natural to study if it possible *to deform* in a suitable way these two algebraic laws so that we obtain a model isomorphic to the conventional Quantum Mechanics. The answer is positive.

2 - HOCHSCHILD COHOMOLOGY AND CHEVALLEY COHOMOLOGY

a) Derivations and deformations of an associative algebra arise from a same cohomology, the so-called *Hochschild cohomology*. Let W be an arbitrary differentiable manifold and ($N = C^\infty(W;\mathbb{R})$, .) the associative algebra defined by the product of functions. A p-cochain ($p \geqslant 1$) C of (N,.) is a p-linear map of N^p into N. The Hochschild coboundary of the p-cochain C is the (p+1)-cochain $\tilde{\partial}C$

defined by :

$$\tilde{\partial}C(u_o,\ldots,u_p) = u_o\, C(u_1,\ldots,u_p) - C(u_o u_1,u_2,\ldots,u_p) +$$

$$(2\text{-}1) \qquad C(u_o,u_1 u_2,\ldots,u_p) - \ldots + (-1)^p\, C(u_o,u_1,\ldots,u_{p-1}\, u_p) +$$

$$(-1)^{p+1}\, C(u_o, \ldots,u_{p-1})\, u_p \quad .$$

We have $\tilde{\partial}^2 = 0$. A 1-cocycle of (N,.) is a derivation of this algebra and so is given by a vector field. A p-cochain C is sayed to be d-differential ($d \geqslant 0$) if it is defined by a multidifferential operator of maximum order d in each argument. If T is an endomorphism of N (1-cochain) which is (d+1)-differential, $\tilde{\partial}T$ is d-differential. Conversely, I have proved :

Proposition - If T is an endomorphism of N such that $C = \tilde{\partial}T$ is d-differential ($d \geqslant 0$), T is (d+1)-differential itself.

We note that if $\tilde{\partial}T$ is null on the constants, it is the same for T. Let $\tilde{H}^p(N;N)$ be the $p^{\underline{th}}$ Hochschild cohomology space for the differential Hochschild cohomology; J. Vey has proved by means of results of Gelfand the following :

Theorem (Vey) - $\tilde{H}^p(N;N)$ is isomorphic to the space of the antisymmetric contravariant p-tensors of W.

Such a tensor defines an alternate multidifferential operator of order 1 and can be identified with this operator. In general, we consider here cochains which are *null on the constants* for which the Hochschild cohomology is not changed.

b) In a symmetric way, derivations and deformations of a Lie algebra arise from a same cohomology, the so-called *Chevalley cohomology* of the Lie algebra corresponding to the adjoint representation. Let (W,F) be a symplectic manifold and (N,P) the corresponding Poisson Lie algebra. A Chevalley p-cochain ($p \geqslant 0$) C is here an *alternate* p-linear map of N^p into N, the 0-cochains being identified with the elements of N. The Chevalley coboundary ∂ is classically defined by the formula :

$$\partial C(u_o,\dots,u_p) = \varepsilon^{\lambda_o\dots\lambda_p}_{o\dots p}\Big(\frac{1}{p!}\{u_{\lambda_o},C(u_{\lambda_1},\dots u_{\lambda_p})\} - \frac{1}{2(p-1)!}C(\{u_{\lambda_o},u_{\lambda_1}\},u_{\lambda_2},\dots,u_{\lambda_p})\Big)$$

where $u_\lambda \in N$ and where ε is the Kronecker skewsymmetrization indicator. A 1-cocycle of (N,P) is a derivation of the Lie algebra, an exact 1-cocycle being an inner derivation. For the d-differential character of a cochain C, we have definitions similar to the definitions concerning the Hochschild cohomology; but we suppose here $d \geqslant 1$: if C is d-differential, ∂C is also d-differential.

Conversely we have :

Proposition - If C is an exact d-differential ($d \geqslant 1$) Chevalley two-cocycle of (N,P), there is a differential operator of order d such that $C = \partial T$.

André Avez and myself have determined all the derivations of (N,P) without a priori differentiability assumption (1972). In particular the derivations which are null on the constants are given by $\mathcal{L}(X)u$, where X is a symplectic vector field. We denote by $H^p(N;N)$ the pth Chevalley cohomology space for the differential Chevalley cohomology *with cochains which are null on the constants.*

3 - FORMAL DEFORMATIONS

I will now recall and extend the main elements of the theory of Gerstenhaber concerning the deformations of the algebraic structures, in particular the associative algebras.

a) Let $E(N;\nu)$ be the space of the formal functions in $\nu \in C$ with coefficients in N; ν is said to be *the deformation parameter*. Consider a bilinear map $N \times N \to E(N;\nu)$ which gives the formal series:

$$(3\text{-}1) \qquad u *_\nu v = \sum_{r=0}^{\infty} \nu^r C_r(u,v) = u.v + \sum_{r=1}^{\infty} \nu^r C_r(u,v)$$

where the C_r are differential cochains of $(N,.)$. These cochains can be extended to $E(N;\nu)$ in a natural way. We say that (3-1) defines a formal deformation of $(N,.)$ if the associativity identity holds formally. If such is the case, (3-1) defines on $E(N;\nu)$ a structure of *formal associative algebra*. If the map (3-1) is arbitrary, we have for $u,v \in N$:

$$(3\text{-}2) \qquad (u *_\nu v) *_\nu w - u *_\nu (v *_\nu w) = \sum_{t=1}^{\infty} \nu^t \tilde{D}_t(u,v,w)$$

where $\tilde{D}_t$ is the 3-cochain :

(3-3) $$\tilde{D}_t(u,v,w) = \sum_{r+s=t} (C_r(C_s(u,v),w) - C_r(u,C_s(v,w))) \quad (r,s \geqslant 0)$$

We are led to put :

(3-4) $$\tilde{E}_t(u,v,w) = \sum_{r+s=t} (C_r(C_s(u,v),w) - C_r(u,C_s(v,w))) \quad (r,s \geqslant 1)$$

and we have the identity :

$$\tilde{D}_t \equiv \tilde{E}_t - \tilde{\partial} C_t$$

If (3-1) is limited at the order q, we have a deformation of order q if the associativity identity is satisfied up to the order (q+1). If such is the case, E_{q+1} is automatically a 3-cocycle of (N,.). We can find a 2-cochain C_{q+1} satisfying $\tilde{D}_{q+1} = \tilde{E}_{q+1} - \tilde{\partial} C_{q+1} = 0$ iff $\tilde{E}_{q+1}$ is exact ; $\tilde{E}_{q+1}$ defines a cohomology class, element of $\tilde{H}^3(N;N)$, which is the obstruction at the order (q+1) to the construction of a deformation. A deformation of order 1 is called infinitesimal. We have $\tilde{E}_1 = 0$ and so only $\tilde{\partial} C_1 = 0$, that is C_1 is a 2-cocycle of (N,.).

b) Consider a formal series in ν :

(3-5) $$T_\nu = \sum_{s=0}^{\infty} \nu^s T_s = Id_N + \sum_{s=1}^{\infty} \nu^s T_s$$

where the T_s $(s > 1)$ are endomorphisms of N ; T_ν acts naturally on $E(N;\nu)$. Consider also another bilinear map $N \times N \to E(N;\nu)$ corresponding to the formal series :

(3-6) $$u *'_\nu v = uv + \sum_{r=1}^{\infty} \nu^r C'_r(u,v)$$

where the C'_r are differential 2-cochains again. Suppose that (3-5) is such that we have formally the identity

$$T_\nu (u \star'_\nu v) = T_\nu u \star_\nu T_\nu v \tag{3-7}$$

By means of universal formulas, we can prove :

Proposition - The deformation (3-1) of (N,.) being given, each formal series (3-5) where the T_s are necessarily differential operators, generates a unique bilinear map (3-6) satisfying (3-7) ; this map is a new deformation which is sayed to be equivalent to (3-1). In particular a deformation is called trivial if it is equivalent to the identity deformation ($C_r = 0$, $r \geqslant 1$).

According to (3-7), T_ν is sayed to transform $\star'_\nu$ in $\star_\nu$. We have :

$$C'_t - C_t + \tilde{G}_t = \tilde{\partial} T_t \tag{3-8}$$

with $\tilde{G}_1 = 0$ and :

$$\tilde{G}_t(u,v) = \sum_{r+s=t} T_s\, C'_r(u,v) - \sum_{s+s'=t} T_s u . T_{s'} v -$$

$$\sum_{r+s=t} (C_r(T_s u, v) + (C_r(u, T_s v)) - \sum_{r+s+s'=t} C_r(T_s u, T_{s'} v) \tag{3-9}$$

$$(r,s,s' \geqslant 1)$$

If two deformations are equivalent at the order q, there appears the 2-cocycle $(C'_{q+1} - C_{q+1} + \tilde{G}_{q+1})$. Its cohomology class, element of $\tilde{H}^2(N;N)$, is *the obstruction to the equivalence for order (q+1).* In particular two infinitesimal deformations defined by the 2-cocycles C_1 and C'_1 are equivalent iff $(C'_1 - C_1)$ is exact.

c) Let $E(N;\lambda)$ be the space of the formal functions of $\lambda \in C$ with coefficients in N. A deformation of the Poisson Lie algebra (N,P) is defined by an *alternate* bilinear map $N \times N \to E(N;\lambda)$ given by :

$$[u,v]_\lambda = P(u,v) + \sum_{r=1}^{\infty} \lambda^r C_{2r+1}(u,v) \tag{3-10}$$

where the C_{2r+1} are differential cochains of (N,P) such that the Jacobi identity holds formally. The Chevalley cohomology plays exactly the same role for the deformations of Lie algebras as the Hochschild cohomology for the deformations of associative algebras.

4 - THE STAR PRODUCTS

a) A Hochschild 2-cochain C(u,v) is said to be *even* if it is symmetric in u,v, *odd* if it is antisymmetric. Similar definition for a Hochschild 3-cochain E(u,v,w) with respect to the symmetry relatively to u,w. It follows from the Vey theorem that if C is an *even* (resp. *odd*) Hochschild 2-cocycle it is *exact*, (resp. *non exact* and *1-differential*). Let E be a Hochschild 3-cocycle ; if E is even, it is exact and $E = \tilde{\partial} C^{(o)}$; if E is odd, $E = T + \tilde{\partial} C^{(\ell)}$, where $C^{(o)}$ (resp. $C^{(\ell)}$) is odd (resp. even) and where T is given by a skewsymmetric 3-tensor.

On the symplectic manifold (W,Λ),P corresponding to the 2-tensor Λ is a *non exact Hochschild 2-cocycle*. We consider only in the following the associative deformations of the form :

$$(4\text{-}1) \qquad u \star_\nu v = u\,v + \nu\, P(u,v) + \sum_{r=2}^{\infty} \nu^r\, C_r(u,v)$$

where the C_r satisfy the following assumptions :

1) *the C_r are null on the constants.*

2) *C_r is even if r is even, odd if r is odd.*

If such is the case, we say that (4-1) defines a $\star_\nu$-product - or star-product - on (W,Λ). We have :

$$u \star_\nu 1 = 1 \star_\nu u = u \qquad\qquad u \star_{-\nu} v = v \star_\nu u$$

A star-product generates by skewsymmetrization *a formal Lie algebra* (3-10) with $\lambda = \nu^2$:

$$(4\text{-}2) \qquad [u,v]_\lambda = (2\nu)^{-1}\,(u \star_\nu v - v \star_\nu u)$$

I have proved the following :

Uniqueness theorem. If a formal Lie algebra (3-10) is generated by a star-product, this star-product is unique.

b) Let $\star_\nu$, $\star'_\nu$ be two equivalent star products, T_ν transforming $\star'_\nu$ in $\star_\nu$. It is possible to prove that we can substitute to T_ν a similar formal series in ν^2. Therefore *two star products which are equivalent are even equivalent with respect to* ν . It follows that two formal Lie algebras generated by two equivalent star products are equivalent with respect to $\lambda = \nu^2$. Conversely, if a formal Lie algebra is generated by a star product, each equivalent formal Lie algebra is generated by an equivalent star-product.

5 - SYMPLECTIC CONNECTIONS AND THE SYMPLECTIC INVARIANT β

a) A symplectic connection Γ is a linear connection without torsion such that $\nabla F = 0$, where ∇ is the operator of covariant differentiation defined by Γ. Let $\bar{\Gamma}$ be an arbitrary linear connection without torsion. Each other connection Γ without torsion differs from $\bar{\Gamma}$ by a tensor T of type (1,2) which is covariantly symmetric. On the domain U of a chart $\{x^i\}$, we have :

$$\nabla_k F_{ij} = \bar{\nabla}_k F_{ij} - T_{ijk} + T_{jik} \qquad (T_{ijk} = F_{ir} T^r_{jk};\ T_{ijk} = T_{ikj})$$

If we choose T defined by :

$$T_{ijk} = \frac{1}{3} (\bar{\nabla}_k F_{ij} + \bar{\nabla}_j F_{ik}) \tag{5-1}$$

we see immediately that $\nabla F = 0$ for Γ and that Γ is thus a symplectic connection. A symplectic manifold admits infinitely many symplectic connections ; such two connections differ by a tensor of type (1,2) deduced from an arbitrary symmetric covariant 3-tensor. Moreover, if G is a Lie group acting on (W,F) by symplectomorphisms and if there is on (W,F) a linear connection invariant under G, *there is on (W,F) a symplectic connection invariant under G* .

b) Suppose that (W,Λ) admits a symplectic connection *without curvature;* if such is the case (W,Λ,Γ) is called *a flat symplectic manifold.* The simplest example is the cotangent bundle of $\mathbb{R}^n$, let $\mathbb{R}^n \times \mathbb{R}^n$. Introduce the bidifferential operators $P^r = P^r_\Gamma$ of maximum order r for each argument, defined by the following expression on each domain U of an arbitrary chart $\{x^i\}$:

(5-2) $$P^r(u,v)|_U = P^r_\Gamma(u,v)|_U = \Lambda^{i_1 j_1} .. \Lambda^{i_r j_r} \nabla_{i_1 .. i_r} u \nabla_{j_1 .. j_s} v \quad (u,v \in N)$$

We set $P^o(u,v) = uv$. For $r = 1$, we obtain the Poisson operator P.

Given a formal function $f(z)$ with constant coefficients such that $f(o) = 1$, substitute P^r to z^r in the expansion of $f(\nu z)$; we obtain a bilinear map $(u,v) \in N \times N \to u \star_\nu v = f(\nu P)(u,v) \in E(N;\nu)$. We wish to choose f so that we define thus a star-product. The answer is given by the following :

Proposition - If (W,Λ,Γ) is a flat symplectic manifold, there is a unique formal function of the Poisson bracket that generates a star-product : it is the exponential function.

We have :

(5-3) $$u \star_\nu v = \sum_{r=0}^{\infty} (\nu^r/r!)\, P^r(u,v) = \exp(\nu P)(u,v)$$

which generates the deformation of the Poisson Lie algebra $(\lambda = \nu^2)$:

(5-4) $$[u,v]_\lambda = \sum_{r=0}^{\infty} (\lambda^r/(2r+1)!)\, P^{2r+1}(u,v) = \nu^{-1} \sinh(\nu P)(u,v)$$

It is remarkable that, for $\nu = \hbar/2i$, we deduce from (5-4) a bracket $\frac{2}{\hbar}\sin(\frac{\hbar}{2}P)$ given in 1949 by Moyal in the context of the Hermann Weyl-Wigner quantization; (5-3) (resp. (5-4) is said to be the Moyal star product (resp. the Moyal bracket).

Consider in (5-4) the term P^3. If this Chevalley 2-cocycle were exact, it would be the coboundary of a 1-cochain that can be supposed 3-differential, according to § 2,b; but it is easy to see that such a coboundary has no term of bidifferential type (3,3). *The deformations (5-3),(5-4) are non trivial even for the order 1.*

c) The Moyal situation can be generalized in the following way : let Γ be an arbitrary symplectic connection; P and $P_\Gamma^2/2$ define always a $\star_\nu$-product at the order 2. If $u \in N$, denote by $\mathcal{L}(X_u)\Gamma$ the symmetric covariant 2-tensor deduced from the Lie derivative of Γ by the hamiltonian vector field X_u. The 2-cochain S_Γ^3 given by :

$$(5\text{-}5) \quad S_\Gamma^3(u,v)|_U = \Lambda^{i_1j_1}\Lambda^{i_2j_2}\Lambda^{i_3j_3}(\mathcal{L}(X_u)\Gamma)_{i_1i_2i_3}(\mathcal{L}(X_v)\Gamma)_{j_1j_2j_3}$$

is a Chevalley 2-cocycle ($\partial S_\Gamma^3 = 0$) of (N,P) which admits the same principal symbol as P^3. The same argument as the argument for P^3 in the flat case, shows that S_Γ^3 is *non exact* ; P, $P_\Gamma^2/2$, $S_\Gamma^3/6$ determine always a $\star_\nu$-product at the order 3. If we change the connection Γ , S_Γ^3 is changed by a coboundary. It follows that the Chevalley cohomology 2-class β of S_Γ^3 is independant of Γ and is an invariant of the symplectic structure. We can prove :

Proposition - The second space $H^2(N;N)$ of Chevalley cohomology admits as generators the class β and the classes defined by the *images by μ^{-1} of the closed 2-forms of W.*

We see that $H^2(N;N)$ depends upon β and the de Rham cohomology of W in dimension 2. We have a similar result for $H^3(N;N)$. In particular, each 1-differential Chevalley cocycle is the image by μ^{-1} of a closed form of W ; it is exact if the form is exact.

6 - THE VEY STAR-PRODUCTS

a) Introduce now the following notation : we denote by Q^r a bi-differential operator of maximum order r in each argument, null on the constants, satisfying the parity assumption and such that its principal symbol coincides with the principal symbol of P^r_Γ. We take in particular $Q^o(u,v) = uv$, $Q^1 = P$. We introduce the following :

Definition - A Vey star-product is a star-product of the form :

$$u *_\nu v = \sum_{r=0}^{\infty} (\nu^r/r!)\, Q^r(u,v) \tag{6-1}$$

A Vey Lie algebra is a formal Lie algebra given by a bracket of the form :

$$[u,v]_\lambda = \sum_{r=0}^{\infty} (\lambda^r/(2r+1)!)\, Q^{2r+1}(u,v) \tag{6-2}$$

My viewpoint differs from the viewpoint of Vey who considers essentially the Lie algebras. Consider a Vey $*_\nu$-product at the order

2; we can show that there is a unique symplectic connection Γ such that :

$$(6\text{-}3) \qquad Q^2 = P^2_\Gamma + \tilde{\partial} H$$

where H is a differential operator of maximum order 2. Suppose that Q^3 is a Chevalley 2-cocycle ($\partial Q^3 = 0$); I have proved similarly that there is a unique symplectic connection Γ such that :

$$(6\text{-}4) \qquad Q^3 = S^3_\Gamma + 3\,\partial H + T$$

where H is a differential operator of maximum order 2 and T a 2-tensor, image by μ^{-1} of a closed 2-form ; P, $Q^2/2$, $Q^3/6$ give a $*_\nu$-product at order 3 iff *the symplectic connections and the operators H of (6-3) and (6-4) coincide.*

b) We have the following on the domain U of an arbitrary chart :

Lemma - *Let C be an even (resp. odd) differential Hochschild 2-cochain which has exactly the bidifferential type (t,s) (with $s \leq t$, $t > 2$). If $s > 1$, the 3-cocycle $\tilde{\partial} C$ has effectively maximum bidifferential type (t,s-1) in u,w ; if s = 1, $\tilde{\partial} C$ has effectively maximum bidifferential type (t-1,1) in u,w .*

It is possible to prove :

Theorem - *Each star-product on a symplectic manifold (W,F) is equivalent to a Vey star-product.*

Sketch of the proof. We can suppose by equivalence that $C_2 = Q^2/2!$. If Q^3 corresponds to Q^2 by (6-3), (6-4), $C_3 - Q^3/3!$ is an odd

Hochschild 2-cocycle of type (1,1) and, changing the notation, we can suppose $C_3 = Q^3/3!$. Proceed by induction and suppose that, by equivalence, our $*_\nu$-product is now a Vey star-product of order $2q-1$ ($q \geqslant 2$). We have :

$$\tilde{\partial} C_{2q} = \tilde{E}_{2q}$$

and it is easy to deduce from the induction assumption that $\tilde{E}_{2q}$ has maximum bidifferential type $(2q,2q-1)$ in u,w.

Suppose that C_{2q} has a maximum bidifferential type (t',s) ; it follows from the lemma that it is impossible that $t' > 2q$ and $s > 1$. We see that either C_{2q} has maximum bidifferential type $(2q,2q)$ or maximum bidifferential type $(t',1)$.

We consider this last case ; $\tilde{\partial} C_{2q}$ has a maximum bidifferential type which is at most $(t'-1,s')$ and we have necessarily $2q < t'$. On the domain U of a natural chart, we set

$$C_{2q|U} = B_U + C_U$$

where B_U has exactly the type $(t',1)$ (with $t' \geqslant 2q+1$) and C_U the maximum order $(t'-1)$. We have :

$$\partial C_{2q+1} = E_q$$

where E_q has the maximum order $(2q+1)$. If C_{2q+1} has an order $> (2q+1)$ it follows from calculus of S. Gutt concerning the Chevalley 2-cocycles that the main part of C_{2q+1} is the main part of a Chevalley 2-cocycle and so of ∂A, where A is a differential operator.

If such is the case, transform the $*_\nu$-product by means of $T_\nu = Id - \nu^{2q} A$, which preserves C_r for $r \leqslant 2q-1$ and changes C_{2q} and C_{2q-1} so that we can suppose that C_{2q+1} has an order $\leqslant 2q+1$ and so that $\tilde{\partial} C_{2q+1}$ has an order $\leqslant 2q+1 \leqslant t'$ in u,w. The relation

$$\tilde{\partial} C_{2q+1} = \tilde{E}_{2q+1}$$

can be written

$$\tilde{\partial} C_{2q+1} = \tilde{F}_{2q+1} + M$$

where

$$M(u,v,w) = P(B_U(u,v),w) + B_U(P(u,v),w) - P(u,B_U(v,w)) - B_U(u,P(v,w))$$

$\tilde{F}_{2q+1}$ has an order $\leqslant t'$ in u,w and if $B_U \neq 0$, M has effectively order $(t'+1)$ in these arguments. If follows that $B_U = 0$. We see that C_{2q} has necessarily maximum bidifferential type $(2q,2q)$. Similarly C_{2q+1} has maximum bidifferential type $(2q+1,2q+1)$.

Introduce on U a flat symplectic connection Γ which defines the $P^r = P^r_\Gamma$. We have on U :

$$\tilde{\partial}(P^{2q}/(2q)!) = \tilde{E}_{2q}(P^r/r!) \qquad r \leqslant 2q-1$$

We have also :

$$\tilde{\partial} C_{2q} = \tilde{E}_{2q}(Q^r/r!)$$

It follows by difference that $\tilde{\partial}(C_{2q} - P^{2q}/(2q)!)$ has a maximum type in u,w inferior to (2q,2q-1), so that $C_{2q} - P^{2q}/(2q)!$ has a maximum type (2q,r) with $r < 2q$. Therefore $C_{2q} = Q^{2q}/(2q)!$ and similarly $C_{2q+1} = Q^{2q+1}/(2q+1)!$. Our theorem is proved.

c) We have obtained a general existence theorem for $*_\nu$-products. But it is also useful to know construction processes for such $*_\nu$-products. I will limit myself to the simplest examples. Consider the flat symplectic manifold defined by the cotangent bundle of the space $\mathbb{R}^n - \{0\}$, that is $E = (\mathbb{R}^n - \{0\}) \times \mathbb{R}^n$. The solvable group K of dimension 2 acts on E in the following way

$$(x,y) \in E = (\mathbb{R}^n - \{0\}) \times \mathbb{R}^n \to (x' = e^{\rho}x, y' = e^{-\rho}(y+\sigma x)) \in E \quad (\rho,\sigma \in \mathbb{R})$$

K preserves the symplectic structure of E and its flat symplectic connection ; it preserves thus the P^r defines by (5-2) and so the Moyal product on E. The space of the orbits of K in E is isomorphic to $T^* S^{n-1}$, where $S^{n-1} = SO(n)/SO(n-1)$ is the sphere of dimension (n-1). We obtain by quotient a natural Vey $*_\nu$-product on $T^* S^{n-1}$ which is invariant under SO(n). It is well-known that the regularized Kepler problem of dimension n admits as phase space the so-called *Moser manifold*, that is $T^*_o S^n$ (the cotangent bundle without the null-section). We have obtained for the Moser manifold an invariant natural $*_\nu$-product. We can deduce from this quotient process the construction of $*_\nu$-products for example for the cotangent bundle of classical groups or of Stiefel or Grassmann manifolds (in the real, complex or quaternion cases).

7 - INTRODUCTION TO A SPECTRAL THEORY AND QUANTIZATION

a) Come back to the flat symplectic manifold $\mathbb{R}^n \times \mathbb{R}^n$. Under suitable assumptions, Hermann Weyl has defined in this case, in terms of Fourier transforms, a map Ω (the Weyl map) which associates with each element u of a large class of classical functions or distributions, an operator $\hat{u}$ of Hilbert space and conversely. The conventional quantization processes in terms of such operators. But the Moyal product corresponds by Ω to the product of operators (for $\nu = \hbar/2i$). If

$$u \star v = \exp(h/2i)P(u,v) \quad \text{we have} \quad \Omega(u \star v) = \Omega(u) \cdot \Omega(v)$$

The Moyal bracket is (up to a constant factor) the image by Ω^{-1} of the natural commutator of these operators. We note that if the intersection of the supports of u,v is compact, we have :

$$\text{(7-1)} \qquad \int_W (u \star v)\eta = \int_W (u.v)\eta$$

where $\eta = F^n/n!$ is the symplectic volume element.

It appears as possible to develop directly Quantum Mechanics in terms of ordinary functions or distributions and star products, without reference to some Ω and to operators, in a complete and autonomous way.

b) Consider a symplectic manifold admitting *a* $\star_\nu$*-product*. We put $N^c = C^\infty(W;\mathbb{C})$ and suppose that ν is purely imaginary. If $u,v \in N^c$, the parity assumption on the star product can be translated by :

$$(7\text{-}2) \qquad \overline{u *_\nu v} = \overline{v} *_\nu \overline{u}$$

Let H be the hamiltonian of our problem. If we consider the value $\hbar/2i$ of the parameter of deformation suggested by the Moyal product, we are led to translate the quantum dynamical equation to

$$(7\text{-}3) \qquad \frac{du_t}{dt} = \frac{i}{\hbar}\, 2\nu\, [H, u_t]_{\nu^2} \qquad (u_t \in E(N^c;\nu) \times \mathbb{R})$$

If we put $\tilde{H} = i\, H/h$, we have :

$$(7\text{-}4) \qquad du_t/dt = \tilde{H} *_\nu u_t - u_t *_\nu \tilde{H}$$

Introduce the $*_\nu$-powers of $\tilde{H}(\tilde{H}^{(*)p} = \tilde{H}^{(*)p-1} *_\nu \tilde{H})$. It follows from the parity assumption that $\tilde{H}^{(*)p}$ depends only upon the even powers on ν. We can define the $*_\nu$-exponential of $\tilde{H}t$ in the following way :

$$(7\text{-}5) \qquad \operatorname{Exp}_*(\tilde{H}\, t) = \sum_{p=0}^{\infty} (t^p/p!)\, \tilde{H}^{(*)p}$$

If $u_o \in E(N^c;\nu)$, define u_t formally by :

$$(7\text{-}6) \qquad u_t = \operatorname{Exp}_*(\tilde{H}\, t) *_\nu u_o *_\nu \operatorname{Exp}_*(-\tilde{H}\, t)$$

(7-6) gives the formal solution of (7-4) taking the value u_o at $t = 0$.

c) Consider now *the viewpoint of the Mathematical Analysis* and give to ν the value $\hbar/2i$. Assume that H is such that, for t in a complex neighborhood of the origin, the right side of (7-5) converges to a distribution on W denoted by $\mathrm{Exp}_{\star}(\tilde{H}t)$ again. Suppose, for simplicity, that, for t in a neighborhood of the origin, $\mathrm{Exp}_{\star}(\tilde{H}t)$ has a unique Fourier-Dirichlet expansion :

(7-7) $$\mathrm{Exp}_{\star}(\tilde{H}t) = \sum_{\lambda \in I} \Pi_{\lambda}\, e^{\frac{i\lambda t}{\hbar}} \qquad (\Pi_{\lambda} \neq 0)$$

where I is a set of $\mathbb{C}$ and $\Pi_{\lambda} \in N^{c}$. This expansion is similar to the spectral expansion of an operator. We deduce from (7-7) :

(7-8) $$\sum_{\lambda \in I} \Pi_{\lambda} = 1 \qquad H \star \Pi_{\lambda} = \Pi_{\lambda} \star H = \lambda\, \Pi_{\lambda}$$

and (7-8) implies :

$$H = \sum \lambda\, \Pi_{\lambda} \qquad \Pi_{\lambda} \star \Pi_{\lambda'} = \delta_{\lambda\lambda'}\, \Pi_{\lambda}$$

where $\delta_{\lambda\lambda'} = 0$ for $\lambda' \neq \lambda$, $\delta_{\lambda\lambda'} = 1$ for $\lambda' = \lambda$. We see that I can be interpreted as the spectrum of H , Π_{λ} being the eigenprojector corresponding to $\lambda \in I$. The spectrum I and the Π_{λ} are characterized by (7-8) in a unique way. A star product is sayed to be *non degenerate* if, for any $u \in N^{c}$, $\bar{u} \star u = 0$ on a domain implies $u = 0$ on this domain. It follows from (7-1) that the Moyal product and the star products deduced by quotient are non degenerate. If such is the case, *the spectrum of each real-valued function admitting a spectral expansion in the sense of (7-7) is real and the corresponding* Π_{λ} *are real-valued.*

Define N_{λ} by :

$$N_{\lambda} = \int_{W} \Pi_{\lambda}\, \tilde{\eta}$$

where $\tilde{\eta} = \eta/(2\pi\hbar)^n$. If N_λ is finite, a normalized state ρ_λ of the system is defined by $\rho_\lambda = \Pi_\lambda/N_\lambda$ and N_λ is the multiplicity of the state in the usual sense of Quantum Mechanics.

d) More generally, we can introduce the Fourier transform in the sense of distributions :

$$\mathrm{Exp}_{\star}(\tilde{H}\,t) = \int e^{\frac{i\lambda t}{\hbar}}\, d\mu(\lambda)$$

and the support of the measure $d\mu$ will be referred as *the spectrum* of H. It is the spectrum of the distribution $\mathrm{Exp}_{\star}(\tilde{H}\,t)$ in the sense of Schwartz (up to the factor $\hbar/i$).

A *state* ρ is here a real-valued distribution (pseudo-probability) on the phase space, normalized by the condition $\int_W \rho\,\tilde{\eta} = 1$, and such that :

$$\rho \star \rho = (1/N)\,\rho$$

where N is the multiplicity. The measured value $< u >_\rho$ of the observable u at time t for the state ρ is :

$$< u >_\rho = \int_W (u_t \star \rho)\,\tilde{\eta} \tag{7-9}$$

This formula gives the Wigner formula in the case of the Moyal product.

e) The previous algorithm directly applied to the flat case gives, *for the n-dimensional harmonic oscillator*, the correct energy levels and multiplicities. *For the Hydrogen Atom*, we consider the Moser manifold $T^{\star}_{o} S_3$ as phase space and introduce , the corresponding natural star-product invariant under SO(4). We obtain then the complete spectrum, that is the negative discrete spectrum and the positive continuous spectrum.

8 - THE CASE WHERE $b_2(W) = 0$

a) For a symplectic manifold (W,F) such that $b_2(W) = 0$, F is exact and $H^2(N;N)$ admits only β as generator. Many general results can be deduced from the study of this particular case. We will show :

Proposition - For a symplectic manifold (W,F) such that $b_2(W) = 0$, all the star-products are equivalent.

Consider on W two star-products :

(8-1) $$u \star_\nu v = uv + \nu P(u,v) + \sum_{r=2}^{\infty} \nu^r C_r(u,v)$$

and

$$u \star'_\nu v = uv + \nu P(u,v) + \sum_{r=2}^{\infty} \nu^r C'_r(u,v)$$

We proceed by induction and suppose that, by transformation of (8-1), we have $C'_r = C_r$ for $r \leqslant 2q-1$ (with $q \geqslant 1$); the even Hochschild 2-cocycle $C'_{2q} - C_{2q}$ is exact so that :

$$C'_{2q} - C_{2q} = \tilde{\partial} A$$

where A is a differential operator. Transforming (8-1) by means of $T_\nu = Id + \nu^{2q}A$, we have $C'_r = C_r$ for $r = 1, \ldots, 2q$. If such is the case, $C'_{2q+1} - C_{2q+1} = T$ where T is an antisymmetric 2-tensor image by μ^{-1} of a closed 2-form that is exact; T is also exact in the Chevalley sense and there is a vector Z such that $T = \partial\mathcal{L}(Z)$. The equivalence given by $T_\nu = Id + \nu^{2q}\ \mathcal{L}(Z)$ preserves the C_r for $r \leqslant 2q$ and transforms C_{2q+1} in $C_{2q+1} + T = C'_{2q+1}$. We see that all star-products are equivalent.

b) Up to now, we have considered only equivalence operators which are defined by differential operators which are null on the constants. For the study of the formal Lie algebras, we are led to introduce equivalence operators which do not satisfy this condition, but for which the differential operators are constant on the constants. We say that we have *a weak equivalence*. Similarly, we consider associative deformations satisfying the parity assumption and for which the cochains of odd rank are null on the constants, the cochains of even rank being non necessarily null on the constants. Such a deformation is called *a weak star-product*. It is possible to show that a weak star-product has the following form :

(8-2) $$u \star_\nu v = k_{\nu^2}(u \,\hat{\star}_\nu\, v) \qquad k_{\nu^2} = 1 + \sum_{r=1}^{\infty} \nu^{2r} k_{2r} \qquad (k_{2r} \in \mathbb{R})$$

where $\hat{\star}_\nu$ is a star-product. Weak star-products generate by skew-symmetrization formal Lie algebras of the form :

(8-3) $$[u,v]_\lambda = P(u,v) + \sum_{r=1}^{\infty} \lambda^r\, C_{2r+1}(u,v)$$

where the C_{2r+1} are *null on the constants*. For such weak star-products, the uniqueness theorem is valid again. Moreover the argument of § 6,b shows that we have :

Theorem - Each weak star-product of (W,F) is equivalent to a weak Vey star-product.

For $b_2(W) = 0$, S. Gutt has proved :

Proposition - If $b_2(W) = 0$, all the formal Lie algebras of the form (8-3) for which $C_3 = Q^3/3!$ are weakly equivalent.

9 - LIE ALGEBRAS GENERATED BY A WEAK TWISTED PRODUCT AND VEY LIE ALGEBRAS

a) Consider, on an arbitrary manifold, a formal Lie algebra :

$$[u,v]_{\nu^2} = P(u,v) + \sum_{r=1}^{\infty} \nu^{2r} C_{2r+1}(u,v) \qquad (9-1)$$

where the C_{2r+1} are null on the constants. *Suppose that $C_3 = Q^3/3!$.* Let U be a contractible domain ($b_2(U) = 0$) of W and introduce the restriction to U of (9-1). This Lie algebra on $(U, F_{|U})$ is weakly equivalent to a Moyal Lie algebra on U generated by a Moyal product. It follows that our Lie algebra $\{C_{2r+1|U}\}$ on U is generated by a weak star-product $\{C_{2r(U)}, C_{2r+1|U}\}$. We set

$$C_{2r(U)}(u,v) = \bar{C}_{2r(U)}(u,v) + k_{2r(U)}uv \qquad k_{2r(U)} \in \mathbb{R}$$

where $\bar{C}_{2r(U)}$ is null on the constants. Consider two contractible domains U,V of W with $U \cap V \neq \emptyset$. We deduce from the uniqueness theorem of § 8 that we have on $U \cap V$:

$$\overline{C}_{2r(U)}(u,v) + k_{2r(U)}uv = \overline{C}_{2r(V)} + k_{2r(V)}uv$$

We see that $k_{2r(U)} = k_{2r(V)} = k_{2r}$ and that $\overline{C}_{2r(U)} = \overline{C}_{2r(V)}$ on $U \cap V$. Therefore there exist bidifferential operators $\overline{C}_{2r}$ null on the constants such that $\overline{C}_{2r(U)} = \overline{C}_{2r|U}$. The k_{2r} and the $\overline{C}_{2r}, \overline{C}_{2r+1}$ define on W a weak star-product that generates the given Lie algebra.

The weak star-product being equivalent to a weak Vey star-product, the Lie algebra (9-1), where $C_3 = Q^3/3!$, is equivalent to a Vey Lie algebra. It is the same for a Lie algebra (9-1) such that

$$C_3 = Q^3/3! + \partial A$$

where A is a differential operator null on the constants. We have

Theorem - 1) A formal Lie algebra (9-1) is equivalent to a Vey Lie algebra iff the Chevalley 2-cocycle C_3 is cohomologous to $Q^3/3!$.
2) A formal Lie algebra (9-1) is generated by a weak star-product iff it is equivalent to a Vey Lie algebra.

It follows from the argument of § 6 ,b that *a Vey Lie algebra is generated by a weak Vey star-product*. A Vey Lie algebra being given, there is a unique Vey Lie algebra deduced by product by a constant k_{ν^2} which is generated by a Vey star-product.

10 - EXISTENCE THEOREM

a) Consider a bilinear map $N \times N \to E(N;\nu)$ given by :

$$u\,v + \nu\, P(u,v) + \sum_{r=2}^{\infty} \nu^r\, C_r(u,v) \qquad (C_1 = P)$$

where the C_r $(r \geqslant 1)$ are differential 2-cochains which are null on the constants and satisfy the parity assumption. We have associated with such a map the 3-cochains :

$$\tilde{E}_t(u,v,w) = \sum_{r+s=t} (C_r(C_s(u,v),w) - C_r(u,C_s(v,w))) \quad (r,s \geqslant 1)$$

A direct calculus shows that we have :

Lemma - If t is odd (resp. even), $\tilde{E}_t$ is an even (resp. odd) Hochschild 3-cochain.

E being a Hochschild 3-cochain, denote by $\hat{\Sigma}E$ the 3-cochain defined by :

$$(10\text{-}1) \qquad (\hat{\Sigma}E)(u,v,w) = E(u,v,w) - E(v,u,w) - E(u,w,v)$$

It is easy to see that, if C is an even Hochschild 2-cochain, we have :

$$(10\text{-}2) \qquad \hat{\Sigma}\,\tilde{\partial}\, C = 0$$

b) Suppose $t = 2q+2$ $(q \geqslant 0)$. In the expression of $\tilde{E}_{2q+2}$, r,s are even together or odd together. In $\hat{\Sigma}\,\tilde{E}_{2q+2}$ the contribution of the terms corresponding to even r,s disappears and we obtain with

$r = 2r'+1,\ s = 2s'+1$:

$$\hat{\Sigma}\,\tilde{E}_{2q+2}(u,v,w) = 2\,S \sum_{r'+s'=q} C_{2r'+1}\,(C_{2s'+1}(u,v),w) \qquad (r',s'\geq 0)$$

where S is the summation after cyclic permutation on u,v,w. We put

(10-3)
$$D_q = S \sum_{r+s=q} C_{2r+1}(C_{2s+1}(u,v),w) \quad (r,s \geq 0)$$

The D_q are similar to the $\tilde{D}_q$ for the deformations of the Poisson Lie algebra : the C_{2r+1} define a deformation of this algebra at order q iff $D_t=0$ for $t = 1,\ldots,q$. We have :

(10-4)
$$\hat{\Sigma}\,\tilde{E}_{2q+2} = 2\,D_q$$

that is, according to (10-2) :

(10-5)
$$\hat{\Sigma}\,\tilde{D}_{2q+2} = 2\,D_q$$

The identities :

$$\hat{\Sigma}\,\tilde{D}_{2t+2} = 2\,D_t \qquad (t = 1,\ldots,q)$$

give the following :

Proposition - By skewsymmetrization of a star product at order 2q+2, we obtain a deformation of the Poisson Lie algebra at order q. In particular D_{q+1} is a Chevalley 3-cocycle.

c) Consider a star product of order 2. It can be written :

(10-6)
$$u\,v + \nu P(u,v) + \nu^2\, C_2(u,v)$$

where we can suppose $C_2 = (1/2)P^2_\Gamma$ for an arbitrary symplectic connection Γ. We will process by recursion. Choose *an odd integer* $t = 2q+1$ $(q \geqslant 1)$ and suppose that there is a star-product of even order $2q$. The Hochschild 3-cocycle E_{2q+1} is even and so exact ; there is an odd differential 2-cochain C_{2q+1}, null on the constants, such that :

(10-7) $$\tilde{E}_{2q+1} = \tilde{\partial} C_{2q+1}$$

Choose the odd cochain C_{2q+1}, *which is defined up to an odd bidifferential operator T of order 1*. We obtain a star-product of odd order $t = 2q+1$. Study the odd Hochschild 3-cocycle :

(10-8) $$\tilde{E}_{2q+2} = T^{(3)}_{2q+2} + \tilde{\partial} C^{(\ell)}$$

where $T^{(3)}_{2q+2}$ is given by an antisymmetric 3-tensor and where $C^{(\ell)}$ is an even two-cochain. *We will show that the alternate 3-cochain $T^{(3)}_{2q+2}$ is a Chevalley 3-cocycle*. It follows from (10-8) that we have :

$$3\, T^{(3)}_{2q+2} = \hat{\Sigma}\, \tilde{E}_{2q+2}$$

let, according to (10-4) :

(10-9) $$T^{(3)}_{2q+2} = (2/3)\, D_q$$

It follows from the proposition of b, for the order $2q$, that D_q is a Chevalley 3-cocycle and it is the same for $T^{(3)}_{2q+2}$. Such a cocycle is the image of a closed 3-form of W by μ^{-1} and it is exact if the 3-form is exact.

d) We search for a modification of C_{2q+1} such that $T^{(3)}_{2q+2}$ vanishs. If we change C_{2q+1} in $C_{2q+1}+T$, where T corresponds to an arbitrary antisymmetric 2-tensor, we have :

$$D_q \to D_q - \partial T \qquad \text{and} \qquad T^{(3)}_{2q+2} \to T^{(3)}_{2q+2} - (2/3)\,\partial T$$

and it is possible to annul $T^{(3)}_{2q+2}$ if this Chevalley 3-cocycle is exact.

It is certainly the case *if we suppose that* $b_3(W) = 0$. If such is the case $\tilde{E}_{2q+2}$ is an exact Hochschild 3-cocycle and we can choose an even 2-cochain $C_{2q+2} = C^{(\ell)}$ such that we obtain a star-product of order (2q+2). We have proved by induction the existence of a star-product.

We have :

Existence theorem - Each symplectic manifold (W,F) such that $b_3(W) = 0$ *admits Vey star-products.*

This theorem has been obtained in 1979, according to one of my suggestions, by O.M. Neroslavsky and A.T. Vlassov.

11 - INVARIANT STAR PRODUCTS

Let (W,F) be a symplectic manifold on which a Lie group acts effectively by symplectomorphisms. We know that if (W,F) admits a G-invariant linear connection, it admits a G-invariant symplectic connection . If such is the case, we have :

Lemma - Let C be a G-invariant exact Hochschild 2-cocycle (resp. 3-cocycle); C is the coboundary of a G-invariant 1-cochain (resp. 2-cochain) with a vanishing part of type 1 (resp. (1,1)) relatively to the connection Γ.

It follows :

Theorem - Let (W,F) be a symplectic manifold on which a compact connected Lie group G acts by symplectomorphisms. If there is a star-product on (W,F), there is a G-invariant star-product.

G being *compact* connected, this result can be applied to the symplectic homogeneous spaces G/H and to the cotangent bundles of a homogeneous space $T^*(G/H)$.

REFERENCES

|1| A. AVEZ et A. LICHNEROWICZ, C.R. Acad. Sci., t.275, A,(1972) p. 113.

|2| F. BAYEN, M. FLATO, C. FRONSDAL, A. LICHNEROWICZ, and D. STERNHEIMER, Ann. of Phys. t 111 (1978), p 61-110 et 111-151.

|3| M. FLATO, A. LICHNEROWICZ, D. STERNHEIMER, C. R. Acad. Sci. t.279, A, (1974) p.877; Compos. Matem. t.31 (1975) p.47-82; C.R. Acad. Sci., t.283, A, (1976), p.19.

|4| M. GERSTENHABER, Ann. of Math. t.79 (1964) p.59-60.

|5| S. GUTT, Lett. in Math. Phys., t.3, (1979), p.297-310.

|6| A. LICHNEROWICZ, Lett. in Math. Phys., t.3 (1979), p 495-502; Ann. di Matem. t.123, (1980), p.287-330.

|7| O.M. NEROSLAVSKY et A.T. VLASSOV, Sur les déformations de l'algèbre des fonctions d'une variété symplectique, (1979) en russe ; C.R. Acad. Sci.(à paraître).

|8| J.E. MOYAL, Proc. Cambridge Philos Soc. t.45 (1949), p.99-124.

|9| S. de GROOT, La transformation de Weyl et la fonction de Wigner : une forme alternative de la mécanique quantique. Presses de l'Université de Montréal (1974).

|10| J. VEY, Comm. Math. Helv. t.50 (1975), p.421-454.

|11| D. ARNAL et J.C. CORTET, J. Math. Phys. t.20 (1979), p.556-563.

LIE GROUP ACTIONS ON POISSON AND CANONICAL MANIFOLDS

C.-M. Marle
Université Pierre et Marie Curie
Paris, France

Lichnerowicz has defined Poisson manifolds (7) and canonical manifolds (5 to 8). Other works on the subject are (2), (3) and (12).

1. POISSON MANIFOLDS.

1.1. Definition

A Poisson structure on a manifold M is a $\mathbb{R}$-bilinear, skew-symmetric map, called a Poisson bracket and noted $(f,g) \mapsto \{f,g\}$, from $C^\infty(M, \mathbb{R}) \times C^\infty(M, \mathbb{R})$ into $C^\infty(M, \mathbb{R})$, such that:

1°) It is a derivation in each of its arguments:

$$\{f_1 f_2, g\} = f_1\{f_2, g\} + f_2\{f_1, g\} ,$$

$$\{f, g_1 g_2\} = \{f, g_1\}g_2 + \{f, g_2\}g_1 .$$

2°) It satisfies the Jacobi identity:

$$\{f,\{g,h\}\} + \{g,\{h,f\}\} + \{h,\{f,g\}\} = 0$$

A manifold M with a Poisson structure is called a Poisson manifold.

1.2. Examples

1°) A symplectic manifold (M,Ω), with the usual Poisson bracket, is a Poisson manifold.

2°) Let $\mathcal{G}$ be a finite-dimensional, real Lie algebra, $\mathcal{G}^*$ its

ISBN 0-12-068720-8

dual space, and Θ a $\mathbb{R}$-valued 2-cocycle of $\mathcal{G}$, i.e. a skew-symmetric bilinear 2-form on $\mathcal{G}$ such that, for all X, Y and Z $\in \mathcal{G}$:

$$\Theta([X,Y], Z) + \Theta([Y, Z], X) + \Theta([Z, X], Y) = 0 \quad .$$

Θ can be seen as a closed, left invariant differential 2-form on any Lie group G with Lie algebra $\mathcal{G}$. We set, for f and $h \in C^\infty(\mathcal{G}^*, \mathbb{R})$, $x \in \mathcal{G}^*$:

$$\{f, h\}_\Theta(x) = \langle x, [df(x), dg(x)]\rangle - \Theta(df(x), dg(x)) \quad .$$

This defines a Poisson structure on $\mathcal{G}^*$(Souriau [11]). For $\Theta = 0$, it is the Kirillov structure [4].

1.3. Proposition

Let M be a Poisson manifold.

1°) For $f \in C^\infty(M, \mathbb{R})$ fixed, there exists a unique vector field $^\# df$, called the Hamiltonian vector field with Hamiltonian f, such that for any $g \in C^\infty(M, \mathbb{R})$:

$$^\# df.g = i(^\# df)\, dg = \{f, g\} \quad .$$

2°) There exists a unique 2-times contravariant skew-symmetric tensor field Λ on M, called the structure tensor, such that:

$$\{f, g\} = \Lambda(df, dg) \quad , \qquad (f \text{ and } g \in C^\infty(M, \mathbb{R}) \;).$$

Proof. These properties are consequences of 1.1, 1°.

Lichnerowicz [7] has shown that property 2° of 1.1 is equivalent to $[\Lambda, \Lambda] = 0$, where $[\ ,\ ]$ is the Schouten bracket for skew-symmetric contravariant tensor fields.

1.4. Proposition

Let (M, Λ) be a Poisson manifold. The map $f \mapsto {}^\# df$ is a Lie algebra homomorphism with, as composition laws, the Poisson bracket on $C^\infty(M, \mathbb{R})$ and the ordinary bracket on the space $\mathcal{C}(M)$ of vector fields on M .

Proof. Using 1.1, 2° and the properties of Lie derivatives:

$^{\#}d\{f, g\}.h = i([^{\#}df, ^{\#}dg])\, dh$, for all f, g, h $\in C^{\infty}(M,\mathbb{R})$,

and therefore $^{\#}d\{f, g\} = [^{\#}df, ^{\#}dg]$.

When M is a symplectic manifold, the values of Hamiltonian vector fields at a given point x of M, fill up the entire tangent space T_xM. This is no longer the case for a Poisson manifold.

1.5. Definition

For each point x of a Poisson manifold (M, Λ), the characteristic direction space $\mathcal{F}_x$ at x is the subspace of T_xM:

$\mathcal{F}_x = \{df(x) \mid f \in C^{\infty}(M,\mathbb{R})\}$.

The map $x \mapsto \mathcal{F}_x$ is called the characteristic direction field. A vector field X is called characteristic if for all $x \in M$, X(x) is in $\mathcal{F}_x$. The space of characteristic vector fields is noted $\Gamma(\mathcal{F})$.

Note that the dimension of $\mathcal{F}_x$ is always even [9], and may depend on x. Therefore $\mathcal{F}$ is not always a vector sub-bundle of TM .

1.6. Proposition

The space $\Gamma(\mathcal{F})$ of characteristic vector fields on a Poisson manifold (M, Λ) is a Lie subalgebra of the space $\mathcal{C}(M)$ of all vector fields on M.

Proof. (See also Ouzilou[10]). In the domain of a chart, characteristic vector fields are linear combinations of the Hamiltonian vector fields $^{\#}dx^i$, which have as Hamiltonians the coordinate functions. One can then check, by using 1.4 and properties of the bracket, that the bracket of two characteristic vector fields is characteristic (see [9] for details).

1.7. Corollary

If the dimension of $\mathcal{F}_x$ is the same for all $x \in M$, then $\mathcal{F}$ is an integrable subbundle of TM. The corresponding foliation of M is called the characteristic foliation.

Proof. This results from the Frobenius theorem.

1.8. Definition

A sub-manifold N of a Poisson manifold (M, Λ) is an integral manifold of the characteristic direction field $\mathcal{F}$, if for all $x \in N$, $T_x N \subset \mathcal{F}_x$. When for all $x \in N$, $T_x N = \mathcal{F}_x$, N is said to be full.

1.9. Example

In Example 1.2, 2°, we define for all $x \in \mathcal{G}^*$, $X \in \mathcal{G}$, $ad^*_X x$ by:

$\langle ad^*_X x, Y\rangle = \langle x, -[X, Y]\rangle$ for all $Y \in \mathcal{G}$.

When considered as a linear function on $\mathcal{G}^*$, any $X \in \mathcal{G}$ is the Hamiltonian of a Hamiltonian vector field given by:

$$^{\#}dX\ (x) = -\ ad^*_X\ x - \Theta(X) ,$$

where $\Theta \in \mathcal{L}(\mathcal{G}, \mathcal{G}^*)$ is such that $\langle\Theta(X), Y\rangle = \Theta(X, Y)$ (X and $Y \in \mathcal{G}$).

Let G be a Lie group having $\mathcal{G}$ as Lie algebra, $\theta: G \to \mathcal{G}^*$ the symplectic 1-cocycle of G associated with Θ (Souriau [11]), and Ad^* the coadjoint representation. We recall that θ and Ad^* are such that, for $g, g_1, g_2 \in G$, $x \in \mathcal{G}^*$, $Y \in \mathcal{G}$, e neutral element:

$$\theta(g_1 g_2) = Ad^*_{g_1}\theta(g_2) + \theta(g_1) \quad ; \quad T_e\theta = \Theta \in \mathcal{L}(\mathcal{G}, \mathcal{G}^*) .$$

$$\langle Ad^*_g\ x, Y\rangle = \langle x, Ad_{g^{-1}}\ Y\rangle .$$

It can then be seen that full integral manifolds of the characteristic direction field of $(\mathcal{G}^*, \{\ \}_\Theta)$ are orbits of the affine action of G on $\mathcal{G}^*$: $(g, x) \longmapsto Ad^*_g\ x + \theta(g)$.

1.10. Proposition

Under conditions of Definition 1.8, let N be an integral manifold of $\mathcal{F}$. There then exists on N a unique closed 2-form Ω_N such that, for all $x \in N$, u and $v \in T_x N$:

$\Omega_N(x)\ (u, v) = \{f, g\}(x)$, if f and $g \in C^\infty(M, \mathbb{R})$ are such that $^\#df(x) = u$, $^\#dg = v$. For N full, Ω_N is a symplectic 2-form.

Proof. We notice that Ω_N is well defined, and by using 1.1,2° we see that Ω_N is closed, and that for N full, it is symplectic.

1.11. The Darboux theorem for Poisson manifolds.

Let (M, Λ) be a m-dimensional Poisson manifold, $x_0 \in M$ such that the dimension of $\mathcal{F}_x$ is constant, equal to 2p ($2p \leqslant m$), for all x in some neighbourhood of x_0. There then exists a chart of M whose domain contains x_0, its coordinates $x^1, \ldots x^m$ being such that:

$$\{f, g\} = \sum_{i=1}^{p} \left(\frac{\partial f}{\partial x^i}\frac{\partial g}{\partial x^{p+i}} - \frac{\partial f}{\partial x^{p+i}}\frac{\partial g}{\partial x^i}\right) \qquad (f \text{ and } g \in C^\infty(M, \mathbb{R})).$$

Proof. Apply the Frobenius theorem and the usual Darboux theorem to the leaves of the characteristic foliation, which by 1.10 are symplectic manifolds. See (7) for details.

The definition of a Hamiltonian vector field $^\#df$ with Hamiltonian f (1.3) is a particular case of the following proposition.

1.12. Proposition

Let (M, Λ) be a Poisson manifold. We define a linear map $\#$ from the space $A^1(M)$ of differential 1-forms on M, in the space $\mathcal{C}(M)$ of vector fields, by setting, whenever β is a 1-form:

$$i(^\#\beta)\ dg = \Lambda(\beta, dg) \qquad \text{for any } g \in C^\infty(M, \mathbb{R}) .$$

Then the kernel of $\#$ is the space $\mathcal{J}^1_{\mathcal{F}}$ of differential 1-forms β such that, for all characteristic vector fields X:

$$i(X)\beta = 0 .$$

The image of $\#$ is the space $\Gamma(\mathcal{F})$ of characteristic vector fields.

Proof. Partitions of unity enable us to study the problem locally. The proof then is similar to that for 1.6 .

1.13. Remark

Contrary to what occurs for a symplectic manifold, the map # is generally no longer an isomorphism. It can be extended into a homomorphism of the graded algebra A(M) of differential forms on M, into the graded algebra of contravariant skew-symmetric tensor fields on M. The kernel of this homomorphism is the ideal $\mathfrak{J}_{\mathfrak{F}}$ generated by $\mathfrak{J}^1_{\mathfrak{F}}$ (1.12). It may be seen that for any $\eta \in \mathfrak{J}_{\mathfrak{F}}$, $d\eta \in \mathfrak{J}_{\mathfrak{F}}$. The De Rham cohomology of M, therefore, induces a cohomology on A(M) quotiented by $\mathfrak{J}_{\mathfrak{F}}$, called the $\mathfrak{F}$-cohomology, which is equivalent to the G-cohomology studied by Lichnerowicz [7].

1.14. Definition

A p-form η on the Poisson manifold (M, Λ) is said $\mathfrak{F}$-closed if $d\eta \in \mathfrak{J}_{\mathfrak{F}}$ (1.13), i. e. if $d\eta(X_1, \ldots X_{p+1}) = 0$ for all $X_1, \ldots X_{p+1}$ $\Gamma(\mathfrak{F})$. A vector field X is said locally Hamiltonian , if there exists a $\mathfrak{F}$-closed 1-form β such that $X = \#\beta$.

2. CANONICAL MANIFOLDS.

2.1. Definition

A canonical manifold (M, Λ, T) is a Poisson manifold (M, Λ), with a map T: M $\to \mathbb{R}$, everywhere regular, such that for all $x \in M$, the kernel of T_xT is the characteristic direction space $\mathfrak{F}_x$.

2.2.Remark

Note that the dimension of a canonical manifold is always odd, 2n + 1, and that for all $x \in M$, the dimension of $\mathfrak{F}_x$ is 2n. For all $t \in \mathbb{R}$, $M_t = T^{-1}(t)$ is a full integral manifold of $\mathfrak{F}$, and, therefore, a symplectic 2n-dimensional manifold. Lichnerowicz [5, 6, 8] has shown that canonical manifolds are the best suited setting for the study of mechanical systems with time-dependent constraints and Hamiltonian, and of canonical transformations. The function T is interpreted as the time.

2.3. Example

Let (N, Ω_N) be a symplectic manifold, $M = \mathbb{R}\times N$, and T the first projection. Any $x \in M$ is a pair (t, y), with $t \in \mathbb{R}$, $y \in N$. We can calculate the Poisson bracket $\{f, g\}$ of two real functions f and g on M, by looking at t as a parameter, and using the Poisson structure of N defined by Ω_N (1.2, 1°). Such a canonical manifold, called a product canonical manifold, is used for the study of mechanical systems whose phase space does not depend on time.

3. AUTOMORPHISMS AND INFINITESIMAL AUTOMORPHISMS.

3.1. Definition

Let (M_1, Λ_1) and (M_2, Λ_2) be two Poisson manifolds. A map $\Phi: M_1 \to M_2$ is a Poisson morphism, if for all f and $g \in C^\infty(M_2, \mathbb{R})$:

$$\{f \circ \Phi, g \circ \Phi\}_{M_1} = \{f, g\}_{M_2} \circ \Phi \quad .$$

When $(M_1, \Lambda_1) = (M_2, \Lambda_2) = (M, \Lambda)$, and when Φ is a diffeomorphism of M onto itself, Φ is a Poisson morphism if and only if Φ^{-1} is a Poisson morphism. In such a case Φ is called an automorphism of the Poisson manifold (M, Λ).

3.2. Definition

An infinitesimal automorphism of a Poisson manifold (M, Λ) is a vector field X on M, such that for all f and $g \in C^\infty(M, \mathbb{R})$:

$$X.\{f, g\} = \{X.f, g\} + \{f, X.g\} \quad .$$

Note that X is an infinitesimal automorphism of (M,Λ) if, and only if, the Lie derivative $\mathcal{L}(X)\Lambda$ is zero. The space of infinitesimal automorphisms, therefore, is a Lie subalgebra of the Lie algebra $\mathcal{V}(M)$ of all vector fields.

3.3. Proposition

1°) A characteristic vector field X of the Poisson manifold (M, Λ) is an infinitesimal automorphism if and only if it is locally Hamiltonian (1.14).

2°) If X and Y are two infinitesimal automorphisms of (M, Λ), and if X is characteristic (and, therefore, by 1° is locally Hamiltonian), then $[X, Y]$ is Hamiltonian (1.3).

Proof. 1°) By 1.12, there exists a 1-form β such that $X = {}^{\#}\beta$. It can then be seen that for all f and $g \in C^{\infty}(M, \mathbb{R})$:

$$X.\{f, g\} - \{X.f, g\} - \{f, X.g\} = d\beta({}^{\#}df, {}^{\#}dg) \quad .$$

2°) If $X = {}^{\#}\beta$, and if Y is an infinitesimal automorphism, a straightforward calculation shows that, for β $\mathcal{F}$-closed, $[X, Y] = -{}^{\#}d\langle\beta, Y\rangle$. $[X,Y]$, therefore, is Hamiltonian.

We observe that on a Poisson manifold, infinitesimal automorphisms may exist, which are not characteristic vector fields, and which, therefore, are not locally Hamiltonian.

3.4. Definition

Let (M, Λ, T) be a canonical manifold.

1°) An automorphism of its canonical structure is a diffeomorphism Φ of M, which is an automorphism of the Poisson structure and moreover satisfies $\Phi \circ T = T$.

2°) An infinitesimal automorphism of its canonical structure is a vector field X, which is an infinitesimal automorphism of the Poisson structure, and which moreover satisfies $X.T = 0$.

Using 3.3, it can be seen that a vector field X on the canonical manifold (M, Λ, T) is an infinitesimal automorphism of the canonical structure if, and only if, its restriction to each symplectic submanifold $M_t = T^{-1}(t)$ is locally Hamiltonian.

4. LIE GROUP ACTIONS.

Let G be a Lie group which acts to the left on a manifold M. The action will be noted $(g, x) \mapsto g.x$. For each element X of the Lie algebra $\mathcal{G}$ of G, the vector field X_M on M, defined by:

$$X_M(x) = \frac{d}{dt} \exp(-tX).x|_{t=0} ,$$

is called the fundamental vector field associated with X.

4.1. Definition

Let G be a Lie group which acts to the left on a Poisson manifold (M, Λ) (resp. on a canonical manifold (M, Λ, T)). We say that this action preserves the Poisson structure (resp. the canonical structure) if each $g \in G$, looked at as a diffeomorphism of M, is an automorphism of the Poisson structure (resp. of the canonical structure).

4.2. Proposition

Under the conditions of Definition 4.1, if the action of G preserves the Poisson (resp. canonical) structure, then for each $X \in \mathcal{G}$, the associated fundamental vector field X_M is an infinitesimal automorphism of this structure. When G is connected, the converse is true.

Proof. It is the same as the corresponding proof for symplectic manifolds (Souriau [11]).

4.3. Proposition. (Symes [12], Ouzilou [10]).

Let G be a Lie group which acts to the left on a Poisson manifold (M, Λ), this action preserving the Poisson structure. We assume that the set P of orbits of this action is a differential manifold, and that the projection $\pi: M \to P$ is a submersion, such that for each $x \in M$, the kernel of $T_x\pi$ is the space $T_x(G.x)$ tangent to the orbit through x. Then there is on P a unique Poisson structure such that π is a Poisson morphism. P is called the reduced Poisson manifold.

Proof. Note that functions on P are in one-to-one correspondence with functions on M which are constant on each G-orbit.

4.4. Example (Symes [12])

Let G be a Lie group, T^*G its cotangent bundle, $q: T^*G \to G$ the canonical projection, α the Liouville 1-form, $\Omega_o = d\alpha$, and Θ a left-invariant closed 2-form on G. We set $\Omega = \Omega_o + q^*\Theta$. Then (T^*G, Ω) is a symplectic manifold, therefore also a Poisson manifold. G acts on itself by left translations, and the canonical lifting of this action to T^*G yields an action of G on T^*G, which preserves the symplectic structure as well as the associated Poisson structure. The space of orbits of this action is $\mathcal{G}^*$, the dual space of the Lie algebra $\mathcal{G}$ of G. The corresponding Poisson structure on $\mathcal{G}^*$ is that given in Example 1.2, 2°.

For a Hamiltonian action of a Lie group G on a symplectic manifold (M, Ω), Souriau [11] has defined the momentum map $J: M \to \mathcal{G}^*$. In the case of a Poisson manifold, a momentum map cannot exist unless the action is such that all fundamental vector fields are Hamiltonian. This restriction is stronger in the present case than in the case of a symplectic manifold, because infinitesimal automorphisms of a Poisson structure may exist, which are not even locally Hamiltonian (see part 3).

4.5. Definition

Under conditions of Definition 4.1, we say that the action of G is Hamiltonian if there exists a map $J: M \to \mathcal{G}^*$ such that, for each $X \in \mathcal{G}$, the corresponding fundamental vector field X_M is the Hamiltonian vector field $\#d\langle J, X\rangle$. The map J is called a momentum map of the group action.

4.6. Remark

Under conditions of Definition 4.5, we have for all X and Y elements of $\mathcal{G}$:

$$d(\{\langle J, X\rangle, \langle J, Y\rangle\} - \langle J, [X, Y]\rangle) \in \mathfrak{I}^1_{\mathcal{F}} .$$

When this quantity is zero for all X and $Y \in \mathcal{G}$, it may be seen that J is Ad^*-equivariant. But in the general case, contrary to what occurs for symplectic manifolds, it is not always possible to find an affine action of G on $\mathcal{G}^*$ which makes J equivariant. Instead we have:

4.7. Proposition

Under conditions of Definition 4.5, we set, for all $x \in M$, $g \in G$:

$$\theta(x, g) = J(g.x) - Ad^*_g J(x) .$$

Then for all fixed $g \in G$, the differential of the function $x \mapsto \theta(x, g)$ is an element of $\mathfrak{I}^1_{\mathcal{F}}$. The action of G on $M \times \mathcal{G}^*$:

$$(g, (x, \xi)) \mapsto (x, Ad^*_g \xi + \theta(x, g))$$

is such that $(Id_M, J) : M \to M \times \mathcal{G}^*$ is equivariant.

Proof. It is similar to the corresponding proof for symplectic manifolds (Souriau [11], Abraham and Marsden [1]).

4.8. Remark

In the case of a canonical manifold, the above proposition can be slightly improved: there exists an action of G on $\mathbb{R} \times \mathcal{G}^*$ which makes (T, J) equivariant [9].

4.9. Noether's theorem. (Souriau [11], Abraham and Marsden [1])

Under conditions of Definition 4.5, let $H: M \to \mathbb{R}$ be a G-invariant function: $H(g.x) = H(x)$ for all x M, g G.

Then J is constant along integral curves of the Hamiltonian vector field $^{\#}dH$.

Proof. It is similar to the corresponding proof for symplectic manifolds. For all $X \in \mathcal{G}$:

$$^{\#}dH.\langle J, X\rangle = \{H, \langle J, X\rangle\} = - {}^{\#}d\langle J, X\rangle.H = - X_M.H = 0 \quad .$$

4.9. Example

We consider again Example 4.4. The action of G on the symplectic manifold (T^*G, Ω) has as momentum map J :

$$J(z) = - Ad^*_{q(z)}(\pi(z)) - \theta(q(z)) \quad ,$$

where $(q, \pi) : T^*G \to G \times \mathcal{G}^*$ is the natural trivialization of T^*G by left translations. θ has been defined in 1.9. As (T^*G, Ω) is symplectic, there is an affine action of G on $\mathcal{G}^*$ which makes J equivariant. This action is:

$$(g, \xi) \mapsto Ad^*_g \xi - \theta(g) \quad .$$

We note that this action is not the same as the one obtained in 1.9, θ being replaced by $-\theta$.

References

1. R. Abraham and J. E. Marsden. Foundations of Mechanics. 2nd edition. The Benjamin/Cummings Publishing Company, Reading, Massachusetts, 1978.
2. M. Flato, A. Lichnerowicz, D. Sternheimer. Algèbres de Lie attachées à une variété canonique. J. Math. Pures et Appl. 54, 1975, p. 445-480.
3. A. Iacob, S. Sternberg. Coadjoint structures, solitons and integrability. Preprint, 1979.
4. A. Kirillov. Eléments de la théorie des représentations. Editions Mir, Moscou 1974.

5. A. Lichnerowicz. Variétés symplectiques et variétés canoniques. Trends in applications of pure mathematics to mechanics. Pages 249-261. Gaetano Fichera, ed. Pitman Publishing, London 1976.

6. A. Lichnerowicz. Variétés symplectiques, variétés canoniques et systèmes dynamiques. Topics in differential geometry. Pages 57-85. Academic Press, New York 1976.

7. A. Lichnerowicz. Les variétés de Poisson et leurs algèbres de Lie associées. J. of Diff. Geom. 12, 1977, p. 253-300.

8. A. Lichnerowicz. La géométrie des transformations canoniques. Preprint, 1979.

9. C.-M. Marle. Lie group actions on a canonical manifold. Journées Fermat, Toulouse, 5-7 mars 1981 (to be published)

10. R. Ouzilou. Actions hamiltoniennes sur les variétés de Poisson Journées Fermat, Toulouse, 5-7 mars 1981 (to be published)

11. J. M. Souriau. Structure des systèmes dynamiques. Dunod, Paris 1969.

12. W. W. Symes. Systems of Toda type, inverse spectral problems and representation theory. Inventiones Math., 59, p. 13-51, 1980.

GAUGE THEORIES AND GRAVITATION

Lecture Notes

by

W. Thirring

Institut für Theoretische Physik
der Universität Wien

Introduction.

In these notes I tried to give a conservative comparison between Maxwell's theory, non abelian gauge theories and general relativity. I concentrated on similarities and differences in the content on concepts related to physics. Even when trying to be mathematically economical and not going to the principal fibre bundle the basic notions and notations of differential geometry have to be used. Thus the first lecture collects the mathematical background, the second generalizes Maxwell's theory to Yang-Mills theories and the last compares this with the tedrad formulation of Einsteins theory. Clearly in these notes I have to concentrate on a few aspects. More details and applications can be found in the volume "Classical field Theory" [1] of my course on mathematical physics, where all of general relativity is developed in the language of differential forms.

DYNAMICAL SYSTEMS
AND MICROPHYSICS

ISBN 0-12-068720-8

1. ANALYSIS OF MANIFOLDS

1.1 Manifolds

A manifold M is a structure which looks locally like the euclidian space R^m, being mapped piecewise into R^m by coordinate systems (charts). Local processes like differentiation can therefore be carried over to M. The principle of general covariance is incorporated in this concept by accepting only notions which do not refer to a particular chart.

1.2 Tangent Bundle

Directions in $q \in M$ are given by trajectories (= mapping of an interval $I \to M$) through q. Representing a trajectory through q by its velocity vector (at this point) in a chart one can endow equivalence classes of trajectories with a linear structure. The latter turns out to be chart independent, its elements $\dot{q}$ form $T_q(M)$, the tangent space at q. $\bigcup_q T_q(M) =: T(M)$, the tangent bundle, has the structure of a manifold with a distinguished projection $\pi: T(M) \to M$, $(q,\dot{q}) \to q$, with $\pi^{-1}(q) = T_q(M)$ a vector space. ($T(M)$ is a vector bundle). A vector field specifies $\forall\, q \in M$ a direction, it is a mapping $X: M \to T(M)$ with $\pi \circ X = 1$ (= identity map on M). Integral curves of X are trajectories which follow at each point the direction prescribed by X, as expressed by the differential equation $\dot{q} = X(q)$. X defines the flow Φ_t, a one parameter group of maps $M \to M$, $q(0) \to q(t)$. For t fixed Φ_t is a diffeomorphism, i.e. a bijection with Φ_t and Φ_t^{-1} differentiable.

1.3 Tensors

The dual space $T_q^*(M)$ is the linear space of linear maps $T_q(M) \to R$. These maps are denoted by scalar products, $p \in T_q^*(M)$ maps $\dot{q}$ into $(p|\dot{q})$. The contangent bundle $T^*(M) = \bigcup_q T_q^*(M)$ has also the canonical projection $T^*(M) \to M$, $(q,p) \to q$ and a covector field ν maps $M \to T^*(M)$ with $\pi \circ \nu = 1$. The sets $\mathcal{T}_0^1(M)$ of vector fields and $\mathcal{T}_0^1(M)$ of covector fields form a module, i.e. a vector space where the scalars are from C(M) = smooth maps $M \to R$. The scalar product extends to the modules, it maps $T_1^0(M) \times T_0^1(M) \to C(M)$, $(\nu,X) \to (\nu|X)$. The tensor product $\nu_1 \otimes \nu_2$ and the wedge product $\nu_1 \wedge \nu_2 = \nu_1 \otimes \nu_2 - \nu_2 \otimes \nu_1$, $\nu_i \in T_1^0(M)$ are bilinear maps $T_0^1(M) \times \times T_0^1(M) \to R$, $(X,Y) \to (\nu_1 \otimes \nu_2 \,|(X,Y)) := (\nu_1|X)(\nu_2|Y)$. Bilinear maps have a natural linear structure, the modules formed from the linear combinations $t = \sum_{ij} t_{ij}\, \nu_i \otimes \nu_j$ (resp. $\nu_i \wedge \nu_j$) are denoted by $T_2^0(M)$ (resp. $E_2(M)$). Products containing T_0^1 or more factors are defined analogously, $\otimes$ and $\wedge$ being distributive and associative but not commutative. The totally antisymmetric tensors of rank p form the module $E_p(M)$ of p-forms, its elements are $\sum_{(j)} t_{j_1 \ldots j_p}\, \nu_{j_1} \wedge \nu_{j_2} \ldots \wedge \nu_{j_p}$ ($E_p(M)$ is C(M) for p = 0, $T_1^0(M)$ for p = 1 and {0} for p < 0 or > m). The scalar product can be extended to the inner product i: $E_p \times T_0^1 \to E_{p-1}$, $(\omega,X) \to i_X\omega$ by the rule

$$i_X(\nu_1 \wedge \nu_2) = (i_X\nu_1) \wedge \nu_2 + (-)^{p_1}\nu_1 \wedge i_X\nu_2,\ \nu \in E_{p_i},$$

and by linearity.

$g \in T_2(M)$ maps $T_0^1(M) \times T_0^1(M) \to C(M)$ but can also be considered as map $T_0^1 \to T_1^0$, $X \to gX$ with $(gX|Y) := (g|X,Y)$. g is called non degenerate if $(g|X,Y) = 0\ \forall\ Y \to X = 0$. In this case the map $T_0^1 \to T_1^0$ becomes a bijection. Let a non degenerate $g \in T_2(M)$ be given, then M is called pseudo-Riemannian (resp. Riemannian, resp. symplectic) if $(g|X,Y) = (g|Y,X)$ (resp. $(g|X,X) \geq 0$, resp. $(g|X,Y) = -(g|Y,X)$ and $dg = 0$). In these cases g extends the scalar product to a map $E_1(M) \times E_1(M) \to C(M)$: $\langle\nu|\omega\rangle := (\nu|g^{-1}\omega)$ and more generally the inner product to a map $E_p \times E_q \to E_{q-p}$ by the rules

1) $i_{\nu_1}\nu_2 = \langle\nu_1|\nu_2\rangle,\ \nu_i \in E_1$,

2) $i_\nu(\omega_1 \wedge \omega_2) = (i_\nu\omega_1) \wedge \omega_2 + (-)^{p_1}\omega_1 \wedge i_\nu\omega_2 \in E_{p_i},\ \nu \in E_1$,

3) $i_{\nu_1 \wedge \nu_2} = i_{\nu_2} \circ i_{\nu_1},\ \nu_i \in E_{p_i}$,

and by linearity.

1.4 Transformation Laws

A diffemorphism $\Phi: M_1 \to M_2$ induces a linear map $T_q(M_1) \to \to T_{\Phi(q)}(M_2)$ by transforming vectors the way Φ changes their representative trajectories. This induces naturally a map Φ': $\mathcal{T}^1_o(M_1) \to \mathcal{T}^1_o(M_2)$ and one defines the transformation Φ' for co-vectors such that the scalar product remains invariant: $(\Phi'\nu|\Phi'X) = = \Phi'(\nu|X) = (\nu|X)\circ\Phi^{-1}$. The definition of Φ' is extended to tensor fields by the rules

$$\Phi'(t_1 \overset{+}{\otimes} t_2) = \Phi'(t_1) \overset{+}{\otimes} \Phi'(t_2), \quad t_i \in \mathcal{T}^{r_i}_{s_i} \quad .$$

The pseudo-Riemannian (resp. symplectic) structure is preserved only for diffeomorphism such that $\Phi'g = g$. They are called isometries (resp. canonical transformation), for them $\Phi'(i_\nu\omega) = = i_{\Phi'\nu}\,\Phi'\omega$, $\nu \in E_p$, $\omega \in E_q$.

1.5 Derivatives

The tangent spaces at different points are not canonically oriented to each other, hence derivatives of tensors t are generally chart dependent. However, given a vector field X one can speak of the derivative of t in the direction of X: X creates a flow Φ_t and Φ'_t transforms tensors accordingly. The infinitesimal change

$$L_X t := \frac{d}{dt}\,\Phi'_{-t}\,t\big|_{t=0}$$

is called the Lie-derivative L_X: $\mathcal{T}^r_s \to \mathcal{T}^r_s$. From Φ' it inherits the properties

1) $L_X(t_1 + t_2) = L_X t_1 + L_X t_2$, $t_i \in \mathcal{T}^r_s$,

2) $L_X(t_1 \otimes t_2) = (L_X t_1) \otimes t_2 + t_1 \otimes L_X t_2, t_i \in \mathcal{T}^r_s$,

3) $L_X i_Y \omega = i_{L_X Y}\omega + i_Y L_X \omega$, $Y \in \mathcal{T}^1_o$, $\omega \in E_p$.

For the inner product as map $E_p \times E_q \to E_{q-p}$ the rule 3) holds only for those vector fields which conserve the pseudo-Riemannian (resp. symplectic) structure, i.e. $L_X g = 0$. They are called Killing (resp. Hamiltonian) vector fields. L_X is chart independent in the sense that for a diffeomorphism Φ one has $\Phi'(L_X t) = L_{\Phi' X}\, \Phi' t$.

In general $(L_X t)(q)$ does not only depend on the value of X at the point q but also on its derivative. Only for a function $f \in E_o$ this does not happen and hence $L_X f$ can be written $(df|X)$, $df \in E_1$. This map d: $E_o \to E_1$ can be generalized to $E_p \to E_{p+1}$ by the rules ($\omega_i \in E_{p_i}$)

1) $d(\omega_1 + \omega_2) = d\omega_1 + d\omega_2$,

2) $d(\omega_1 \wedge \omega_2) = (d\omega_1) \wedge \omega_2 + (-)^{p_1} \omega_1 \wedge d\omega_2$,

3) $d \circ d = 0$,

and is called the exterior derivative. It is chart independent, $\Phi' \circ d = d \circ \Phi'$ and hence $L_X \circ d = d \circ L_X$ also $L_{X|E_p} = i_X \circ d + d \circ i_X$.

1.6 The Hodge Duality Map

m independent covector fields $e^i \in E_1$, $i = 1 \ldots m$ are called a basis. They generate a nowhere vanishing m-form $\varepsilon := e^1 \wedge e^2 \wedge \ldots e^m$. Generally, a basis and thus this ε exist only locally. If there exists globally a nowhere vanishing $\varepsilon \in E_m$ the manifold is called orientable. If it also has a pseudo-Riemannian structure we may normalize ε by $i_\varepsilon \varepsilon = 1$ and define the *-isomorphism $E_p \to E_{m-p}$: $^*\nu = i_\nu \varepsilon$. $^*\nu$ is called the dual of ν and $\wedge$ is the dual of i in the sense that $^*(\omega \wedge \nu) = i_\omega {}^*\nu$.

1.7 Integration

The exterior derivative dx^i of local coordinates x^i provides a local basis and any $\omega \in E_m$ can be written $\omega = f\, dx^1 \wedge \ldots \wedge dx^m$, $f \in C(M)$. Defining for ω's with support in the domain of the chart $\int \omega := \int dx^1 \ldots dx^m f(x)$ one finds that this linear functional is invariant under diffeomorphisms $\Phi: x \to \bar{x}$ as Φ' multiplies f by the Jacobian Det $\partial x/\partial \bar{x}$. $\int \nu$, $\nu \in E_n$, over an n-dimensional submanifold N is defined by integrating as above the restriction $\nu_{|N}$ of ν to N. The latter is given in a chart where the image of N is $\{x: x^{n+1} = \ldots = x^m = 0\}$ by putting $dx^{n+1} = \ldots = dx^m = 0$ in

$$\nu = \sum_{(j)} \nu_{j_1 \ldots j_n} dx^{j_1} \wedge \ldots \wedge dx^{j_n} .$$

With this notation $\int$ inverts d in the sense of Stokes law

$$\int_N d\omega = \int_{\partial N} \omega, \qquad \omega \in E_{n-1}, \quad \partial N = \text{boundary of } N.$$

2. HAMILTONIAN SYSTEMS

2.1 Hamiltonian Vector Fields

Symplectic geometry has a simpler local structure than Riemannian geometry. There always is a chart (q_i, p_i) where the 2-form, which shall be called ω, is of the form $\omega = \sum_{i=1}^{m} dq_i \wedge dp_i$. In particular, on $T^*(M)$ this 2-form is canonically given in terms of the coordinates introduced in 1.3.

$$\Omega = \underbrace{\omega \wedge \omega \wedge \ldots \wedge \omega}_{\text{m-times}}$$

is a nowhere vanishing 2m-form and defines a volume measure on $T^*(M) \equiv$ phase space. Ω is invariant under canonical transformations too. Hamiltonian vector fields are also simpler than Killing vector fields, they are all of the form $X_H = i_{dH}\omega$, $H \in C(T^*(M))$. Their integral curves satisfy Hamilton's equations of motion with H(q,p) the Hamiltonian.

The flow generated by X_H is conveniently expressed in terms of the Poisson bracket $\{G,H\} = \{\omega|X_G,X_H\} = -\{H,G\} = L_{X_H} G$, $G,H \in C^1(T^*(M))$. It satisfies the Jacobi-identity $\{F,\{G,H\}\} + \{G,\{H,F\}\} + \{H,\{F,G\}\} = 0$; thus $\{F,H\} = \{G,H\} = 0 \Rightarrow \{\{F,G\}H\}=0$. Formally the flow of H can be written $\Phi_t = \exp tL_{X_H}$. From the volume preserving nature of Φ_t one draws some general conclusions. If a region of finite volume is time invariant then almost every orbit in it returns infinitely often to each neighbourhood of its points (Poincaré's Recurrence Theorem). Furthermore the set of orbits which come from infinity but then remain in a finite region is of measure zero (Schwarzschild's Capture Theorem).

2.2 Global Constants of Motion

If $\exists\, K_i \in C(M)$, $i = 1 \ldots n$, $\{K_i,H\} = 0$ and dK_i are independent on $\{N:(q,p) \in T^*(M),\ K_i(q,p) = \alpha_i \in R\}$ then the 2m-n-dimensional submanifold N is invariant under Φ_t. If $n = 2m-1$ the orbits are one-dimensional submanifolds. If $n = m$ and $\{K_i,K_j\} = 0$ the system is called integrable. If, furthermore, N is compact and connected then it is diffeomorphic to the m-dimensional torus and is in general covered densely by one trajectory. A small perturbation of H for an integrable system will generically destroy the constants but there remain enough time invariant tori to fill a large fraction of the volume of phase space (KAM-theorem).

2.3 Scattering Theory

If a flow Φ_t approaches for $t \to \pm\infty$ another flow Φ_t^0 in the sense that the Möller-transformation $\Omega_\pm = \lim_{t\to\pm\infty} \Phi_{-t} \circ \Phi_t^0$ exists then the two flows are diffeomorphic $\Omega_\pm \circ \Phi_t^0 = \Phi_t \circ \Omega_\pm$ and $H\circ\Omega_\pm = H_o$. In particular they have the same number of constants of the motion. If Φ_t^0 is the free motion then $\Omega_\pm$ maps the free trajectories into the trajectories of Φ_t and the latter has also the maximal number of constants. The scattering transformation $S = \Omega_+^{-1} \circ \Omega_-$ transforms the straight line which is tangent to a trajectory for $t \to -\infty$. It commutes with the free flow, $S\circ\Phi_t^0 = \Phi_t^0\circ S \leftrightarrow H_o\circ S = H_o$.

S transforms in particular the momenta for $t \to -\infty$ into those for $t \to +\infty$ and thus contains all the information about scattering.

3. MAXWELL'S EQUATIONS

Maxwell's equations restrict the electromagnetic fieldstrength $F \in E_2$ by

$$dF = 0, \quad d^*F = - {}^*J \tag{3.1}$$

or in integral form

$$\int_{\partial N_3} F = 0, \quad \int_{\partial N_3} {}^*F = - \int_{N_3} {}^*J \quad \forall\, N_3 = \text{compact, 3-dimensional} \tag{3.2}$$

Here $J \in E_1$ is the current and, because of 1.5, (3.1) implies the conservation law

$$d^*J = 0 \quad \text{or} \quad \int_{\partial N_4} {}^*J = 0 \quad \forall\, N_4 = \text{compact, 4-dimensional}\ . \tag{3.3}$$

Actually $d^*F = - {}^*J$ implies not only the conservation of the charge but also its vanishing in certain circumstances. For instance, if F is a periodic 2-Form (example: plane waves) and N_3 its periodicity volume then the total charge in it $\int_{N_3} {}^*J$ is zero since the contributions of opposite sides in $\int_{\partial N_3} F$ cancel. Furthermore, if we define a closed universe by the existence of N_3 = spacelike, compact 3-dimensional and without boundary then it has total charge zero as $\int_{\partial N_3} {}^*F = \int_{\emptyset} {}^*F = 0$.

$dF = 0$ implies locally $F = dA$ where $A \in E_1$ is determined up to the gauge transformation

$$A \to A + d\Lambda\ , \quad \Lambda \in E_0\ . \tag{3.4}$$

A allows the construction of a Lagrangian $L \in E_4$ such that (3.1) result from the requirement

$$\delta \int_{N_4} L = 0 \quad \forall\ \delta A|_{\partial N_4} = 0:$$

$$L = -\frac{1}{2}\, dA \wedge {}^*dA - A \wedge {}^*J \qquad (3.5)$$

If $d^*J \neq 0$ the functional $W(A) := \int_{N_4} L$ has no stationary point since under $A \to A + d\Lambda$, $d\Lambda|_{\partial N_4} = 0$ it changes linearly $W \to W + \int_{N_4} \Lambda \cdot d^*J$. To get consistent field equations one has to construct Lagrangians with additional symmetries which guarantee current conservation. For instance, if J is generated by 2 scalar fields $\phi_\alpha \in E_0$ then the covariant exterior derivative

$$(D\phi)_\alpha := d\phi_\alpha - \tau_{\alpha\beta} A \phi_\beta , \quad \tau = \begin{pmatrix} 0 & 1 \\ -1 & 0 \end{pmatrix} \qquad (3.6)$$

is invariant under the (space dependent) isospin rotation $\phi \to e^{\tau\Lambda}\phi$ when combined with the gauge transformation $A \to A + d\Lambda$. Therefore, also the Lagrangian

$$L = -\frac{1}{2}\, dA \wedge {}^*dA - \frac{1}{2}\, D\phi \wedge {}^*D\phi - \frac{1}{2}\, m^2\ \phi \wedge {}^*\phi \qquad (3.7)$$

has this invariance. This prevents W to increase linearly with Λ and thus $d^*J = 0$ follows also from second of the Euler equations:

$$d^*F = -{}^*J, \quad {}^*J = -\frac{\delta L}{\delta A} = \tau_{\alpha\beta}\ \phi_\alpha \wedge {}^*(D\phi)_\beta, \quad D^*D\phi = m^2\, {}^*\phi. \qquad (3.8)$$

The simplest way to see this formally is to consider the variation of L with ϕ

$$\delta L = \sum_\alpha \delta\phi_\alpha \left[\frac{\delta L}{\partial \phi_\alpha} - d\,\frac{\delta L}{\partial (d\phi_\alpha)}\right] + d\Big(\sum_\alpha \delta\phi_\alpha \frac{\delta L}{\partial d\phi_\alpha}\Big) \qquad (3.9)$$

The Euler equations require [...] = 0 and since L is invariant under $\phi \to e^{\Lambda\tau}\phi$ with $d\Lambda = 0$ we get to first order in Λ:
$d(\Lambda\phi_\alpha \, \tau_{\alpha\beta}(D\phi)_\beta) = \Lambda d^*J = 0$.

4. YANG-MILLS THEORIES

To generalize the O(2)-group of the previous example to more general Lie-groups G we formulate it abstractly as follows: Let G be represented (in matrix notation) by $\phi \to g\phi$, $\phi \in E_0$ and g a matrix valued 0-form, i.e. the group representative g being space-time dependent. Then

$$d\phi \to dg.\phi + g.d\phi \tag{4.1}$$

and the "covariant exterior derivative"

$$D\phi := d\phi + A\phi \to g\, D\phi \tag{4.2}$$

provided the Lie-algebra (= matrix) valued 1-Form A transforms as

$$A \to gAg^{-1} - dg \,.\, g^{-1} \tag{4.3}$$

However, in the non abelian case the exterior derivative of the inhomogenous member in (4.3) does not vanish. To get a "co-variant exterior derivative" DA which transforms according to the adjoint representation

$$DA \to g\,DA\,g^{-1} \tag{4.4}$$

we have to add to dA a term $A \wedge A$. This Lie-algebra valued 2-Form is to be understood as follows. Expand A in a basis λ_σ of the Lie-algebra, $A = A_\sigma \lambda_\sigma$, $A_\sigma \in E_1$, then $A \wedge A = A_\sigma \wedge A_\tau \; \lambda_\sigma \lambda_\tau =$
$= A_\sigma \wedge A_\tau \frac{1}{2} [\lambda_\sigma \lambda_\tau]$ and thus $A \wedge A$ vanishes only in the abelian case.

Now

$$dA \to dg \wedge Ag^{-1} - gA \wedge dg^{-1} + g\,dA\,g^{-1} + dg \wedge dg^{-1} \tag{4.5}$$

$$A \wedge A \to g\,A \wedge A\,g^{-1} - gA\,g^{-1} \wedge dg.g^{-1} - dg \wedge Ag^{-1} + dg\,g^{-1} \wedge dg\,g^{-1}$$

and since $g.g^{-1} = 1$ implies $dg^{-1} = -\,g^{-1}\,dg\,g^{-1}$ we see that

$$DA := dA + A \wedge A \tag{4.6}$$

has the transformation property (4.4). We can now generalize electrodynamics by forming the gauge invariant Lagrangian

$$L = -\frac{1}{2}\,\mathrm{tr}\,DA \wedge {}^*DA - \frac{1}{2}\,(D\phi,\ {}^*D\phi) - \frac{1}{2}\,m^2(\phi,{}^*\phi) =$$

$$= L^{\text{field}} + L^{\text{matter}} \tag{4.7}$$

where $(\phi,\phi) = (g\phi,\ g\phi)$ is a bilinear form invariant under the group action and tr the trace in the Lie-algebra. The field equations are again of the general form

$$d^*F = -\,{}^*J, \quad F = DA, \quad -{}^*J = \frac{\partial L}{\partial A}, \quad D^*D\phi = m^2\,{}^*\phi\ . \tag{4.8}$$

But to look at it more in detail we have to employ suitable bases. For the Lie-algebra we shall use λ's which diagonalize tr, $\mathrm{tr}\,\lambda_\sigma\lambda_{\sigma'} = \delta_{\sigma\sigma'}$ and let C be the structure constants in this basis: $[\lambda_\sigma,\lambda_{\sigma'}] = C_{\sigma\sigma'\tau}\lambda_\tau$. Furthermore, choose components of ϕ such that $(\phi,\phi) = \sum\limits_\alpha \phi_\alpha{}^2$. Then

$$L^{\text{field}} = -\frac{1}{2}\,F_\sigma \wedge {}^*F_\sigma$$

$$F_\sigma = DA_\sigma = dA_\sigma + \frac{1}{2}\,C_{\alpha\beta\sigma}\,A_\alpha \wedge A_\beta \tag{4.9}$$

and if we call J the colour current and A the gluon field then there is a gluon contribution

$$J_\alpha^{\text{field}} = -\frac{\partial L^{\text{field}}}{\partial A_\alpha} = C_{\alpha\beta\sigma}\,A_\beta \wedge {}^*F_\sigma \tag{4.10}$$

to the total colour current

$$^*J_\alpha = C_{\alpha\beta\sigma} A_\beta \wedge {}^*F_\sigma + (\lambda_\alpha)_{\beta\sigma}\, \phi_\beta \wedge {}^*(D\phi)_\sigma \tag{4.11}$$

The Yang-Mills-equation $d^*F_\alpha = -{}^*J_\alpha$ requires again not only current conservation $d^*J_\alpha = 0$ but also that the total colour charge vanishes in a closed universe or for plane wave F's. In the same way as in §3 we see that (4.8) is consistent in the sense that $d^*J = 0$ also results from the equations of the matter field ϕ. For this to become evident consider in (4.1) an infinitesimal constant gauge transformation $g = 1 + \Lambda_\sigma \lambda_\sigma$, $\Lambda_\sigma \to 0$ and $d\Lambda_\sigma = 0$. According to (4.3) A changes by $\delta A_\sigma = \Lambda_\rho A_\tau\, C_{\rho\tau\sigma}$ and since $\delta L = 0$ the generalisation of (3.9) tells us that $d(\Lambda_\rho (C_{\rho\tau\sigma} A_\tau \wedge {}^*F_\sigma + (\lambda_\rho)_{\alpha\beta}\, \phi_\alpha \wedge {}^*(D\phi)_\beta)) = 0$ provided (4.8) is satisfied. However, there is an essential difference to Maxwells theory inasmuch as J_σ does not transform according to the adjoint representation of G, i.e. under a gauge transformation we do not have $J \to g\, J g^{-1}$ but rather an inhomogenous transformation law. The reason is that the gluon contribution contains A. As a consequence, at any point the colour current can be transformed to zero. Actually, a homogenous transformation law would not be consistent with (4.8): Although ${}^*F \to g\, {}^*F g^{-1}$, $d^*F \not\to g\, d^*F\, g^{-1}$ but only the covariant exterior derivative

$$D^*F_\sigma = d^*F_\sigma + C_{\alpha\beta\sigma}\, A_\alpha \wedge {}^*F_\beta \tag{4.12}$$

transforms like $D^*F \to g\, D^*F\, g^{-1}$. We note that the additional part in D^*F is just the gluon current. Thus the Yang-Mills equation (4.8) can be written alternatively

$$D^*F = -{}^*J^{\text{matter}}, \qquad {}^*J^{\text{matter}} = -\frac{\delta L^{\text{matter}}}{\delta A} \tag{4.13}$$

where now both sides transform homogenously. However, J^{matter} alone is not conserved, $d^*J^{\text{matter}} \neq 0$. Now we generalize (4.12) with $[A \wedge V] = A \wedge V - (-)^p V \wedge A$ to

$$DV = dV + [A \wedge V], \ [A \wedge V]_\gamma := A_\alpha \wedge V_\beta \ C_{\alpha\beta\gamma} \tag{4.14}$$

for the covariant derivative of p-forms V which transform according to the adjoint representation of G. Note that DA had to be defined differently since A has the transformation law (4.6). But for F (4.14) is applicable and we have

$$DF = d(dA + A \wedge A) + A \wedge (dA + A \wedge A) - (dA + A \wedge A) \wedge A = 0 \tag{4.15}$$

Furthermore we deduce from (4.14)

$$DDV = [F \wedge V], \quad [F \wedge V]_\gamma = F_\alpha \wedge V_\beta \ C_{\alpha\beta\gamma} \tag{4.16}$$

and in particular $DD^*F_\gamma = F_\alpha \wedge {}^*F_\beta \ C_{\alpha\beta\gamma} = 0$ since according to (1.6) $-{}^*(F_\sigma \wedge {}^*F_{\sigma'}) = i_{F_\sigma} F_{\sigma'} = i_{F_{\sigma'}} F_\sigma$. Thus we have the gauge covariant version of (4.8)

$$DF = 0, \quad D^*F = -{}^*J^{matter} \Rightarrow D^*J^{matter} = 0 \tag{4.17}$$

as counter part to Maxwell's equation.

5. GRAVITATION

In gauge theories of gravitation one gauges a space-time group rather than an internal group as in §3. Therefore the analogy to §3 is not perfect and various groups, like the translation, Lorentz, Poincaré and de-Sitter groups have been tried [2] out. We shall follow here a proposal by Cho[3] and Freund and work with the translations. The physical reason for this choice is that one knows that gravity is coupled to energy and momentum and these are the generators of translations. One could think that this leads to a trivial theory since translations are an abelian group and the field we consider transform trivially, i.e. for $x \to x' = x + \Lambda$ one has $\phi(x') = \phi(x + \Lambda)$. However, by gauging the group and thus allowing for an x-dependent Λ one gets the group of all diffeomorphism. In particular the basis dx^i of the cotangent space of R^4 does no longer stay invariant but $dx^i \to dx^i + d\Lambda^i$. One now considers a basis $e^i \in T^0_1(M)$ as the gauge potentials of this theory and requires invariance under local rotations $e^i \to L^i{}_k\, e^k$ of the basis, $dL^i{}_k \neq 0$. So far we did not mention a Riemannian structure and it is economical to incorporate the information given by the metric into the e^i's by using an orthogonal basis, i.e. $g = e^i \otimes e_i$, $e_i := \pm e^i$, - for $i = 0$, + for $i = 1,2,3$. L is then restricted to be a local Lorentz transformation, $L^i{}_k\, L^k{}_j = \delta^i_j$. We now have the task of constructing an action which is invariant under both, general coordinate transformations and $e^i \to L^i{}_k\, e^k$. Actually, $\int L_m$ with $L_m = L$ from (3.7) has already this invariance and the only problem is the gravitational part L_g involving the e^i's. It turns out that

$$L_g = \frac{1}{2}\,(de^i \wedge e^k) \wedge {}^*(de_k \wedge e_i) - \frac{1}{4}\,(de^i \wedge e_i) \wedge {}^*(de^k \wedge e_k) \qquad (5.1)$$

leads to an invariant action since $L_g + d(e_k \wedge {}^*de^k)$ is invariant under local Lorentz transformations. The verification of this claim is somewhat lengthy and is given in an appendix. The Euler equations resulting from $L_m + L_g$ have the same structure as (3.8)

$$\begin{aligned} d{}^*F_k &= -\,{}^*J_k \\ {}^*F_k &= \frac{\delta L_g}{\delta de^k} = e^i \wedge {}^*(de_i \wedge e_k) - \frac{1}{2}\, e_k \wedge {}^*(de^i \wedge e_i) \\ -\,{}^*J_k &= \frac{\delta L_g}{\delta e^k} + \frac{\delta L_m}{\delta e^k} \end{aligned} \qquad (5.2)$$

$$\frac{\delta L_g}{\delta e^k} = de^j \wedge i_{e_k}\, {}^*F_j - i_{e_k}\, L_g$$

$$\frac{\delta L_m}{\delta e^k} = dA \wedge i_{e_k}\, {}^*dA + d\phi \wedge i_{e_k}\, {}^*d\phi - i_{e_k}\, L_m$$

Apart from the trivial difference that we now have 4 equations since we have 4 e^i's, there are 2 features not encountered in Maxwell's theory but in non abelian gauge theories.

1) There is a contribution to J_k from L_g
2) Although the action and hence (5.2) are invariant under $e^i \to L^i{}_k e^k$ if $dL^i{}_k \neq 0$ J_k does not transform as $J^i \to L^i{}_k J^k$ but inhomogenously.

The physical interpretation of these facts is as follows. J_k are the currents of energy and momentum, in fact the matter part is just the sum of the usual energy-momentum tensors of a scalar and vector field. They are constructed like the Hamiltonian $H = \dot{q}p - L$ if $\dot{q} \to dA$ (or $d\phi$) and $p \to {}^*dA$ (or ${}^*d\phi$). There is a gravitational contribution of the same structure. (In the natural basis it would be the Landau-Lifshitz pseudotensor). $d^*F_k = -{}^*J_k$ does not only require the conservation law $d^*J_k = 0$ but again that the total energy and momentum of a closed universe are zero. Furthermore, if there were a strictly periodic solution for the ${}^*F_k$ (plane gravitational wave) then it would not carry energy or momentum. The fact 2) is usually expressed by saying that the gravitational contribution of energy cannot be localized, it can be transformed to zero at a given point.

The equations (5.2) are an alternative way of writing Einsteins equation. They bear the same relation to them as (4.8) to (4.17): Since the matter part of J_k transforms homogenously under local Lorentz transformations one obtains an invariant version by putting the gravitational contribution to the left hand side. In this form gravitation does not contribute to energy and momentum but the latter are not conserved but a covariant exterior derivative of them vanishes (Bianchi identity).

APPENDIX

To show that L_g in (5.1) leads to an invariant action under local Lorentz transformations, it is sufficient to prove the infinitesimal version $\delta L_g \equiv$ exact (δe) for $\delta e_i := \varepsilon_{ij}(x)e^j$, $\varepsilon_{ij} = -\varepsilon_{ji}$ and $\delta \langle e_i, e_j\rangle \equiv \delta\eta_{ij} \equiv 0$. To this end, use (5.2) to write L_g as $L_g = \frac{1}{2}\, de^i \wedge {}^*F_i$ and $\delta L_g \equiv \delta e_i \wedge \delta L_g/\delta e_i + d(de_i \wedge {}^*F^i)$ for <u>any</u> variation of the basis. Obviously, L_g is invariant under <u>global</u> Lorentz transformations ε_{ij} = const. $\forall$ i,j, thus $e_{[i} \wedge \delta L_g/\delta e_{j]} \equiv -d(e_{[i} \wedge {}^*F_{j]}$ and therefore $\delta L_g \equiv -d\,\varepsilon_{ij}\wedge e^{[i}\wedge{}^*F^{j]}$ if $\varepsilon_{ij} = \varepsilon_{ij}(x)$. A simple calculation using (5.2) leads to $e^{[i} \wedge {}^*F^{j]} \equiv -\frac{1}{2}\, d^*e^{ij}$, wherefore $\delta L_g \equiv \frac{1}{2}\, d\varepsilon_{ij} \wedge d^*e^{ij} \equiv$ $\equiv -\frac{1}{2}\, d(d\varepsilon_{ij} \wedge {}^*e^{ij})$ which completes the proof. Note that $\delta(e_i \wedge {}^*de^i) \equiv e_i \wedge {}^*(d\varepsilon^{ij} \wedge e_j) \equiv \frac{1}{2}\, d\varepsilon^{ij} \wedge {}^*e_{ij}$ and therefore $\delta[L_g + d(e_i \wedge {}^*de^i)] \equiv 0$.

REFERENCES

[1] W. Thirring, A Course in Mathematical Physics, Springer New York, Wien (1978).

Y. Choquet-Bruhat, C. DeWitt-Morette, M. Dillard-Bleick: Analysis, Manifolds and Physics, North Holland 1978.

[2] For further Review see

W. Thirring, Acta Phys. Austr. Suppl. XIX, 1978.

[3] Y.M. Cho, Einstein Lagrangian as the Translational Yang-Mills Lagrangian, Phys. Rev. <u>D14</u> (1976) 2521.

Y.M. Cho, Gauge Theory of Poincaré Symmetry, Phys. Rev. <u>D14</u> (1976) 3335.

RIEMANNIAN GEOMETRY AND MECHANICS : THE KEPLER PROBLEM.

Pham Mau Quan
Département de Mathématiques
Université Paris Nord

1. THE JACOBI SYSTEM.

Many problems of classical mechanics can be solved within the framework of Riemannian geometry. This is the case for Jacobi systems defined by any triplet (M,g,V) where M is the configuration space, g the quadratic form associated to the kinetic energy and V the potential of forces. The motion of such a system derives from the Lagrangian $L = \frac{1}{2}|\dot{x}|^2 - V$. L is independent of the time, one has the energy integral $E = \frac{1}{2}|\dot{x}|^2 + V$. It is known that the motions of constant energy $E = h$ are defined by the Lagrangian

$$L_h = \sqrt{2(h-V)}\ |\dot{x}| \qquad \dot{x} = \frac{dx}{du} \tag{1.1}$$

u is any arbitrary parameter. The time $x^o = t$ will be given by

$$\dot{x}^o = \frac{\partial L_h}{\partial h} = \frac{L_h}{2(h-V)} . \tag{1.2}$$

(1.1) leads one to introduce the Riemannian manifold (M_h , g_h) where $M_h = \{x \in M \mid h - V > 0\}$ and $g_h = 2(h - V)g$. The trajectories of energy value h are then the geodesics of (M_h , g_h). The

DYNAMICAL SYSTEMS
AND MICROPHYSICS

ISBN 0-12-068720-8

Maupertuisian action defined by L_h is equal to the length of the geodesic and the time is given by the dynamical path (1.2).

Actually, by investigating the geometry of the Riemannian manifold (M_h, g_h), one can obtain the properties of the motion of any Jacobi system. In this paper, we make such a study of the Kepler problem.

2. THE KEPLER PROBLEM

By the Kepler problem we mean the study of the motion of one particle of mass 1 under the action of the potential of forces $V = -\frac{1}{|x|}$ in the Euclidean space $\mathbb{R}^n$. For n = 3, this is the reduced form of the two bodies problem in celestial mechanics or of the hydrogen atom problem in quantum mechanics.

The motion on each energy surface $E = \frac{1}{2}|\dot{x}|^2 - \frac{1}{|x|} = h$ is defined by the geometry of the Riemannian manifold (M_h, g_h) where $M_h = \{ x \in \mathbb{R}^n \mid x \neq 0,\ h + \frac{1}{|x|} > 0 \}$ and $g_h = 2(h + \frac{1}{|x|})g$, g is the Euclidean metric of R^n. The orbits on each energy surface are the sole geodesics of this Riemannian manifold.

In order to interpret the collisions, one must include in M_h the point x = 0 and extend the metric g_h to this point. Nevertheless, the extended metric $\bar{g}_h$ has one singularity in x = 0. Fortunately, this singularity is integrable in the sense that the length of any curve joining any point of M_h to x = 0 is defined and bounded. For h > 0 *(hyperbolic case)* and h = 0 *(parabolic case)* we obtain one Riemannian manifold $(\bar{M}_h, \bar{g}_h)$ homeomorphic to R^n, the metric of which presents one integrable singularity in $x = 0$. For $h<0$ *(elliptic case)* M_h has, in addition to this singularity $x=0$, a boundary ∂M_h on which $g_h = 0$. We can identify ∂M_h (the sphere of radius $a = \frac{1}{|h|}$ in the chart) with one unique point ω', so that we obtain a compactification of M_h. Thus we have one Riemannian manifold $(\bar{M}_h, \bar{g}_h)$ homeomorphic to S^n , the metric $\bar{g}_h$ of which presents two isolated integrable singularities at 0 and ω'.

One can point out that in all the cases, $(\bar{M}_h, \bar{g}_h)$ is complete with respect to the Riemannian distance, hence geodesically complete. In the hyperbolic and parabolic cases, the geodesics are not closed. In the elliptic case, the geodesics are closed and have the same length equal to the Maupertuisian action $W_h = \oint L_h$ $= \pi\sqrt{2}|h|^{-1/2}$ and the motion is periodic with the period $T = \frac{dW}{dh} = \frac{\pi}{\sqrt{2}}|h|^{-3/2}$.

3. CURVATURE AND MOTION

The Riemannian metric $\bar{g}_h$ is conform with the Euclidean metric g . A standard calculation results in the Riemannian curvature

$$R_{ij,kl} = e^{-4f}\{C_{ik}g_{jl} - C_{il}g_{jk} + g_{ik}C_{jl} - g_{il}C_{jk} - \Delta_1 f(g_{ik}g_{jl} - g_{il}g_{jk})\}$$

where $e^{2f} = 2(h+\frac{1}{|x|})$ and $C_{ij} = \nabla_i\partial_j f - \partial_i f\partial_j f$, $\Delta_1 f = g^{ij}\partial_i f\partial_j f$ are evaluated in the Euclidean metric of R^n. This permits the deduction of the Ricci curvature R_{ij} and the scalar curvature R

$$R_{jk} = \frac{e^{-6f}}{r^4}\{[3(n-2) + 2(2n-5)hr]g_{jk} - 3(n-2)(1+2hr)n_j n_k\}$$

$$R = \frac{n-1}{r^4}e^{-6f}[3(n-2) + 4(n-3)hr]$$

where $r = |x|$ and $n_i = x_i/|x|$.

The Riemannian curvature $R_{ij,kl}$ defines at each point $x \in \bar{M}_h$ one symmetric bilinear form $S_x : \Lambda^2 T_x\bar{M}_h \times \Lambda^2 T_x\bar{M}_h \to \mathbb{R}$, $\Lambda^2 T_x\bar{M}_h$ is given with the induced metric $g_{ij,kl} = g_{ik}g_{jl} - g_{il}g_{jk}$. It is the sectional curvature K which influences the behaviour of the geodesics, hence of the motion. At $x \in M$, the sectional curvature is given in the direction of the 2-plane P defined by a pair of orthonormal vectors (u,v)

$$K(P) = S_x(u\wedge v, u\wedge v) = R_{ij,kl}(u^i v^j - u^j v^i)(u^k v^l - u^l v^k) .$$

With a view towards determining its signature, we choose one orthonormal frame $(u_1,\ldots,u_n)$ such that $u_1 = \frac{x}{|x|}$ and research the eigen-values of $(R_{ij,kl})$ relatively to $(g_{ij,kl})$. It is easy to show that $u_1 \wedge u_\alpha$, $u_\alpha \wedge u_\beta$, $\alpha,\beta = 2,\ldots,n$, are eigen-vectors associated to the eigen-values

$$\rho_{1\alpha} = -\frac{4}{r^4}\, e^{-6f} hr \qquad \rho_{\alpha\beta} = -\frac{1}{r^4}\, e^{-6f}\left[1 + 4(1+hr)\right]$$

One deduces

Theorem 3.1 :

1) $h < 0 \Leftrightarrow$ S is semi-definite with $\rho_{1\alpha} > 0$, $\rho_{\alpha\beta} < 0$

2) $h = 0 \Leftrightarrow$ S is degenerate with $\rho_{1\alpha} = 0$, $\rho_{\alpha\beta} < 0$

3) $h > 0 \Leftrightarrow$ S is negative definite with $\rho_{1\alpha} < 0$, $\rho_{\alpha\beta} < 0$

For $n = 3$, one can make use of the Ricci curvature and for $n = 2$, the scalar curvature. So for $n = 3$, the motion is elliptic if R_{ij} is positive definite, parabolic if R_{ij} is degenerate and hyperbolic if R_{ij} is semi-definite. For $n = 2$, the motion is elliptic if $R > 0$, parabolic if $R = 0$ and hyperbolic if $R < 0$.

4. GEODESIC FLOW AND PHASE SPACE

Putting $x = |h|^{-1}x'$, we obtain

Lemma 4.0 : *For $h < 0$ (resp. $h > 0$), all the riemannian manifolds $(\bar{M}_h, \bar{g}_h)$ are homothetic.*

Hence, it suffices to consider three values of h, for instance $h = -\frac{1}{2}, 0, +\frac{1}{2}$.

The Keplerian orbits on the energy surface $E = h$ are geodesics of the Riemannian manifold $(\bar{M}_h, \bar{g}_h)$. This leads to study the geodesic flow on the tangent bundle $T\bar{M}_h$, which also is defined by the Hamiltonian flow of the Hamiltonian $H_h = \frac{1}{2}|y|^2/(h+\frac{1}{|x|})$ for $H_h = 1$ on the cotangent bundle $T^*\bar{M}_h$. Instead of the normal parameter σ, one can take a new parameter u defined by $du = \dfrac{d\sigma}{(h+\frac{1}{|x|})|x|}$ (by virtue of (1.2) $du = \dfrac{dx^o}{|x|}$). The geodesic flow is then defined by the Hamiltonian flow of $H = \frac{1}{2}|y|^2 - \frac{1}{|x|}$, which is the solution of the differential equations

$$\frac{dx}{du} = |x|y \qquad \frac{dy}{du} = -\frac{x}{|x|^2} \tag{4.1}$$

for $H = h$. We have to consider three cases.

<u>$h < 0$: elliptic case</u>

Theorem 4.1 : *In the elliptic case, $\bar{M}_h$ is diffeomorphic to S^n and there exists one analytic symplectomorphism $\mu : T^*\bar{M}_h \to T^*S^n$ such that*

1) μ maps the geodesic flow of $(\bar{M}_h, \bar{g}_h)$ onto that of S^n.

2) the collision orbits are mapped into the geodesics of S^n

passing through the image of the singularity 0 .

By virtue of the lemma 4.0, it suffices to prove the theorem for $h = -\frac{1}{2}$. It is known that the geodesic flow of the sphere $S^n \subset R^{1+n} : \xi_o^2 + \xi_1^2 + \ldots + \xi_n^2 = 1$ is defined in the stereographic chart $S^n - \{(1,0,\ldots,0)\} \to \mathbb{R}^n$ by the Hamiltonian flow of the function $K = \frac{1}{2}(1+|x|^2)|y|$ for $K = 1$, which is solution of the differential equations

$$\frac{dx}{ds} = \frac{y}{|y|^2} \qquad\qquad \frac{dy}{ds} = -\ |y|x \tag{4.2}$$

where s is the normal parameter (arc length).

One sees that (4.1) reduces to (4.2) by the substitution $(u,x,y) \to (s,y,-x)$ and conversely, (4.2) reduces to (4.1) by inverse substitution. From that, one deduces $s = u$ and the mapping $\mu : (x,y) \to (y,-x)$ is clearly a canonical diffeomorphism i.e. a symplectomorphism of $T^*\bar{M}_h$ onto T^*S^n . This allows one to transfer the analytic structure of S^n on $\bar{M}_h$ such that μ becomes analytic. Hence, the proof of 1) and 2) is immediate.

Remarks.- Property 2 gives rise to a regularization of the collision orbits in the cotangent bundle with the parameter u (the same variable is present in the Kepler's equation in Celestial Mechanics). This is the regularization found by J. Moser.

Theorem 4.1 leads to one representation of elliptic orbits onto large circles of the S^n sphere. The stereographic mapping gives us then a geometric solution to the motion equations (4.1).

<u>$h > 0$: hyperbolic case</u>

Theorem 4.2 : In the hyperbolic case, $\bar{M}_h$ *is diffeomorphic to the hyperbolic space* H^n *and there exists one analytic symplectomorphism* $\mu : T^*\bar{M}_h \to T^*H^n$ *such that*

1) μ *maps the geodesic flow of* $(\bar{M}_h, \bar{g}_h)$ *onto that of* H^n

2) the collision orbits are mapped into the geodesics of H^n *passing through the image of the singularity* 0.

H^n denotes here any sheet of the two-sheeted hyperboloid $-\xi_o^2 + \xi_1^2 + \ldots + \xi_n^2 = -1$ of $\mathbb{R}^{1,n}$ endowed with the metric induced by the pseudo-Euclidean metric of Minkowski space. If ω is the point $(1,0,\ldots,0)$, one has a chart $H^n - \{\omega\} \to \mathbb{R}^n$ by making one "stereographic" projection of center ω with the pseudo-distance $d(\xi,\xi') = \{|-(\xi_o-\xi_o')^2 + \sum_{k=1}^{n} (\xi_k-\xi_k')^2|\}^{1/2}$. H_-^n $(\xi_o < 0)$ will be mapped onto the open ball $B^n : |x| < 1$, endowed with the Riemannian metric $\frac{g}{(1-|x|^2)^2}$. H_+^n $(\xi_o > 0)$ will be mapped onto the open set $|x| > 1$ of $\mathbb{R}^n$ completed with the infinite point and endowed with the Riemannian metric $\frac{g}{(|x|^2-1)^2}$. It can be proved that these two Riemannian manifolds are isometric by the inversion of $\mathbb{R}^n$ with center 0 and power 1.

H^n has constant negative curvature -1.

The proof of this theorem is then analogous to that of the theorem 4.1, by pointing out that $K = \frac{1}{2}(1-|x|^2)|y|$ is the Hamiltonian function of the geodesic flow of H_-^n and $K = \frac{1}{2}(|x|^2-1)|y|$ that of H_+^n. One can make the same remarks as in the elliptic case.

h = 0 : parabolic case

Theorem 4.3 : In the parabolic case, $\bar{M}_o$ *is diffeomorphic to a manifold* P^n *diffeomorphic to* R^n *and there exists an analytic symplectomorphism* $\mu : T^*\bar{M}_o \to T^*P^n$ *such that*

1) μ *maps the geodesic flow of* $(\bar{M}_o, \bar{g}_o)$ *onto that of* P^n *endowed with a Riemannian metric of constant curvature 0*

2) the collision orbits are mapped into the geodesics of P^n *passing through the image of the singularity* 0 .

One can note that the geodesic flow of the Riemannian manifold $(\mathbb{R}^n, \frac{g}{|x|^4})$ is defined by the Hamiltonian flow of the function $K = \frac{1}{2}|x|^2|y|$ for $K = 1$. The proof of this theorem is then analogous to that of the theorem (4.1) in the elliptic case. Thus, P^n is a Riemannian manifold which admits the representation $(\mathbb{R}^n, \frac{g}{|x|^4})$, $x = 0$ corresponds to the collisions. Elsewhere, one can verify that $(\mathbb{R}^n, \frac{g}{|x|^4})$ is isometric to the Euclidean space $(\mathbb{R}^n, g)$ by the inversion ρ of center 0 and power 1, so that it is of null constant curvature and its geodesics are circles ending at the origin 0 as limit point, inasmuch as its geodesics are the transformed by ρ of the straight lines of $(\mathbb{R}^n, g)$.

We can also obtain the previous result by making $h \to 0$ in the elliptic or hyperbolic case. We will see that P^n is the paraboloid $2\xi_o - \xi_1^2 - \dots - \xi_n^2 = 0$ embedded in $\mathbb{R}^{1,n+1}$.

5. THE SYMMETRY GROUP

When one studies the Kepler motion, the symmetry group that appears first is SO(3). Though, it can be larger and containing SO(3) as a subgroup. Actually, W. Pauli in 1926 and

V. Fock in 1935 had shown relation between the spectrum of the hydrogen atom and a certain Lie algebra structure isomorphic to that of SO(4). Recently, in the 1970s, J. Moser, J.M. Souriau, E. Onofri have investigated symmetry groups larger than SO(3) and acting on the phase space. We formalize this problem here.

From the initial definition of the Riemannian manifold $(\bar{M}_h, \bar{g}_h)$ of the Kepler problem, one sees that SO(n) operates as an isometry group on $(\bar{M}_h, \bar{g}_h)$ leaving invariant the point 0 (and ω' in the case $h < 0$). The previous study of the geodesic flows allows the enlarging of the action group SO(n). Nevertheless, it leads to three different groups. Actually, it is known that S^n, H^n, P^n admit SO(n+1), SO(1,n) and $SO(n)\times\mathbb{R}^n$ respectively as greatest connected isometry groups. The action of $SO(n)\times\mathbb{R}^n$ on P^n will be deduced from its action on $\mathbb{R}^n$ euclidean space. These groups operate naturally on the corresponding phase-spaces T^*S^n, T^*H^n, T^*P^n as symmetry groups.

For purposes of unifying these three different aspect of the Kepler problem, according to the sign of h, one can introduce the projective quadric defined by the set of isotropic lines of $R^{1,n+1}$

$$- x_0^2 + x_1^2 + \dots + x_n^2 + x_{n+1}^2 = 0 \tag{5.1}$$

This projective quadric has three different affine forms : the S^n-sphere $x_1^2 + x_2^2 + \dots + x_{n+1}^2 = 1$ obtained for $x_0 = 1$, the two sheeted hyperboloid $- x_0^2 + x_1^2 + \dots + x_n^2 = - 1$ obtained for $x_{n+1} = 1$ and the elliptic paraboloid $- 2z + x_1^2 + \dots + x_n^2 = 0$ obtained for $w = x_0 + x_{n+1} = 2$, $z = x_0 - x_{n+1}$.

Using the pseudo-distance $d(x,y) = \left[\left|-(x_0-y_o)^2+\sum_{k=1}^{n+1}(x_k-y_k)\right|\right]^{1/2}$ to make an inversion of center $\omega = (1,0,\ldots,0,1)$ and power 2, one defines a "stereographic" chart of $S^n-\{\omega\}$, $H^n-\{\omega\}$ and $P^n-\{\omega\}$ into the hyperplanes $x_{n+1} = 0$, $x_0 = 0$ and $w = x_0 + x_{n+1} = 2$ respectively. In these charts, S^n, H^n_-, P^n are represented by

$$(\mathbb{R}^n, \frac{g}{(1+|x|^2)^2}) \quad , \quad (\mathbb{B}^n, \frac{g}{(1-|x|^2)^2}) \quad , \quad (\mathbb{R}^n, \frac{g}{|x|^4})$$

We find again the results of §.4 .

S^n, H^n, P^n are embedded in the isotropic cone (5.1) of $\mathbb{R}^{1,n+1}$. We can give to their equations the matrix form

$$^tX\eta X = 0 \tag{5.2}$$

where η is the matrix associated to the pseudo-euclidean metric of $\mathbb{R}^{1,n+1}$ and X the column matrix associated to the vector such that $x_0 = 1$, $x_{n+1} = 1$ and $w = x_0 + x_{n+1} = 2$ respectively. Their cotangent bundles T^*S^n , T^*H^n , T^*P^n are embedded in $\mathbb{R}^{1,n+1}\times\mathbb{R}^{1,n+1}$, having equations

$$^tX\eta X = 0 \qquad ^tX\eta Y = \varepsilon \tag{5.3}$$

where Y is defined as X and $\varepsilon = -1$, $+1$, 0 respectively.

It ensues from the definition of S^n, H^n, P^n as affine forms of the projective quadric (5.1) that $SO(1,n+1)$ leaves the equations (5.3) invariant. Any element $s \in SO(1,n+1)$ transforms X into one isotropic vector X' which is normalized as X , and one defines thus the action of $SO(1,n+1)$ on the cotangent bundles T^*S^n, T^*H^n, T^*P^n . Thus we have

<u>*Theorem 5.1*</u> *:* SO(1,n+1) *is the smallest group which contains as subgroups* SO(1+n), SO(1,n) *and* $SO(n)\times \mathbb{R}^n$ *and which operates transitively on the phase spaces* T^*S^n, T^*H^n, T^*P^n .

So SO(1,n+1) is the symmetry group of the Kepler problem. It preserves the Hamiltonian flow.

6. THE CASE n = 3

In the elliptic case, the phase space is diffeomorphic to the cotangent bundle T^*S^3. Since S^3 is a Lie group, this bundle is trivial. The phase space is hence diffeomorphic to $S^3\times \mathbb{R}^3$. The unit bundle $T_1^*S^3$, associated to the normal parametrization of the geodesics, is then diffeomorphic to $S^3\times S^2$.

Otherwise, if one keeps away the constant solutions $x = x_o$ $y = 0$ of the equations (4.1), which correspond to the constant geodesics $x = x_o$ of the S^3 sphere, one must carry off the null section of T^*S^3. The space $T^*S^3\setminus S^3$ is then diffeomorphic to $S^3\times S^2\times \mathbb{R}_+$ and one finds again the Kepler manifold of J. Moser and J.M. Souriau. In fact, by virtue of the lemma 4.0, an affine parametrization of the geodesics introduces one parameter s' proportional to s ; it corresponds to the energy level h if $s'=|h|s$.

In the hyperbolic case, the phase space is T^*H^3 and one has analogous results : $T^*H^3 \simeq H^3\times \mathbb{R}^3$, $T_1^*H^3 \simeq H^3\times S^2$, $T^*H^3\setminus H^3 \simeq H^3\times S^2\times \mathbb{R}_+$. In the parabolic case, the phase space is T^*P^3 and one has : $T^*P^3 \simeq P^3\times \mathbb{R}^3$, $T_1^*P^3 \simeq P^3\times S^2$, $T^*P^3\setminus P^3 \simeq P^3\times S^2\times \mathbb{R}_+$.

Finally, the group SO(1,4) operates as the symmetry group

on the phase spaces T^*S^3, T^*H^3, T^*P^3. It contains as subgroups $SO(4)$, $SO(1,3)$, $SO(3)\times \mathbb{R}^3$ which act on these phase spaces respectively as isometry groups. It also contains $SO(3)$ as a subgroup, which appears at first sight and is a common subgroup of the three groups $SO(4)$, $SO(1,3)$, $SO(3)\times \mathbb{R}^3$.

If the orbits are not oriented, on must consider the groups $O(1,4)$, $O(4)$, $O(1,3)$, $O(3)\times \mathbb{R}^3$ and $O(3)$.

7. THE COADJOINT REPRESENTATION

Now, the Lie algebra $\mathfrak{g} = \mathfrak{so}(1,n+1)$ of $G = SO(1,n+1)$ is compound with matrices A such that ${}^tA\eta + \eta A = 0$. This means the matrix $A' = \eta A$ is skew-symmetric and conversely $A = \eta A'$. Each element $A \in \mathfrak{g}$ can be identified with a 2-form $A' \in \Lambda^2 R^{1,n+1}$ and one can define in a natural way the following scalar product on

$$\langle A,B\rangle = \sum \eta_{ij,kl} a'_{ij} b'_{kl} \tag{7.1}$$

where $\eta_{ij,kl} = \eta_{ik}\eta_{jl} - \eta_{il}\eta_{jk}$ is the natural metric on $\Lambda^2 R^{1,n+1}$ and a'_{ij}, b'_{kl} are components of A', B'. Hence $\langle A,B\rangle = - \mathrm{Tr}(AB)$, where the trace of a matrix is defined relatively to $\eta_{ij,kl}$.

<u>Proposition 7.1</u> : *The pseudo-Euclidean metric* $\tilde{\eta}$ *defined on* $\mathfrak{so}(1,n+1)$ *by (7.1) is bi-invariant.*

It is easy to see that $\mathrm{Ad}_s^* \tilde{\eta} = \tilde{\eta}$.

Hence, one can identify the dual $\mathfrak{g}^*$ to $\mathfrak{g}$ and bring back the study of the coadjoint representation to that of the adjoint representation. One thus obtains :

Theorem 7.2 : The set $Q = \{Z \in \mathfrak{g}^* \mid Z = (X^tY-Y^tX)\eta,\ X,Y$ *defined as in* (5.3)$\}$ *is a 2n-dimensional orbit of the coadjoint representation of* SO(1,n+1) *into* $\mathfrak{g}^* = \mathfrak{so}^*(1,n+1)$. *Moreover*

1) *Q is symplectomorphic to the phase space in each case*
2) *the adjoint action of* G = SO(1,n+1) *on Q is isomorphic to the action of* SO(1,n+1) *defined previously.*

For the proof, one can first verify that each element of the form $X'\eta$ where X' is a skew-symmetric matrix, is contained in $\mathfrak{so}(1,n+1)$ i.e. $Z = (X^tY-Y^tX)\eta$ belongs to $\mathfrak{so}(1,n+1)$. It must be proved that for every $s \in SO(1,n+1)$, $Ad_s\, Q \subset Q$. Presently s verifies ${}^ts\eta s = \eta$, consequently $s^{-1} = \eta\,{}^ts\eta$. Now

$$\begin{aligned} sZs^{-1} &= s(X^tY-Y^tX)\eta\eta\,{}^ts\eta \\ &= s(A^tY-Y^tX)\,{}^ts\eta \\ &= (sX^tY^ts - sY^tX^ts)\eta \\ &= [(sX)^t(sY) - (sY)^t(sX)]\eta \end{aligned}$$

So if $Z \in Q$, $Ad_sZ \in Q$.

Moreover, by definition, Q is a 2n-dimensional submanifold of $\mathbb{R}^{1,n+1} \times \mathbb{R}^{1,n+1}$. Consequently, Q is a 2n-dimensional orbit of the coadjoint representation of SO(1,n+1).

It is then easy to verify the properties 1) and 2).

8. ORBIT MANIFOLDS

The existence of symmetry groups in the Kepler problem offers a method for finding the orbit manifolds. Its knowledge is useful for solving periodic orbits in the perturbated problem.

<u>*Theorem 8.1*</u> *: In the elliptic case, the manifold of oriented orbits is a symmetric space diffeomorphic to the homogeneous space* $SO(n+1)/SO(2)\times SO(n-1)$, *and in particular for* $n = 2$ *to* S^2 *and for* $n = 3$ *to* $S^2\times S^2$.

In fact, $SO(n+1)$ acts transitively on S^n as greatest connected isometry group. It acts naturally and transitively on $T_1^*S^n$ and on the set of oriented orbits of its geodesic flow. The isotropy group which leaves one orbit invariant is $SO(2)\times SO(n-1)$. One deduces from this property the theorem. For n = 2, the orbit manifold is diffeomorphic to $SO(3)/SO(2) \simeq S^2$ and for n = 3, it is diffeomorphic to $SO(4)/SO(2)\times SO(2) \simeq S^2\times S^2$ since one has $SO(4) \simeq SO(3)\times SO(3)$.

These two last results are obtained with a direct method by J. Moser.

<u>*Theorem 8.2*</u> *: In the hyperbolic case, the manifold of oriented orbits is a symmetric space diffeomorphic to the homogeneous space* $SO^+(1,n)/SO^+(1,1)\times SO(n-1)$, *and in particular for* $n = 2$ *to* $H^{1,1}$ *and for* $n = 3$ *to* $H^{1,1}\times S^2$.

$H^{1,1}$ is the one sheeted hyperboloid $-x_0^2 + x_1^2 + x_2^2 = +1$ in R^3 . $SO^+(1,p)$ is the greatest connected isometry group of the Minkowski space $R^{1,p}$ which preserves both the partial space and time orientations.

The proof is analogous to that of theorem 8.1.

One can note that $H^{1,1}$ is diffeomorphic to $S^1\times R$.

Theorem 8.3 : In the parabolic case, the manifold of oriented orbits is a symmetric space diffeomorphic to the homogeneous space $SO(n)\times \mathbb{R}^n / \mathbb{R}\times(SO(n-2)\times \mathbb{R}^{n-2})$, *and in particular for n = 2 to* $S^1\times \mathbb{R}$ *and for n = 3 to* $S^2\times S^1\times \mathbb{R}$.

$SO(n)\times \mathbb{R}^n$ is the greatest connected isometry group acting on the Euclidean space $\mathbb{R}^n$, hence on P^n and $T_1^*P^n$. It is the semi-direct product of SO(n) by the translation subgroup identified with $\mathbb{R}^n$.

The proof is analogous to that of the previous cases.

9. STABILITY OF PERIODIC ORBITS

Proposition 9.1 : $T_1^*S^n$ *is a fiber bundle over* $G_{2,n-1}$ *with fiber* S^1.

The geodesic flow on $T_1^*S^n$ is indeed a periodic one. It defines a 1-parameter group isomorphic to S^1, which acts freely on $T_1^*S^n$. The quotient space $T_1^*S^n/S^1$ is exactly the orbit manifold which, by virtue of theorem 8.1, is diffeomorphic to the homogeneous space $SO(n+1)/SO(2)\times SO(n-1)$ which is diffeomorphic to the Grassmann manifold $G_{2,n-1}$.

Now, the perturbated Kepler problem is equivalent to this one : " Given a small variation of the metric g on S^n, are there periodic geodesics of the unperturbated metric from which periodic geodesics of the perturbated metric $\tilde{g} = g + \varepsilon 1 + o(\varepsilon)$ branch off ? ".

Any perturbation modifies the Hamiltonian vector field F

associated with the geodesic flow Φ_t on $T_1^*S^n$. For the study of the stability of the periodic orbits, one can use the averaging method.

Let more generally a periodic flow Φ_t with period 1 on a compact manifold N. Let $\hat{N} = N/S^1$ the quotient space and $\pi : N \to \hat{N}$ the projection. For any vector field v on N, define the averaged vector field

$$\bar{v} = \int_0^1 \Phi_{t*} v \, dt$$

where $\Phi_{t*} v = T\Phi_t . v o \Phi_t$. $\bar{v}$ is invariant along each orbit of the flow Φ_t. We obtain by projection a well defined vector field $\hat{v}$ on $\hat{N}$. We have the following main lemma :

Lemma 9.2 : Let $\tilde{F} = F + \varepsilon f + o(\varepsilon)$ *be the perturbated vector field where* F *is the generator of the flow* Φ_t. *If the averaged vector field* $\hat{f}$ *possesses a non degenerate singular point at* $\hat{x}$ *then the flow* $\tilde{\Phi}_t$ *of* $\tilde{F}$ *admits a periodic orbit of period closed to* 1 *in a preassigned neighbourhood of* $\pi^{-1}(\hat{x})$, *provided* ε *is small enough.*

For proof, see J. Moser [5].

From the topological properties of $\hat{N}$, one can then infer how many singular points $\hat{f}$ will have in general.

For the perturbated Kepler problem we have

Theorem 9.3 : For small perturbations of the Kepler problem, the number of stable periodic orbits is at least equal to n if n is even and n+1 if n is odd.

Indeed, the Euler-Poincaré characteristic of a compact manifold $\hat{N}$ gives a lower bound for the number of singular points, provided they are non degenerate, of any vector field on $\hat{N}$ and it is defined by the formula

$$\chi(\hat{N}) = \sum_{i=0}^{m} (-1)^i b_i(\hat{N})$$

where $m = \dim \hat{N}$ and $b_i(\hat{N})$ is the i-th Betti number of $\hat{N}$. Taking $\hat{N} = G_{2,n-1}$ one has the theorem.

Elsewhere, the geodesic flow on $T_1^*S^n$ is a Hamiltonian one. To the perturbated metric $\tilde{g} = g + \varepsilon l + o(\varepsilon)$ corresponds the perturbated Hamiltonian $\tilde{K} = K + \varepsilon L + o(\varepsilon)$. For small perturbations, the singular points of the average perturbation vector field $\hat{f}$ will appear as the critical points of the average perturbation function $\hat{L}$ defined by

$$\bar{L} = \int_0^1 \Phi_t^* L \, dt \ .$$

The number of critical points of a function on a manifold can now be estimated by the Morse theory provided only non degenerate critical points occur. It is given by the formula

$$\Gamma(\hat{N}) = \sum_{i=0}^{m} b_i(\hat{N}).$$

One can show that $\chi(G_{2,n-1})$ is equal to $\Gamma(G_{2,n-1})$. Thus, we do not get a better result because of the Hamiltonian character of the geodesic flow.

REFERENCES

[1] H. BACRY , Nuovo Cimento, 41, 1966.

[2] V. FOCK , Z. Phys., 98, 1935.

[3] G. GYORGYI , Nuovo Cimento, 53, 1968.

[4] E. ONOFRI , Dynamical quantization of the Kepler manifold, J. Math. Phys., 17,401-408, 1976.

[5] J. MOSER , Regularization of Kepler's problem and the averaging method on a manifold, Com. Pure Appl. Math., XXIII, 609-636, 1970.

[6] M. PAULI, Z. Phys., 36, 1926.

[7] PHAM MAU QUAN, C.R.A.S., 291, 219-221 et 291-302, 1980.

[8] J.M. SOURIAU , Sur la variété de Képler, Proc. Convegno di Geometria Simplettica, Roma 1973.

[9] S. SMALE , Topology and mechanics, Inv. Math., 10, 305-331, 1970 and 11, 45-64, 1970.

CONFINEMENT PROBLEMS IN MATHEMATICAL PHYSICS, CLASSICAL AND MODERN.

M.S. Berger
Center for Applied Mathematics and Mathematical Science
University of Massachusetts
Amherst, MA 01003, U.S.A.

1. INTRODUCTION.

The confinement problem arises in many areas of nonlinear physical science both classical and modern. The common mathematical problem consists in finding a piecewise smooth vector field A (i.e. one form) on a given domain Ω satisfying (after reduction) certain canonical nonlinear elliptic partial differential equation and boundary conditions (generally involving a free boundary) such that the solenoidal part of A is focussed by nonlinearity into a bounded subdomain Ω_1 of Ω (Ω_1 is referred to as a confinement domain). In contemporary gauge theories one is given a nonquadratic action functional S(A), a gauge group G, and one requires (in addition to the above) that the confinement field A arise as the absolute minimum of S(A) subject to appropriate boundary conditions. Standard examples from classical mathematical physics include steady vortices in ideal fluids and equilibrium configurations in plasma physics governed by Lundquist equations, an area of contemporary importance due to its connec-

DYNAMICAL SYSTEMS
AND MICROPHYSICS

ISBN 0-12-068720-8

tion with fusion energy. A relatively simple contemporary gauge theory in which confinement problems occur is vortices in type II superconductivity associated with the Ginzberg-Landau action functional. A very contemporary question of great importance is the quark confinement problem in non Abelian gauge theory. Recently Stephen Adler has reformulated a coherent theory for this problem in which, (a) a major aspect of the quark confinement problem can be reduced to an exact confinement problem in the sense defined above and, (b) the highly successful MIT bag model (a highly phenomenological approach) emerges as a logical consequence of the theory.

In the sequel we shall discuss a range of ideas paying careful attention to the mathematical structures involved.

2. THE CLASSICAL EXAMPLES.

(i) Let A denote the velocity vector of an ideal incompressible fluid. The steady vortices of permanent shape moving with a constant speed S satisfy the Euler equations

$$A \times \operatorname{curl} A = \nabla H \tag{1a}$$

$$\operatorname{div} A = 0 \tag{1b}$$

with associated boundary condition at infinity related to the speed S . The vector $\omega = \operatorname{curl} A$ is called the vorticity of the fluid motion. This confinement problem then requires the determination of a fluid motion in which the vorticity is confined to bounded domain Ω . Such domains are called "vortex rings" since solid tori are generally the observed confinement domain, although limiting forms of solid tori (viz sphere - like figures and circular filaments) are also possible.

(ii) The equilibria of plasma fusion physics (e.g. Tokomak fusion) are governed by the Lundquist equations

$$B \times \operatorname{curl} B = \nabla p \tag{2a}$$

$$\operatorname{div} B = 0 \tag{2b}$$

$$j = \operatorname{curl} B \tag{2c}$$

Here B denotes the magnetic vector, p the associated pressure and $j = \operatorname{curl} B$ is the current associated with the magnetic field.

The confinement problem in this case is associated with finding bounded subdomains strictly contained in a torus held in equilibrium by a externally imposed magnetic field.

An interesting and informative variant occurs in the plasma physics of astrophysics. In the simplest case we ask if force free confined smooth vector fields for the Lundquist equations (i.e., $\nabla p = 0$) exist on R^3 with zero boundary conditions at infinity. It can be proved that the answer is negative (this fact will be useful later on).

In all these examples, it is useful to exploit the fact that the desired confinement figures are associated with vector fields whose divergence vanishes. Thus, assuming axial asymmetry with respect to appropriate axes, we can reduce the confinement problems for equations (1) and (2) to a single equation for a single scalar function usually termed by "stream function ψ" with $u(r,z) = (u_r, u_\theta, r_z)$ and $u_r = -\frac{1}{r}(\partial\psi/\partial z)$, $u_z = \frac{1}{r}(\partial\psi/\partial r)$.

Moreover, all these problems involve the determination of the exact shape of the confinement domain in question (a so-called free boundary problem), since this is an unknown in each problem.

3. VORTICES IN TYPE II SUPERCONDUCTIVITY.

The vortices in this case are determined as (finite action) critical points of the Ginzberg-Landau action functional $S_\lambda(A,\phi)$ (for $\lambda > 1$)

$$S_\lambda(A,\phi) = \int_{R^2} \{\tfrac{1}{2}|\operatorname{curl} A|^2 + \sum_{j=1}^{2} |(\partial_j - iA_j)\phi|^2 + \tfrac{\lambda}{4}(|\phi|^2 - 1)^2\}dx \tag{3a}$$

associated with the topological invariant the "vortex number" $N \neq 0$ given by

$$N = \frac{1}{2\pi} \int_{R^2} \text{curl } A \, dx \tag{3b}$$

Again the confinement problem here involves the vector A, the vector potential of the magnetic field h (so that $h = \text{curl } A$). We do not demand in this case that the magnetic field be totally confined to a bounded domain but rather that this be true apart from an exponentially decaying remainder.

As is well known this problem of vortices is one of Abelian gauge theory with $G = U(1)$, and the Gibbs free energy functional $G = S_\lambda(A,\phi) - NH$ is minimal, where H is the external magnetic field strength. In this case the vortex with $N = 1$ is stable.

4. ADLER'S APPROACH TO QUARK CONFINEMENT.

In any nonlinear gauge theory we minimize a nonquadratic action function $S(A)$ over a class of smooth fields satisfying appropriate boundary conditions. The functional $S(A)$ is invariant under a gauge group G . In Adler's approach we also require invariance with respect to the renormalization group.

This leads us to consider the function 1

$$S_\varepsilon(A) = \int_{R^3} \left\{ |F(A)|^2 \{1 + \varepsilon \log \left[\frac{|F(A)|^2}{\mu^4}\right] \} \right\}$$

where $\varepsilon = 1/4 \, b_o g^2$, and b_o is the "asymptotic freedom constraint" related to the gauge group SU(3).

Relative to this action functional we consider a source term j, a renormalization constant $S_\varepsilon(K)$ and an amended functional

$$W(A) = [\, S_\varepsilon(A) = S_\varepsilon(K) - j\phi \,]$$

The function $S_\varepsilon(A)$ written above contains the radiative correction term ($\varepsilon \neq 0$) based on renormalization group invariance. The pure action functional with $\varepsilon = 0$ does not imply quark invariance as was observed by Adler after numerous computer studies.

Now it is important to analyze just how this new minimization problem leads to the kind of confinement problem described above. To this end we let j be pair of classical sources (quarks) of equal magnitude at points x_1, x_2 .

$$j = Q\,[\,\delta(x - x_1) - \delta(x - x_2)\,]$$

and set $F^2 = E^2 + B^2$

$$E = -\nabla\phi \qquad B = \text{curl } A$$

Moreover we suppose that $|E|$ is axisymmetric around the line joining the two quarks sources.

Now the effect of the functional $S_\varepsilon(A)$ with $\varepsilon \neq 0$ is to move its minimum to a nonzero value $F^2 = K^2$ from the usual zero value, and this has a key effect on the minimization problem. Indeed, if $|E|^2 < K^2$ (we use this fact by taking the minimum as close as possible to K^2) we have $|B|^2 = (K^2 - |E|^2)$, whereas if $|E|^2 > K^2$ we minimize by taking $B = 0$. Thus we find 2 domains Ω_1 and Ω_2 , depending if $|\nabla\phi| > K$ or $|\nabla\phi| < K$.

Now the Euler-Lagrange equation of the resulting problem can be written

$$\nabla(\tilde{\varepsilon} E) = j \tag{4}$$

where

$$\tilde{\varepsilon}(|\nabla\phi|^2) = \begin{cases} 1/4 \log \{|\nabla\phi|^2/K^2\} & \Omega_1 \\ 0 & \Omega_2 \end{cases}$$

Now to convert this to a confinement problem we set $D = \tilde{\varepsilon} E$ and note that, because of axial symmetry and the fact that $\text{div } D = 0$ at points off the axis of symmetry and at points on the axis other than x_1 and x_2 ,

$$D = \text{curl}\,[\frac{\hat{\theta}}{2\pi}\,\Phi]$$

where $\Phi = \Phi(R,z)$ is the analogue of the stream function in Section 1. The resulting Euler equation can be written

$$\text{div}\,[\,\sigma(R,\ |\nabla\Phi|\,)\nabla\phi\,] = 0 \quad \text{in} \quad \Omega_1 \tag{5}$$

where

$$\sigma(R,\ |\nabla\Phi|\,) = \frac{1}{R^2\,\tilde{\varepsilon}\,(D)}$$

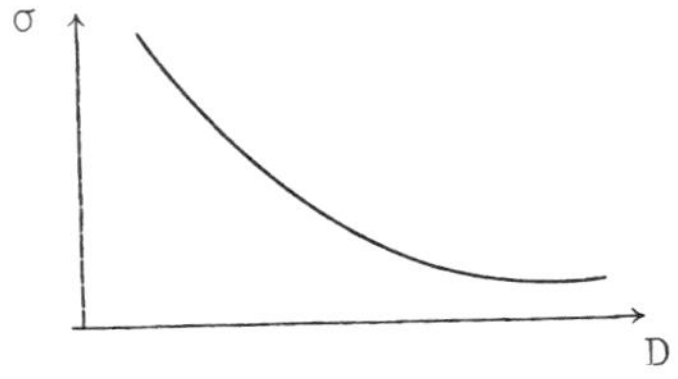

and $\tilde{\varepsilon} = \tilde{\varepsilon}(D)$ has the behavior as shown.
For small D, σ behaves like D^{-1}.
For large D, σ behaves like $[\log D]^{-1}$.

Now the confinement problem consists in finding a C^1 function Φ such that Φ satisfies (5) in Ω_1, $\Delta\Phi = 0$ in Ω_2, $\Phi = |Q|$ on $\partial\Omega_1$ the free boundary. This results in an exact confinement problem in our sense for the vector field $\underline{D}$. The expected solution graph in the positive quadrant of the (r,z) plane is a connected convex subset joining both axes. The MIT bag model of quark confinement results by approximating the σ, D graph above by a step function.

The proof of the existence of a solution for this confinement problem and its physical consequences will appear in a later publication.

Acknowledgement.

The contents of this chapter are based on research supported by the National Science Foundation.

BIBLIOGRAPHY

[1] Adler, S., The Mechanism for Confinement in Massive Quark QCD. (to appear in "Current Problems in Particle Theory").

[2] Berger, M.S., *Nonlinearity and Functional Analysis*, Academic Press, New York, N.Y. 1977 (Chapter 6.4).

[3] Berger, M.S., Vortex Phenomena (research article to appear in volume on *Global Analysis*. North Holland Publishers, 1982).

PART II

SYSTEM THEORY APPROACHES TO MECHANICS

OPTIMALITY AND REACHABILITY WITH FEEDBACK CONTROL

G. Leitmann

Department of Mechanical Engineering
University of California
Berkeley, California

PREFACE

This chapter contains a discussion of conditions which must be met if a dynamical system is "controlled" either in an "optimal" fashion or such that a trajectory of the system belongs to the boundary of the "reachable set", that is, of the region containing all trajectories emanating from a given set. By "optimality" we mean the attainment of the minimum value of a given functional. In either case, the choice of the control value at given instant is based on the system's current state; thus, we are dealing with feedback control.

1. DYNAMICAL SYSTEM

We are concerned with a subset of the universe that is defined by n real numbers referred to as the <u>state</u> of the system, $x = (x_1, x_2, \ldots, x_n)^T \in R^n$, which may change with the passing of time $t \in (-\infty, \infty)$. The evolution of the state is governed by a <u>controller</u>. Such a system is called a <u>dynamical system</u>. Here

ISBN 0-12-068720-8

we are concerned with a dynamical system whose evolution is determined by an ordinary differential equation, the <u>state equation</u>.

Given a <u>terminal</u> state x^1, we are interested in the motion of the system's state to that terminal state due to the controller's decisions. It is convenient to let time be one of the state components, say $x_n \triangleq t$, and to introduce time-to-go $\tau \triangleq t_1 - t$, where $t_1 = x_n^1$ is the <u>terminal time</u>. Thus, in the ensuing discussion we must keep in mind that what is initial with respect to time t is terminal with respect to time-to-go τ, and conversely.

Now consider a Lebesgue measurable function

$\hat{u}(\cdot) : [0,\tau_f] \to R^m$

and a prescribed C^1 - function

$\hat{f}(\cdot) : R^n \times R^m \to R^n$.

Then the state equation is

$$\hat{x}'(\tau) = f(\hat{x}(\tau),\hat{u}(\tau)) \tag{1.1}$$

where prime denotes $d/d\tau$. The evolution of the state is given by the absolutely continuous function

$$\hat{x}(\cdot) : [0,\tau_f] \to R^n , \hat{x}(0) = x^1 , \tag{1.2}$$

satisfying (1.1) for almost all $\tau \in [0,\tau_f]$.

2. FEEDBACK CONTROL

The controller governs the state evolution through the choice of $\hat{u}(\tau)$ for $\tau \in [0,\tau_f]$. This control decision is based on information. If time t, or time-to-go τ, is used as the only information, we speak of <u>open-loop control</u>. Here we shall consider control choices based on the knowledge of the system's state, $\hat{x}(\tau)$; we speak of <u>closed-loop</u> or <u>feedback control</u>. Thus, a feedback control is a function

$$p(\cdot) : R^n \to R^m .$$

With such a choice by the controller, the state equation (1.1) becomes

$$\hat{x}'(\tau) = f(\hat{x}(\tau), p(\hat{x}(\tau))) \tag{2.1}$$

where, provided it exists, $\hat{x}(\cdot) : [0,\tau_f] \to R^n$, $\hat{x}(0) = x^1$, is a solution of (2.1). Indeed, without further restrictions on the class of <u>admissible</u> feedback controls, a solution need not exist, or may be not unique. We shall not insist on uniqueness, but we shall assume that there exists at least one solution (possibly in a generalized sense)[1-5].

Given a feedback control $p(\cdot)$, suppose there exists a solution $\hat{x}(\cdot) : [0,\tau_f] \to R^n$ with $\tau_f > 0$ and $\hat{x}(0) = x^1$. Recall here that x^1 is the prescribed <u>terminal</u> state with respect to the time t. Then at $\tau \in [0,\tau_f]$ there is a control value

$$\hat{u}(\tau) = p(\hat{x}(\tau)) . \tag{2.2}$$

Let $K_{\hat{x}}(x^1)$ denote the set of all solutions generated by $p(\cdot)$ from state x^1, and let $K_{\hat{u}}(x^1)$ denote the set of corresponding control functions $\hat{u}(\cdot) : [0,\tau_f] \to R^m$ generated via relation (2.2); that is,

$K_{\hat{u}}(x^1) \triangleq \{\hat{u}(\cdot) \mid \hat{x}(\cdot) \in K_{\hat{x}}(x^1)\}$.

Now we restrict feedback controls to a set of admissible ones.

Definition 2.1. The set P is the set of admissible feedback controls if and only if

(i) for all $x^1 \in R^n$ and all $p(\cdot) \in P$, $\hat{u}(\cdot) \in K_{\hat{u}}(x^1)$ is bounded and Lebesgue measurable,

(ii) given $p'(\cdot), p''(\cdot) \in P$, $t' = x'_n \in (-\infty,\infty)$ and $p(\cdot)$ such that

$$p(x) = p'(x) \quad \text{for} \quad x_n < x'_n$$

$$p(x) = p''(x) \quad \text{for} \quad x_n \geqslant x'_n ,$$

implies that $p(\cdot) \in P$,

and

(iii) given the set-valued function

$$U(\cdot) : R^n \to \text{nonempty subsets of } R^m ,$$

$p(\cdot) \in P$ implies that

$$p(x) \in U(x) \qquad \forall x \in R^n .$$

The condition (i) is imposed to assure the existence of integrals introduced subsequently in the definition of a performance index. The condition (ii) is a "group property" of admissible feedback controls that permits one to "join" feedback controls; this property is of basic importance in the derivation of necessary conditions for both "optimality" and "reachability". Finally, the condition (iii) is imposed to allow for state-dependent control constraints.

To assure the existence of a solution of (2.1) for every initial value of the state and of the control, we shall assume henceforth that the set of admissible controls, P, is sufficiently "rich".

<u>Assumption 2.1.</u> For every $x \in R^n$ and for every $u \in U(x)$ there exists a control $p_{x,u}(\cdot) \in P$ whose restriction to some open neighborhood of x is of class C^1 and for which $p_{x,u}(x) = u$.

Finally we note that all conditions involving R^n can be restricted by replacing R^n with a domain (open connected set) of R^n.

3. PLAYABILITY

In terms of time t, the controller wishes to "steer" the state from one belonging to a prescribed <u>initial set</u> $\theta \subset R^n$ to the given <u>terminal state</u> x^1.

<u>Definition 3.1.</u> A feedback control $p(\cdot)$ is playable at x^1 if and only if

(i) $p(\cdot) \in P$,

and

(ii) there is at least one solution $\hat{x}(\cdot) : [0,\tau_f] \to R^n$ of (2.1) such that $\hat{x}(0) = x^1$, $\hat{x}(\tau) \notin \theta$ for $\tau < \tau_f$, and $\hat{x}(\tau_f) \in \theta$. In terms of time-to-go τ, such a solution is a terminating one (on θ) and τ_f is the corresponding terminal value of τ; a corresponding triplet $\{x^1, p(\cdot), \hat{x}(\cdot)\}$ is also called terminating.

4. PERFORMANCE INDEX

In order to define "optimality", we introduce a performance index or cost. Let

$f_o(\cdot) : R^n \times R^m \to R$

be a prescribed C^1-function. Then for given $x^1 \in R^n$, $p(\cdot) \in P$ and corresponding solution $\hat{x}(\cdot) : [0, \tau_f] \to R^n$, $\hat{x}(0) = x^1$, the cost is defined by

$$V(x^1, p(\cdot), \hat{x}(\cdot)) \triangleq \int_0^{\tau_f} f_o(\hat{x}(\tau), p(\hat{x}(\tau)))\, d\tau \quad . \tag{4.1}$$

5. OPTIMALITY

In addition to "termination" on θ, it is desired to minimize the value of the cost (4.1).

Definition 5.1. The feedback control $p^*(\cdot)$ is optimal on $X \subseteq R^n$ if and only if

(i) it is playable at all $x^1 \in X \setminus \theta$,

and

(ii) for all $x^1 \in X \setminus \theta$

$$V(x^1, p^*(\cdot), \hat{x}^*(\cdot)) \leqslant V(x^1, p(\cdot), \hat{x}(\cdot))$$

for all terminating $\{x^1, p^*(\cdot), \hat{x}^*(\cdot)\}$ and $\{x^1, p(\cdot), \hat{x}(\cdot)\}$.

We note that the condition (ii) of Definition 3.1, together with (4.1), implies that

$$\lim_{\tau \to \tau_f} V(\hat{x}(\tau), p(\cdot), \hat{x}(\cdot) \big|_{[\tau, \tau_f]}) = 0$$

and, in particular,

$$\lim_{\tau \to \tau_f^*} V(\hat{x}^*(\tau), p^*(\cdot), \hat{x}^*(\cdot) \big|_{[\tau, \tau_f^*]}) = 0 \quad . \tag{5.1}$$

Hence, in view of the uniqueness of the minimum, we can define a function

$$V^*(\cdot) : X \cup \theta \to R$$

such that

$$V^*(x^1) = \begin{cases} V(x^1, p^*(\cdot), \hat{x}^*(\cdot)) & \forall\, x^1 \in X \setminus \theta \\ 0 & \forall\, x^1 \in \theta \end{cases} \tag{5.2}$$

for all "terminating" $\{x^1, p^*(\cdot), \hat{x}^*(\cdot)\}$. Note that $V^*(\cdot)$ is defined on $X \cup \theta$; that is, it is defined for those feedback controls which are optimal on X .

In the following three sections we state necessary conditions for optimality of a feedback control. This derivation is based on a geometric approach that was used earlier[6] for problems in which the terminal state set (target) and the initial state are prescribed rather than the initial state set and the terminal state (in terms of time). Since the derivation is entirely analogous to that of Ref. 6, we shall not present the details of the derivation but rather refer the reader to that reference.

Before stating some necessary conditions for optimality, we shall restrict the class of optimal feedback controls.

Definition 5.2. Let $p^*(\cdot)$ be a feedback control that is optimal on X and generates a "terminating" solution $\hat{x}^*(\cdot) : [0,\tau_f^*] \to R^n$. Such a solution is regular if and only if

(i) $\hat{x}^*(\tau) \in \text{int } X$ for $\tau \in [0,\tau_f^*)$;

(ii) every $\hat{x}^*(\tau)$, $\tau \in [0,\tau_f^*)$, possesses a neighborhood on which $V^*(\cdot)$ is of class C^2 ;

(iii) every $\hat{x}^*(\cdot)$, $\tau \in [0,\tau_f^*]$, possesses a neighborhood on which $p^*(\cdot)$ is of class C^1 .

Unless stated otherwise, the subsequent discussion is restricted to regular solutions.

6. NECESSARY CONDITIONS -- A FUNCTIONAL EQUATION

For the first of our necessary conditions we may relax the condition (ii) of Definition 5.2 by restricting $V^*(\cdot)$ to be of class C^1 on a neighborhood of $\hat{x}^*(\tau)$. Indeed, since $p^*(\cdot)$ must generate a "terminating" solution from every $x^1 \in X$, this amounts to assuming that $V^*(\cdot)$ is of class C^1 in a neighborhood of every $x \in \text{int } X$.

Now, as in Ref. 6, it can be shown that, if $p^*(\cdot)$ is optimal on X, then for all $x \in \text{int } X$

$$\min_{u \in U(x)} [f_o(x,u) + \sum_{i=1}^{n} \frac{\partial V^*(x)}{\partial x_i} \hat{f}_i(x,u)]$$

$$= f_o(x,p^*(x)) + \sum_{i=1}^{n} \frac{\partial V^*(x)}{\partial x_i} \hat{f}_i(x,p^*(x))$$

$$= 0$$

subject to the boundary condition $V^*(x) = 0$ for $x \in \theta$.

7. NECESSARY CONDITIONS -- A MINIMUM PRINCIPLE

Next we turn to another necessary condition that is a consequence of (6.1).

Let

$$\frac{\partial \hat{f}}{\partial x} \triangleq \begin{bmatrix} 0 & \frac{\partial f_o}{\partial x_1} & \cdots & \frac{\partial f_o}{\partial x_n} \\ \vdots & & & \vdots \\ 0 & \frac{\partial \hat{f}_n}{\partial x_1} & \cdots & \frac{\partial \hat{f}_n}{\partial x_n} \end{bmatrix}$$

$$\frac{\partial \hat{f}}{\partial u} \triangleq \begin{bmatrix} \frac{\partial f_o}{\partial u_1} & \cdots & \frac{\partial f_o}{\partial u_m} \\ \vdots & & \vdots \\ \frac{\partial \hat{f}_n}{\partial u_1} & \cdots & \frac{\partial \hat{f}_n}{\partial u_m} \end{bmatrix}$$

$$\frac{dp^*}{dx} \triangleq \begin{bmatrix} 0 & \frac{\partial p_1^*}{\partial x_1} & \cdots & \frac{\partial p^*{}_1}{\partial x_n} \\ \vdots & & & \vdots \\ 0 & \frac{\partial p_m^*}{\partial x_1} & \cdots & \frac{\partial p_m^*}{\partial x_n} \end{bmatrix}$$

all evaluated at $x = \hat{x}^*(\tau)$, $u = p^*(\hat{x}^*(\tau))$.

Consider also the function $H(\cdot) : R^{n+1} \times R^n \times R^m \to R$ defined by

$$H(\lambda,x,u) \triangleq \sum_{i=0}^{n} \lambda_i f_i(x,u) . \tag{7.1}$$

Then, as in Ref. 6, it can be shown that there exists an absolutely continuous function $\hat{\lambda}(\cdot) : [0,\tau_f^*] \to R^{n+1}$ satisfying

$$\hat{\lambda}'(\tau) = - \left[\frac{\partial \hat{f}}{\partial x} + \frac{\partial \hat{f}}{\partial u} \frac{dp^*}{dx} \right]^T \hat{\lambda}(\tau) \tag{7.2}$$

for almost all $\tau \in [0,\tau_f^*]$, such that

$$\begin{aligned} \min_{u \in U(x^*(\tau))} & H(\hat{\lambda}(\tau),\hat{x}^*(\tau),u) \\ &= H(\hat{\lambda}(\tau),\hat{x}^*(\tau),p^*(\hat{x}^*(\tau))) \\ &= 0 . \end{aligned} \tag{7.3}$$

Furthermore, the zeroth component of $\hat{\lambda}(\tau)$ is

$$\hat{\lambda}_o(\tau) \equiv 1 . \tag{7.4}$$

If the control constraint set is state-independent, that is, if

$$U(x) \equiv \text{constant} , \tag{7.5}$$

then (7.2) reduces to

$$\hat{\lambda}'(\tau) = - \left[\frac{\partial \hat{f}}{\partial x} \right]^T \hat{\lambda}(\tau) . \tag{7.6}$$

The "initial" state, $\hat{x}^*(0)$, is prescribed; namely, $\hat{x}^*(0) = x^1$. The "terminal" state, $\hat{x}^*(\tau_f^*)$, must belong to the given set θ. If θ is a point in R^n, $\theta = \{ x^o \}$, then -- leaving aside the equation for $\hat{\lambda}_o(\tau)$ -- there are 2n boundary conditions for the integration of (2.1) and (7.2) or (7.6). If θ is not a point in R^n, one can give additional necessary conditions at $\hat{x}^*(\tau_f^*) \in \theta$.

8. NECESSARY CONDITIONS -- TRANSVERSALITY

We suppose now that the set θ is closed in R^n and that its boundary, $\partial\theta$, is a smooth manifold of dimension $n - r$, $r \leqslant n$; that is,

$$\partial\theta \triangleq \{ x \in R^n \mid \theta_i(x) = 0 \quad , \quad i = 1,2,\ldots,r \} \tag{8.1}$$

where the $\theta_i(\cdot) : R^n \to R$ are of Class C^1 , and the grad $\theta_i(x)$, $i = 1,2,\ldots,r,$ are linearly independent for $x \in \partial\theta$. For $r = n$, $\partial\theta$ reduces to a point; thus, suppose that $r < n$.

Then, as in Ref. 6, it can be shown that

$$\sum_{i=1}^{n} \hat{\lambda}_i(\tau_f^*) \, \eta_i = 0 \tag{8.2}$$

for all $\eta = (\eta_1,\eta_2,\ldots,\eta_n)^T$ such that

$$\sum_{i=1}^{n} \frac{\partial\theta_j(\hat{x}^*(\tau_f^*))}{\partial x_i} \eta_i = 0 \ , \quad j = 1,2,\ldots,r \ . \tag{8.3}$$

The relation (8.2), together with (8.3), yields $n - r$ conditions at $\tau = \tau_f^*$, in addition to the r conditions

$\theta_i(\hat{x}^*(\tau_f^*)) = 0 \quad , \quad i = 1,2,\ldots,r$.

Before turning to a discussion of "reachability", let us summarize the aforegoing necessary conditions for optimality in terms of time t , rather than time-to-go τ .

9. SUMMARY OF NECESSARY CONDITIONS

Given a feedback control $p(\cdot) \in P$, the state equation (2.1) in terms of τ has the counterpart

$$\dot{x}(t) = f(x(t),p(x(t))) \tag{9.1}$$

where dot denotes d/dt, and

$$f(x,u) = -\hat{f}(x,u) \quad \forall\ (x,u) \in R^n \times R^m . \tag{9.2}$$

The solution of (9.1), $x(\cdot) : [t_o,t_1] \to R^n$, $x(t_1) = x^1$, is such that

$$x(t) = \hat{x}(t_1-t) \qquad \forall\ t \in [t_o,t_1] . \tag{9.3}$$

In terms of τ, playability at x^1 is defined in Definition 3.1. Now we state the analogous definition in terms of t.

<u>Definition 9.1</u>. A feedback control $p(\cdot)$ is playable at x^1 if and only if

(i) $p(\cdot) \in P$,

and

(ii) there is at least one solution $x(\cdot) : [t_o,t_1] \to R^n$ of (9.1) such that $x(t_o) \in \theta$, $x(t) \notin \theta$ for $t > t_o$, and $x(t_1) = x^1$.

The performance index (4.1) is unchanged; namely,

$$\int_0^{\tau_f} f_o(\hat{x}(\tau),p(\hat{x}(\tau)))\, d\tau = \int_{t_o}^{t_1} f_o(x(t),p(x(t)))\, dt , \tag{9.4}$$

so that

$$V(x^1,p(\cdot),\hat{x}(\cdot)) = V(x^1,p(\cdot),x(\cdot)) .$$

We can now restate the necessary conditions, utilizing (9.2) and (9.4). We recall first the relation (6.1); namely, if

$p^*(\cdot)$ is optimal on X , then for all $x \in \text{int } X$

$$\max_{u \in U(x)} [-f_o(x,u) + \sum_{i=1}^{n} \frac{\partial V^*(x)}{\partial x_i} f_i(x,u)]$$

$$= -f_o(x,p^*(x)) + \sum_{i=1}^{n} \frac{\partial V^*(x)}{\partial x_i} f_i(x,p^*(x))$$

$$= 0 \quad , \tag{9.5}$$

subject to the boundary condition $V^*(x) = 0$ for $x \in \theta$.

Next we consider the conditions of Sections 7-8. Let

$$\frac{\partial f}{\partial x} \triangleq \begin{bmatrix} 0 & -\frac{\partial f_o}{\partial x_1} & \cdots & -\frac{\partial f_o}{\partial x_n} \\ \cdot & & & \cdot \\ \cdot & & & \cdot \\ \cdot & & & \cdot \\ 0 & \frac{\partial f_n}{\partial x_1} & \cdots & \frac{\partial f_n}{\partial x_n} \end{bmatrix} ,$$

$$\frac{\partial f}{\partial u} \triangleq \begin{bmatrix} -\frac{\partial f_o}{\partial u_1} & \cdots & -\frac{\partial f_o}{\partial u_m} \\ \cdot & & \cdot \\ \cdot & & \cdot \\ \cdot & & \cdot \\ \frac{\partial f_n}{\partial u_1} & \cdots & \frac{\partial f_n}{\partial u_m} \end{bmatrix} ,$$

and $\frac{dp^*}{dx}$ as defined earlier, all evaluated at $x = x^*(t)$, $u = p^*(x^*(t))$, where the solution $x^*(\cdot) : [t_o^*,t_1] \to R^n$ generated by $p^*(\cdot)$ is such that $x^*(t) = \hat{x}^*(t_1-t)$. Then there exists an absolutely continuous function $\lambda(\cdot) : [t_o^*,t_1] \to R^{n+1}$

satisfying

$$\dot{\lambda}(t) = - \left[\frac{\partial f}{\partial x} + \frac{\partial f}{\partial u} \frac{dp^*}{dx} \right]^T \lambda(t) \tag{9.6}$$

for almost all $t \in [t_o^*, t_1]$, such that

$$\max_{u \in U(x^*(t))} H(\lambda(t), x^*(t), u)$$

$$= H(\lambda(t), x^*(t), p^*(x^*(t)))$$

$$= 0 \ . \tag{9.7}$$

Furthermore, the zeroth component of $\lambda(t)$ is

$$\lambda_o(t) \equiv - 1 \ . \tag{9.8}$$

Finally, the transversality condition becomes

$$\sum_{i=1}^{n} \lambda_i(t_o^*) \, \eta_i = 0 \tag{9.9}$$

for all η such that

$$\sum_{i=1}^{n} \frac{\partial \theta_j(x^*(t_o^*))}{\partial x_i} \eta_i = 0 \ , \quad j = 1,2,\ldots,r \ . \tag{9.10}$$

10. REACHABILITY

Now let us turn to the question of <u>reachability</u>, and recall the definition of <u>playability</u>, Definition 9.1. If $p(\cdot)$ is playable at x^1 , then the state x^1 can be <u>reached</u> from some state in θ by use of the feedback control $p(\cdot)$. We are concerned with the set of all states which can be reached from states in θ by means of all admissible feedback controls.

<u>Definition 10.1</u>. The reachable set

$$X \triangleq \bigcup_{x^o \in \theta} X(x^o)$$

where

$$X(x^o) \triangleq \{ x^1 \in R^n \mid \exists\ p(\cdot) \in P \text{ which generates a solution } x(\cdot) : [t_o,t_1] \to R^n \text{ of (9.1) such that } x(t_o) = x^o ,\ x(t) \neq x^o \text{ for } t > t_o \text{ and } x(t_1) = x^1 \} .$$

Before considering a condition that must be met at the point of a trajectory belonging to the boundary of the reachable set, ∂X, we state an assumption.

Assumption 10.1. Given $\bar{x} \in \partial X$, $\bar{x} \notin \theta$, there are a neighborhood $N(\bar{x})$ of $\bar{x}$ and a C^1-function $W_{\bar{x}}(\cdot) : N(\bar{x}) \to R$ such that

$$W_{\bar{x}}(\bar{x}) = 0 \quad \text{and}$$

$$W_{\bar{x}}(x) \leqslant 0 \qquad \forall\ x \in N(\bar{x}) \cap \bar{X} .$$

Then one can establish readily the following result.

Lemma 10.1. Let $\bar{p}(\cdot) \in P$ be a feedback control that generates a solution $\bar{x}(\cdot) : [t_o,t_1] \to R^n$ of (9.1) such that $\bar{x}(t_o) \in \theta$, $\bar{x}(t) \notin \theta$ for $t > t_o$, and $\bar{x}(t) \in \partial X$ for all $t \in [t',t''] \subset [t_o,t_1]$. Let $\bar{x} \triangleq \bar{x}(\bar{t})$, $\bar{t} \in [t',t'')$. Then, provided Assumption 10.1 is satisfied,

$$\max_{u \in U(\bar{x})} \operatorname{grad}^T W_{\bar{x}}(\bar{x})\ f(\bar{x},u)$$
$$= \operatorname{grad}^T W_{\bar{x}}(\bar{x})\ f(\bar{x},\bar{p}(\bar{x}))$$
$$= 0 .$$

Proof. Consider a $\bar{\bar{p}}(\cdot) \in P$ and the control $p(\cdot)$ such that

$$p(x) = \bar{p}(x) \quad \text{for} \quad x_n < \bar{x}_n$$

$$p(x) = \bar{\bar{p}}(x) \quad \text{for} \quad x_n \geqslant \bar{x}_n \quad .$$

In view of Definition 2.1, $p(\cdot) \in P$. Now let $\bar{\bar{x}}(\cdot) : [\bar{t}, t_2] \to R^n$, $\bar{\bar{x}}(\bar{t}) = \bar{x}$, denote a solution of (9.1) generated by $\bar{\bar{p}}(\cdot)$. Then $p(\cdot)$ generates the solution $x(\cdot) : [t_0, t_2] \to R^n$ such that

$$x(t) = \bar{x}(t) \quad \text{for} \quad t \in [t_0, \bar{t}]$$

$$x(t) = \bar{\bar{x}}(t) \quad \text{for} \quad t \in [\bar{t}, t_2] \quad .$$

Finally, as a consequence of Definition (10.1), it follows that $x(t) \in X$ for all $t \in [\bar{t}, t_2]$.

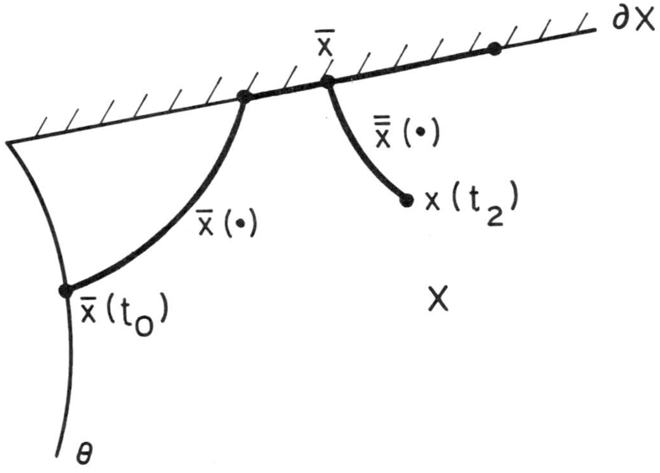

Figure 10.1. Trajectories in the Proof of Lemma 10.1

Then, for sufficiently small $\varepsilon > 0$,

$$W_{\bar{x}}(x(\bar{t}+\varepsilon)) = W_{\bar{x}}(\bar{x}) + \varepsilon \, \text{grad}^T W_{\bar{x}}(\bar{x}) \, f(\bar{x},p(\bar{x})) + o(\varepsilon) \quad .$$

Hence, as a consequence of Assumption 10.1,

$$\text{grad}^T W_{\bar{x}}(\bar{x}) \, f(\bar{x},p(\bar{x})) \leqslant 0 \; .$$

But $\bar{\bar{p}}(\cdot) \in P$ is arbitrary; that is, one may choose any $\bar{\bar{p}}(\bar{x}) = p(\bar{x}) \in U(\bar{x})$. Thus,

$$\text{grad}^T W_{\bar{x}}(\bar{x}) \, f(\bar{x},u) \leqslant 0 \quad \forall \, u \in U(\bar{x}) \quad .$$

In an entirely analogous manner one can show that

$$\text{grad}^T W_{\bar{x}}(\bar{x}) \, f(\bar{x},\bar{p}(\bar{x})) = 0 \; .$$

This concludes the proof.

11. EXAMPLE: TIME-OPTIMAL NAVIGATION

In order to illustrate the results of the aforegoing sections we consider a version of the classical navigation problem of Zermelo[7]. Consider a boat moving with velocity $\vec{v}$ of constant magnitude v relative to a stream that flows with constant velocity $\vec{s}$ relative to the shore. Taking $v = 1$, and letting x_1 and x_2 be the position coordinates (with the x_2-axis in the direction of $\vec{s}$), the equations of motion of the boat are

$$\dot{x}_1(t) = s + u_1(t) \quad , \quad \dot{x}_2(t) = u_2(t) \quad , \qquad (11.1)$$

where s is the constant magnitude of $\vec{s}$, and $u_1 = \cos\phi$, $u_2 = \sin\phi$; see Figure 11.1. Thus, the control constraint set

$$U(x) = \{u \in R^2 \mid u_1^2 + u_2^2 = 1\} \equiv \text{const.} \overset{\Delta}{=} U \quad . \tag{11.2}$$

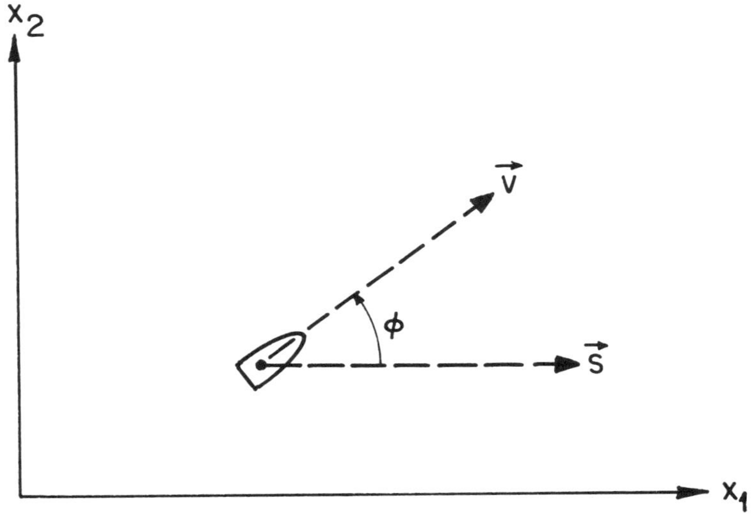

Figure 11.1. Navigation Geometry

In terms of feedback control $p(x) \in U \subset R^2$, with $x \overset{\Delta}{=} (x_1, x_2, t)^T$, the state equation (9.1) becomes

$$\dot{x}_1(t) = s + p_1(x(t))$$

$$\dot{x}_2(t) = p_2(x(t)) \tag{11.3}$$

$$\dot{x}_3(t) = 1 \quad .$$

It is desired to steer the boat from a given initial position, (0,0), to a given terminal one, (x_1^1, x_2^1), in minimum time.

Thus,

$$\theta = \{x \in R^3 \mid x_1 = 0 \ , \ x_2 = 0\} \tag{11.4}$$

and the terminal state $x(t_1) = (x_1^1, x_2^1, t_1) \overset{\Delta}{=} x^1$ is prescribed.

The performance index is the transfer time; that is,

$$V(x^1, p(\cdot), x(\cdot)) = \int_{t_0}^{t_1} dt \ . \tag{11.5}$$

In general, three possibilities must be considered: $s < v$, $s = v$, and $s > v$. Hereafter we shall restrict the discussion to the latter case, $s > v = 1$.

Let us turn first to the reachable set defined in Definition 10.1. As a consequence of the control constraint (11.2), it is

$$X = \{x \in R^3 \mid x_2^2 \leqslant \frac{x_1^2}{s^2-1} \} \tag{11.6}$$

The projection of the reachable set onto R^2 is shown in Figure 11.2.

Thus, for

$\bar{x} \in \{x \in \partial X \mid x_2 > 0\}$

there is a $W_{\bar{x}}(\cdot)$ such that

$$W_{\bar{x}}(x) = x_2 - \frac{x_1}{\sqrt{s^2-1}} \overset{\Delta}{=} W_+(x) \ . \tag{11.7}$$

Similarly, for

$\bar{x} \in \{x \in \partial X \mid x_2 < 0\}$

there is a $W_{\bar{x}}(\cdot)$ such that

$$W_{\underline{x}}(x) = -x_2 - \frac{x_1}{\sqrt{s^2-1}} \stackrel{\Delta}{=} W_-(x) \quad . \tag{11.8}$$

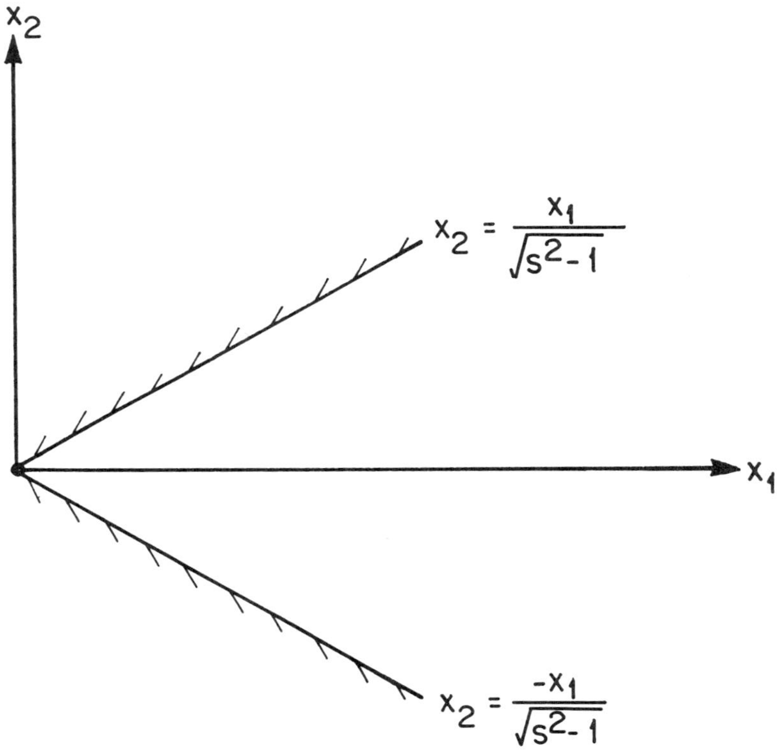

Figure 11.2. Projection of X onto R^2

Before taking up a discussion of Lemma 10.1, we turn to the question of optimality. In particular, we shall invoke necessary conditions to deduce a feedback control that is a candidate for one that is optimal on X . To do this, we employ (9.5); namely, if $p^*(\cdot)$ is optimal on X , then for all $x \in \text{int } X$

$$\max_{u \in U} \left[-1 + \frac{\partial V^*(x)}{\partial x_1}(s+u_1) + \frac{\partial V^*(x)}{\partial x_2} u_2 + \frac{\partial V^*(x)}{\partial t}\right]$$

$$= \left[-1 + \frac{\partial V^*(x)}{\partial x_1}(s+p_1^*(x)) + \frac{\partial V^*(x)}{\partial x_2} p_2^*(x) + \frac{\partial V^*(x)}{\partial t}\right]$$

$$= 0 \quad . \tag{11.9}$$

The first part of this relation yields

$$p_i^*(x) = \frac{\partial V^*(x)}{\partial x_i}\left[\left(\frac{\partial V^*(x)}{\partial x_1}\right)^2 + \left(\frac{\partial V^*(x)}{\partial x_2}\right)^2\right]^{-1/2} \tag{11.10}$$

for $i = 1,2$. The second part of (11.9) then becomes

$$-1 + \left[\left(\frac{\partial V^*(x)}{\partial x_1}\right)^2 + \left(\frac{\partial V^*(x)}{\partial x_2}\right)^2\right]^{1/2} + s\,\frac{\partial V^*(x)}{\partial x_1} + \frac{\partial V^*(x)}{\partial t} = 0. \tag{11.11}$$

subject to $V^*(x) = 0$ for $x_1 = x_2 = 0$ and $t \in R$.

Now one can verify by direct substitution that

$$V^*(x) = \left[s\,x_1 - \sqrt{x_1^2-(s^2-1)\,x_2^2}\,\right](s^2-1)^{-1} \tag{11.12}$$

so that, in view of (11.10), the candidate optimal control is such that

$$\begin{aligned} p_1^*(x) &= \frac{x_1}{V^*(x)} - s \\ p_2^*(x) &= \frac{x_2}{V^*(x)} \end{aligned} \tag{11.13}$$

for $x \in \operatorname{int} X$.

Now we note that for $\bar{x} \in \partial X$

$$\begin{aligned} \hat{p}_1(\bar{x}) &\triangleq \lim_{x\to\bar{x}} p_1^*(x) = -\frac{1}{s} \\ \hat{p}_2(\bar{x}) &\triangleq \lim_{x\to\bar{x}} p_2^*(x) = \begin{cases} \sqrt{s^2-1}\,/s & \text{for } \bar{x}_2 > 0 \\ -\sqrt{s^2-1}\,/s & \text{for } \bar{x}_2 < 0 \end{cases} \end{aligned} \tag{11.14}$$

Finally, upon substituting (11.13) in the first two of state equations (11.3), one is led to conclude that the projections onto R^2 of corresponding trajectories are straight lines.

It is left to the interested reader to verify that the feedback control (11.13) also satisfies the necessary conditions (9.6)-(9.10) of the minimum principle; see also Ref. 8.

Since any $\hat{p}(\cdot) \in P$ satisfying (11.14) results in a trajectory that belongs entirely to ∂X provided $x^1 \in \partial X$, we wish to verify Lemma 10.1; namely,

$$\max_{u \in U} \left[\frac{\partial W_{\bar{x}}(\bar{x})}{\partial x_1}(s+u_1) + \frac{\partial W_{\bar{x}}(\bar{x})}{\partial x_2} u_2 + \frac{\partial W_{\bar{x}}(\bar{x})}{\partial t} \right]$$

$$= \frac{\partial W_{\bar{x}}(\bar{x})}{\partial x_1}(s+\hat{p}_1(\bar{x})) + \frac{\partial W_{\bar{x}}(\bar{x})}{\partial x_2} \hat{p}_2(\bar{x}) + \frac{\partial W_{\bar{x}}(\bar{x})}{\partial t}$$

$$= 0 \ . \qquad (11.15)$$

For example, if $\bar{x} \in \partial X$ such that $\bar{x}_2 > 0$, then it follows from (11.7) that $W_{\bar{x}}(x) = W_+(x)$ so that

$$\frac{\partial W_+(x)}{\partial x_1} = -(s^2-1)^{-1/2} \ , \ \frac{\partial W_+(x)}{\partial x_2} = 1 \ , \ \frac{\partial W_+(x)}{\partial t} = 0 \ . \qquad (11.16)$$

Now it is a consequence of the first part of (11.15) that $\hat{p}_i(x)$ is given by (11.14). The second part of (11.15) is then verified by direct substitution.

Acknowledgement

The contents of this chapter are based on research supported by the National Science Foundation under Grant ECS-7813931.

REFERENCES

1. Filippov, A. F., Differential equations with many-valued discontinuous right-hand side, Soviet Mathematics, Vol. 4, No. 4, 1963.
2. Roxin, E., On generalized dynamical systems defined by contingent equations, Journal of Differential Equations, Vol. 1, No. 2, 1965.
3. Hermes, H., Discontinuous vector fields and feedback control, in Differential Equations and Dynamical Systems, Hale, J. K. and LaSalle, J. P., Eds, Academic Press, N.Y., 1967.
4. Davy, J. L., Properties of the solution set of a generalized differential equation, Bulletin of the Australian Mathematical Society, Vol. 6, 1972.
5. Hajek, O., Discontinuous differential equations, Journal of Differential Equations, Vol. 32, No. 2, 1979.
6. Leitmann, G., Optimal feedback control for dynamical systems with one or two controllers, in Dynamical Systems and Microphysics, Blaquière, A., Fer, F., and Marzollo, A., Eds., Springer-Verlag, Vienna, 1980.
7. Zermelo, E., Über das Navigationsproblem bei ruhender oder veränderlicher Windverteilung, Zeitschrift für Angewandte Mathematik und Mechanik, Vol. 11, No. 2, 1931.
8. Leitmann, G., The Calculus of Variations and Optimal Control, Plenum Press, N.Y., 1981.

BIBLIOGRAPHY

1. Blaquière, A. and Leitmann, G., On the geometry of optimal processes, Parts I, II, III, University of California, Berkeley, IER Reports AM-64-10, AM-65-11, AM-66-1.
2. Blaquière, A., and Leitmann, G., On the geometry of optimal processes, in Topics in Optimization, Leitmann, G., Ed., Academic Press, New York, 1967.
3. Blaquière, A., and Gérard F., On the geometry of optimal strategies in two-person games of kind, Journal of Computer Science, Vol. 2, No. 3, 1968.
4. Blaquière, A., Gérard, F., and Leitmann, G., Quantitative and Qualitative Games, Academic Press, New York, 1969.

OPTIMIZATION AND CONTROLLABILITY IN PROBLEMS OF RELATIVISTIC DYNAMICS AND OF GEOMETRICAL OPTICS

A. Blaquière

Laboratoire d'Automatique Théorique
Université de Paris 7
Paris, France

INTRODUCTION.

The main purpose of this article is to discuss the problem of finding geodesics in the Riemannian space-time of General Relativity, from the point of view of the theory of optimal control.

As usual we shall consider a mass-point, with strictly positive proper mass, which moves along a path in a given Riemannian space-time. An integral of action of Hamilton, or as we shall say an integral cost, is associated with each possible path. Then the problem of finding geodesics is stated as the problem of finding paths of minimum cost joining two points.

One of the reasons for relying on the modern theory of optimal control is that, in the theory of General Relativity, the set to which the velocity vectors belong, or, as we shall say, the control constraint set, is *closed*. It follows that a correct treatment of the problem cannot be based on the calculus of variations, in which *open* control constraint sets only are considered. Correlatively

ISBN 0-12-068720-8

we cannot make use of the Euler-Lagrange equations unless we have proved that we are entitled to do so.

In general, up to our best knowledge, the literature specialized in the area of General Relativity seems not to worry about this need, as historical reasons have conferred the freedom of the city upon variational methods. Why worry as long as Mechanics is in order ?

This is one question we shall answer, in the frame of certain assumptions. By giving the expected proof, we shall make precise the conditions according to which variational methods can be safely used. In addition we shall make precise the situations in which they are forbidden.

By saying a few words about this delimitation, in this introduction, we enter the very kernel of the matter.

In the optimal control problem we shall consider, the mass-point is supposed to emanate from a given point-source. An important feature of this problem is that it separates into two sub-problems, namely

- first *find the reachable set,* that is the set of all points in space-time which can be reached from the given point-source, along a path; and
- secondly, for each arbitrarily given end point in the reachable set, *find an optimal path,* that is one which emanates from the source, ends at that point, and along which the integral of action of Hamilton is minimum.

The first of these problems is a problem of controllability or equivalently a (one-player) qualitative game, whereas the second one is a problem of optimal control or equivalently a (one-player) quantitative game.

Associated with these two problems are two classes of paths, namely

- paths all of whose points are interior points of the reachable set; and

- paths all of whose points lie in the boundary of the reachable set.

A basic result brought in this article is expressed by Proposition 2. It is a separation theorem according to which, under proper assumption, there exist no other path than those which range in one or the other of these two classes.

Moreover for paths ranging in the first set, we have shown that the control belongs to the interior of the control constraint set. It follows that for such paths the variational methods are valid, and that in particular the use of the Euler-Lagrange equations is permitted.

On the other hand, for paths ranging in the second set, the control belongs to the boundary of the control constraint set. It follows that for such paths the variational methods are non valid, and that the use of the Euler-Lagrange equations is prohibited.

The first case lies in the domain of classical relativistic Mechanics in which traditional methods apply, whereas in the second case traditional methods of Mechanics do not apply.

Does the second case correspond to any real situation in Mechanics ?

As we will see, the cost of boundary paths - those ranging in the second set - is zero. Such paths are geodesics of null length of the given Riemannian space-time. In the theory of General Relativity, they are supposed to be the trajectories of photons in the given gravitational field.

In other words, the set of boundary paths emanating from the given point-source is the same as the set of light rays of geometrical optics emanating from the same source.

Hence the second case is concerned with the domain of geometrical optics.

The conclusion is that the modern theory of optimal control provides a theoretical frame in which relativistic dynamics of a mass-point and geometrical optics are on the same footing.

In addition to the fact that this treatment removes a good deal of the inconsistencies which are found in the literature concerned with these two areas (e.g., though it is claimed that there are connections between them, the dynamics of a mass-point and geometrical optics are treated as two problems of completely different nature) it brings to light an unusual fact; that is, contrary to ideas currently accepted,

1. in a problem of relativistic dynamics of a mass-point, with proper mass *strictly positive,* there can exist cases in which the speed of the point is *equal* to the speed of light in the vacuum; and

2. the proper mass of a photon, when the photon is modelized as a mass-point whose trajectory is a geodesic of null-length along the light-cone, needs not be zero.

However, by our separation theorem (namely Proposition 2) there exist no path along which the mass-point behaves as a regular mass-point of Mechanics during part of the time, and as a photon during part of the time.

That is, for an observer the two domains, relativistic Mechanics and geometrical Optics, appear as completely separated.

In this introduction we have restrained our account to the basic ideas which legitimate the splitting of this article into two parts, namely Part I and Part II. In particular we have left out of our scope some connections with Wave Mechanics, pointed out in the text, which will be analyzed in detail elsewhere.

PART I

AN OPTIMAL CONTROL PROBLEM IN RELATIVISTIC DYNAMICS OF A MASS-POINT

In Part I we shall discuss a problem which is concerned with finding geodesics in the Riemannian space-time of General Relativity. In the literature this problem is sometimes stated as a variational problem and the equations for the geodesics are the Euler equations associated with it. In view of further developments, we shall put this problem in the form of a (closed-loop) optimal control problem.

1. PROBLEM STATEMENT, AND FIRST ASSUMPTIONS.

Let the *system* under consideration be a mass point with (constant) *proper mass* $m_o > 0$ and consider the *one-player differential game* (i.e., the problem of *closed-loop control)* defined by

(a) the set $\mathcal{E} \times T$ of *states* $y = (x,t)$ of the system, where $\mathcal{E}$ is a domain of a 3-dimensional Euclidean space R^3, containing state $(0,0)$, and $T = (-\infty, +\infty)$;

(b) the set U of *strategies* $u(\cdot)$, which are functions defined on $\mathcal{E} \times T^+$, where $T^+ = (0, +\infty)$, of class C^1 on $\mathcal{E}$ and piece-wise continuous on T^+, such that

$$u(x,t) \in K(x,t) \subset R^3 \quad \text{for each } (x,t) \in \mathcal{E} \times T^+$$

where, for each $(x,t) \in \mathcal{E} \times T^+$, $K(x,t)$ is a given subset of R^3;

(c) the class of *state equations*

$$\dot{x} = u(x,t), \qquad u(\cdot) \in U \tag{1}$$

where "$\cdot$" means d/dt ;

(d) the *initial set* Θ, or *source*

$$\Theta = \{(0,0)\} \subset \mathcal{E} \times T$$

Further let there be given a *metric*

$$ds^2 = \Sigma\, g_{ij} dx_i dx_j, \quad i,j = 1,2,3,4, \quad x_4 \equiv t$$

where the g_{ij} are functions defined and of class C^1 on $\mathcal{E} \times T$, and the determinant g of the g_{ij} is non-zero. We shall suppose that $g_{ij} = g_{ji}$ for all i,j = 1,2,3,4 .

We shall denote by g^{ij} the coefficients of the inverse matrix $(g_{ij})^{-1}$.

Let the *constraint set* K(x,t) be given by

$$G(x,t,u) \triangleq (ds^2/dt^2) \geqslant 0$$

A strategy $u(\cdot) \in U$ is termed *admissible*, or *playable at* $(x^1,t_1) \in \mathcal{E} \times T^+$ if it generates a solution $x(\cdot)$ of (1) on the interval $[0,t_1]$, such that $x(0) = 0$ and $x(t_1) = x^1$. The ordered set of states

$$\pi = \{(x,t): \ x = x(t), \quad t \in [0,t_1]\}$$

is called a *path*. In this definition the ordering is the "natural" one induced by time $x_4 \equiv t$. We shall denote by J(x,t) the set of all strategies admissible at $(x,t) \in \mathcal{E} \times T^+$.

We shall define the *cost* of path π generated by $u(\cdot)$ by

$$V(x^1,t_1,u(\cdot)) = \int_0^{t_1} L(x(t),t,u(x(t),t))dt$$

where $L(x,t,u) = -m_0 c(ds/dt) = -m_0 c\sqrt{G(x,t,u)}$; c is the speed of light in vacuum in the absence of gravitational field.

The *reachable set* is $Y = \Theta \cup Y^+$, where

$$Y^+ = \{y = (x,t): \ (x,t) \in \mathcal{E} \times T^+, \ \exists u(\cdot) \in J(x,t)\}$$

Then a classical problem in (closed-loop) optimal control is the following: find a strategy $u^*(\cdot) \in U$ such that

$$u^*(\cdot) \in J(x,t) \quad \text{and} \quad V(x,t,u^*(\cdot)) \leqslant V(x,t,u(\cdot))$$

for all $u(\cdot) \in J(x,t)$, and for all $(x,t) \in Y^+$.

Such a strategy, if it exists, is called an *optimal strategy on* Y.

We shall let $V^*(x,t) = V(x,t,u^*(\cdot))$ for all $(x,t) \in Y^+$, and

$V^*(0,0) = 0$. Function $V^*(\cdot)$ is defined on Y with range in R. It is termed *optimal value function.*

The problem of finding geodesics, i.e., shortest curves joining two points on some manifold can be put in the form of a control problem of the type above. By an extension of the notion of "distance", the minimum value of the cost at (x,t) is termed the *geodetic distance* between Θ and (x,t)) [1].

Before we proceed, we shall draw up a list of some of our assumptions.

Assumption 1. For all $(x,t) \in Y^+$, for all $u \in K(x,t)$, there exists a strategy $u(\cdot) \in U$, which is continuous, and continuously differentiable with respect to x, on some open neighbourhood of (x,t) in $\mathcal{E} \times T^+$, and such that $u(x,t) = u$.

Assumption 2. Y is the closure(1) of a domain $D \subset \mathcal{E} \times T^+$, whose boundary ∂Y is given by

$$W(x,t) = 0$$

where function $W(\cdot)$ is defined and continuous on $\mathcal{E} \times T$, $\operatorname{grad} W(x,t)$ is defined, continuous and non-zero on $(\mathcal{E} - \{0\}) \times T$, and $\partial W(x,t)/\partial t$ is defined, continuously differentiable and strictly negative on $\mathcal{E} \times T$.

We shall denote by $\mathring{Y}$ the *interior* of Y, that is $\mathring{Y} = D$, and by ∂Y the *boundary* of Y.

Assumption 3. There exists an optimal strategy on Y, namely $u^*(\cdot) \in U$, which is continuously differentiable on $\mathcal{E} \times T^+$.

Assumption 4. For all $(x,t) \in \mathcal{E} \times T^+$, the constraint set K(x,t) is the closure of a domain $\Omega(x,t) \subset R^3$, whose boundary $\partial K(x,t)$ is given by

$$G(x,t,u) = 0$$

where function $G(\cdot)$ is defined and of class C^1 on $\mathcal{E} \times T \times R^3$, and $G(x,t,0) > 0$.

We shall denote by $\mathring{K}(x,t)$ the interior of K(x,t).

---

(1) In the topology induced by $R^3 \times T$ on $\mathcal{E} \times T^+$.

Assumption 5. Function $V^*(\cdot)$ is continuous on Y and twice continuously differentiable on $\mathring{Y}$.

2. A FIRST BASIC PROPOSITION.

In fact we have two problems, namely: first *find the reachable set* Y, which is a problem of controllability or equivalently a (one-player) qualitative game(2); and secondly *find an optimal strategy on* Y, which is a problem of optimal control or equivalently a (one-player) quantitative game(3). Indeed these problems are not independent from one another.

Since the problem of optimal control is more classical than the problem of controllability, we shall consider it first; and we shall discuss the other one in Part II.

In order to prepare further discussions, we shall state here without proof a first basic proposition. Its proof is similar to the proof of Theorem 3.1 of Ref.(2) except for minor changes due to the fact that the *terminal set* or *target* of Ref.(2) is replaced here by an initial set or source. Note that Theorem 3.1 of Ref.(2) which has been derived for two-player games applies as well to one-player games, as a special case, if one "freezes" the maximizing player(4) .

Proposition 1. If Assumptions 1 - 5 are satisfied , then at each state $y = (x,t) \in \mathring{Y}$

$$\text{(i)} \qquad \max_{u \in K(y)} H(\lambda(y),\lambda_t(y),y,u) = H(\lambda(y),\lambda_t(y),y,u^*(y)),$$

$$\text{(ii)} \qquad H(\lambda(y),\lambda_t(y),y,u^*(y)) = 0$$

---

(2) According to the terminology of Ref.(2).

(3) According to the terminology of Ref.(2).

(4) Closed-loop control problems cannot be handled in a similar fashion as the ones of open-loop control. They require appropriate techniques.

where

$$H(\lambda, \lambda_t, y, u) = \lambda_o L(x,t,u) + \lambda_t + \lambda \cdot u ,$$

and

$$\lambda_o = -1, \quad \lambda(y) = \text{grad } V^*(x,t),^{(5)} \quad \lambda_t(y) = \partial V^*(x,t)/\partial t .$$

Further along an optimal path represented by $x^*(\cdot): [0,t_1] \to \mathcal{E}$ generated by $u^*(\cdot)$, such that $(x^*(t),t) \in \overset{\circ}{Y}$ and $u^*(x^*(t),t) \in \overset{\circ}{K}(x^*(t),t)$ for all $t \in (0,t_1]$, we have the *adjoint equations*

$$\dot{\lambda}_\alpha = - \partial H(\lambda,\lambda_t,y,u)/\partial x_\alpha - \sum_{\beta=1}^{3} (\partial H(\lambda,\lambda_t,y,u)/\partial u_\beta)(\partial u^*_\beta(y)/\partial x_\alpha) \qquad \alpha = 1,2,3 \tag{2}$$

$$\dot{\lambda}_t = - \partial H(\lambda,\lambda_t,y,u)/\partial t - \sum_{\beta=1}^{3} (\partial H(\lambda,\lambda_t,y,u)/\partial u_\beta)(\partial u^*_\beta(y)/\partial t) \tag{3}$$

where the partial derivatives are computed for $u = u^*(y)$, and the derivatives $\dot{\lambda}_\alpha$, $\dot{\lambda}_t$ are computed at state y, along the optimal path generated by $u^*(\cdot)$ through y .

3. PATHS IN AUGMENTED STATE SPACE.

By letting $V^o(x,t) = (- 1/m_0 c)V^*(x,t)$, conditions (i) and (ii) of Proposition 1 are replaced by

$$\min_{u \in K(x,t)} (- \sqrt{G(x,t,u)} + (\partial V^o/\partial t) + (\partial V^o/\partial x) \cdot u) \tag{4}$$

$$= - \sqrt{G(x,t,u^*(x,t))} + (\partial V^o/\partial t) + (\partial V^o/\partial x) \cdot u^*(x,t) = 0$$

where $\partial V^o/\partial x = (\partial V^o/\partial x_1, \partial V^o/\partial x_2, \partial V^o/\partial x_3)$ and $\partial V^o/\partial t$ are computed at state $(x,t) \in \overset{\circ}{Y}$.

---

(5) grad $V^*(x,t) = \partial V^*(x,t)/\partial x = (\partial V^*(x,t)/\partial x_1, \partial V^*(x,t)/\partial x_2, \partial V^*(x,t)/\partial x_3)$.

This minor change in the notation is equivalent to replacing the cost function in the problem statement of paragraph 1 by

$$L(x,t,u) = \sqrt{G(x,t,u)} \qquad (5)$$

Then, because of the change of the sign, optimality means maximizing the new integral cost.

In other words, to our first version which is a problem of *least action* there corresponds the present version which is a problem of *maximum length*.

In order to make use of condition (4) we need know which one of the following cases is met, namely

Case 1 : $u^*(x,t) \in \overset{\circ}{K}(x,t)$; or

Case 2 : $u^*(x,t) \in \partial K(x,t)$.

We will see later that the condition $(x,t) \in \overset{\circ}{Y}$ rules out Case 2. However, for the time being, let us consider these two possibilities.

Case 1.

In Case 1 we have $G(x,t,u^*(x,t)) \neq 0$, and since $G(x,t,u) \neq 0$ for all $u \in \overset{\circ}{K}(x,t)$, condition (4) implies

$$-1 + (\partial V^o/\partial t)(1/\sqrt{G(x,t,u)}) + (\partial V^o/\partial x) \cdot (u/\sqrt{G(x,t,u)}) \geqslant 0 \qquad (6)$$

for all $u \in \overset{\circ}{K}(x,t)$; and

$$-1 + (\partial V^o/\partial t)(1/\sqrt{G(x,t,u^*)}) + (\partial V^o/\partial x) \cdot (u^*/\sqrt{G(x,t,u^*)}) = 0 \qquad (7)$$

where we have written for brevity u^* in the place of $u^*(x,t)$.

A new step is taken by letting

$$\begin{aligned} \tilde{u}_4 &= dt/ds = 1/\sqrt{G(x,t,u)} \\ \tilde{u} &= dx/ds = (dx/dt)(dt/ds) = u/\sqrt{G(x,t,u)} \\ \tilde{u}_4^* &= 1/\sqrt{G(x,t,u^*)} \\ \tilde{u}^* &= u^*/\sqrt{G(x,t,u^*)} \end{aligned} \qquad (8^*)$$

which means replacing the state equations (1) by

$$\begin{aligned} dx/ds &= \tilde{u}(x,t) \\ dt/ds &= \tilde{u}_4(x,t) \end{aligned} \qquad (8)$$

with s as a new parameter.

We have thus introduced a 5-dimensional space whose points are (x_1, x_2, x_3, t, s), Fig.(2).

An important remark is that, associated with equations (8) is the new constraint

$$G_s(x,t,\tilde{u},\tilde{u}_4) = 1 \tag{9}$$

where

$$G_s(x,t,\tilde{u},\tilde{u}_4) = (ds^2/ds^2) = \sum_{i,j} g_{ij}(x,t)\tilde{u}_i\tilde{u}_j, \qquad i,j = 1,2,3,4. \tag{10}$$

That is, $(\tilde{u},\tilde{u}_4)$ belongs to a *constraint set* $K_s(x,t)$ defined by (9) and (8*).

In other words, conditions (6) and (7) are now replaced by

$$\begin{aligned} \min_{(\tilde{u},\tilde{u}_4)\in K_s(x,t)} & (-1 + (\partial V^o/\partial t)\tilde{u}_4 + (\partial V^o/\partial x)\cdot\tilde{u}) = \\ & = (-1 + (\partial V^o/\partial t)\tilde{u}_4^* + (\partial V^o/\partial x)\cdot\tilde{u}^*) = 0 \end{aligned} \tag{11}$$

In view of the differentiability of $G_s(x,t,\tilde{u},\tilde{u}_4)$ with respect to the $\tilde{u}_\alpha$, $\alpha = 1,2,3,4$, one can prove easily that

$$\begin{aligned} \frac{\partial V^o(x,t)}{\partial x_\alpha} &= k \left.\frac{\partial G_s(x,t,\tilde{u},\tilde{u}_4)}{\partial \tilde{u}_\alpha}\right|_{\substack{\tilde{u} = \tilde{u}^* \\ \tilde{u}_4 = \tilde{u}_4^*}} \\ \frac{\partial V^o(x,t)}{\partial t} &= k \left.\frac{\partial G_s(x,t,\tilde{u},\tilde{u}_4)}{\partial \tilde{u}_4}\right|_{\substack{\tilde{u} = \tilde{u}^* \\ \tilde{u}_4 = \tilde{u}_4^*}} \end{aligned} \tag{12}$$

where k is a non-zero real number (Fig.1).

From (11) and (12) we deduce

$$-1 + k((\partial G_s/\partial\tilde{u}_4)\tilde{u}_4^* + (\partial G_s/\partial\tilde{u})\cdot\tilde{u}^*) = 0$$

where $G_s = G_s(x,t,\tilde{u},\tilde{u}_4)$, $\partial G_s/\partial\tilde{u} = (\partial G_s/\partial\tilde{u}_1, \partial G_s/\partial\tilde{u}_2, \partial G_s/\partial\tilde{u}_3)$, and the partial derivatives with respect to the $\tilde{u}_\alpha$, $\alpha = 1,2,3,4$,

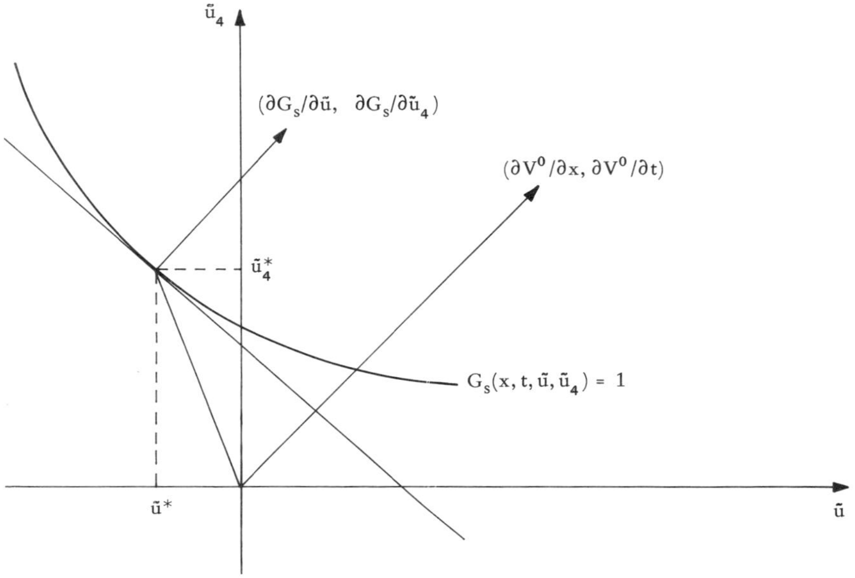

Fig. 1

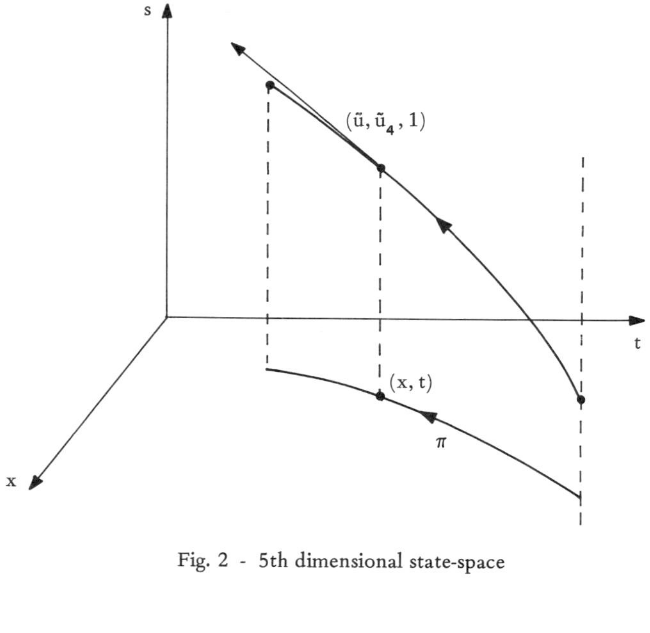

Fig. 2 - 5th dimensional state-space

are computed for $\tilde{u} = \tilde{u}^*(x,t)$ and $\tilde{u}_4 = \tilde{u}_4^*(x,t)$.

At last, since

$$(\partial G_s/\partial \tilde{u}_4)\tilde{u}_4^* + (\partial G_s/\partial \tilde{u}) \cdot \tilde{u}^* = 2G_s(x,t,\tilde{u}^*,\tilde{u}_4^*) = 2$$

we obtain the value of k, namely

$$k = 1/2 \qquad (13)$$

We recognize the method we have used as being a special case of the *Lagrange multipliers rule*, according to which

$$\frac{\partial}{\partial \tilde{u}_\alpha}(-k(G_s(x,t,\tilde{u},\tilde{u}_4)-1) - 1 + (\partial V^o/\partial t)\tilde{u}_4 + (\partial V^o/\partial x) \cdot \tilde{u}) = 0 \qquad (14)$$

$$\alpha = 1,2,3,4,$$

the partial derivatives with respect to the $\tilde{u}_\alpha$, $\alpha = 1,2,3,4$ being computed for $\tilde{u} = \tilde{u}^*(x,t)$ and $\tilde{u}_4 = \tilde{u}_4^*(x,t)$.

Since from (10)

$$\partial G_s/\partial \tilde{u}_i = 2 \sum_{j=1}^{4} g_{ij}\, \tilde{u}_j, \qquad i = 1,2,3,4$$

we have

$$\tilde{u}_j = (1/2) \sum_{i=1}^{4} g^{ij}(\partial G_s/\partial \tilde{u}_i), \qquad j = 1,2,3,4$$

and from (12) and (13) it follows that

$$\tilde{u}_j^* = \sum_{i=1}^{4} g^{ij}(\partial V^o/\partial x_i), \qquad j = 1,2,3,4$$

By substituting in (11) we obtain

$$\sum_{i,j} g^{ij}(\partial V^o/\partial x_i)(\partial V^o/\partial x_j) = 1, \qquad i,j = 1,2,3,4 \qquad (15)$$

which is the equation of Hamilton-Jacobi associated with (11).

Case 2.

In Case 2 we have $G(x,t,u^*(x,t)) = 0$, so that we can no longer make use of the parameter s in a similar way as in the discussion above.

Here, let us introduce a parameter τ such that $dt/d\tau > 0$, and let[6]

$$\begin{aligned} \tilde{u}_4 &= dt/d\tau \\ \tilde{u} = dx/d\tau = (dx/dt)(dt/d\tau) &= u\tilde{u}_4 \\ \tilde{u}_4^* &= 1 \\ \tilde{u}^* &= u^* \end{aligned} \tag{16*}$$

which means replacing the state equations (1) by

$$\begin{aligned} dx/d\tau &= \tilde{u}(x,t) \\ dt/d\tau &= \tilde{u}_4(x,t) > 0 \end{aligned} \tag{16}$$

with τ as a new parameter.

Let us associate with equations (16) the new constraint

$$G_\tau(x,t,\tilde{u},\tilde{u}_4) = 0 \tag{17}$$

where

$$\begin{aligned} G_\tau(x,t,\tilde{u},\tilde{u}_4) &= (ds^2/d\tau^2) = (ds^2/dt^2)(dt^2/d\tau^2) \\ &= G(x,t,u)\tilde{u}_4^2 = \sum_{i,j} g_{ij}(x,t)\tilde{u}_i\tilde{u}_j, \qquad i,j = 1,2,3,4. \end{aligned} \tag{18}$$

That is, let $u \in \partial K(x,t)$, so that $(\tilde{u},\tilde{u}_4)$ belongs to the *constraint set* $K_\tau(x,t)$ defined by (17) and (16*).

From condition (4), since $G(x,t,u) = 0$ for all $u \in \partial K(x,t)$, we deduce

$$\begin{aligned} &\min_{(\tilde{u},\tilde{u}_4) \in K_\tau(x,t)} ((\partial V^o/\partial t)\tilde{u}_4 + (\partial V^o/\partial x) \cdot \tilde{u}) \\ &= (\partial V^o/\partial t)\tilde{u}_4^* + (\partial V^o/\partial x) \cdot \tilde{u}^* = 0 \end{aligned} \tag{19}$$

In view of the differentiability of $G_\tau(x,t,\tilde{u},\tilde{u}_4)$ with respect to the $\tilde{u}_\alpha$, $\alpha = 1,2,3,4$, again one can prove easily that[7]

(6) We shall keep the same notation $\tilde{u},\tilde{u}_4,\tilde{u}^*,\tilde{u}_4^*$, as in Case 1, but with a different meaning, as there is no possible confusion.

(7) Note that, since the determinant g of the g_{ij} is different from zero, $\partial G_\tau/\partial\tilde{u}$ and $\partial G_\tau/\partial\tilde{u}_4$ both vanish for $\tilde{u}=0$, $\tilde{u}_4=0$ only.

$$\frac{\partial V^o(x,t)}{\partial x_\alpha} = k \left. \frac{\partial G_\tau(x,t,\tilde{u},\tilde{u}_4)}{\partial \tilde{u}_\alpha} \right|_{\substack{\tilde{u} = \tilde{u}^* \\ \tilde{u}_4 = \tilde{u}_4^*}}$$

$$\frac{\partial V^o(x,t)}{\partial t} = k \left. \frac{\partial G_\tau(x,t,\tilde{u},\tilde{u}_4)}{\partial \tilde{u}_4} \right|_{\substack{\tilde{u} = \tilde{u}^* \\ \tilde{u}_4 = \tilde{u}_4^*}} \qquad (20)$$

where k is a non-zero real number.

Again, the method we have used in a special case of the Lagrange multipliers rule, according to which

$$\frac{\partial}{\partial \tilde{u}_\alpha} (-k\, G_\tau(x,t,\tilde{u},\tilde{u}_4) + (\partial V^o/\partial t)\tilde{u}_4 + (\partial V^o/\partial x) \cdot \tilde{u} = 0 \qquad \alpha = 1,2,3,4, \qquad (21)$$

the partial derivatives with respect to the $\tilde{u}_\alpha$, $\alpha = 1,2,3,4$, being computed for $\tilde{u} = \tilde{u}^*(x,t)$ and $\tilde{u}_4 = \tilde{u}_4^*(x,t)$.

The following computations are similar to the ones of Case 1. It leads to

$$\sum_{i,j} g^{ij} (\partial V^o/\partial x_i)(\partial V^o/\partial x_j) = 0, \qquad i,j = 1,2,3,4 \qquad (22)$$

which is the equation of Hamilton-Jacobi associated with (19).

Concluding comments.

1. As we have said before, we will see later that, at interior points of the reachable set, Case 2 is ruled out. Accordingly *parameter* s *is the only one we will have to consider when dealing with paths all of whose points, except the initial one, are interior points of the reachable set* Y .
2. If, according to the classical formulation of optimal control problems, we introduce the *cost-variable* x_o, such that

$$dx_o/dt = L(x,t,u), \qquad u = u(x,t)$$

we see that, in our problem where $L(x,t,u)$ is given by (5), we have

$$dx_o/dt = ds/dt \; .$$

Accordingly *parameter* s *can be identified with the usual cost-variable* x_0 .
Note that the *addition to* s *of an arbitrary constant* C *is irrelevant.*

3. By introducing parameter s, we have reduced our problem to an *autonomous* one; that is, one for which s does not enter explicitly in the control, in the right hand side of state equations (8).
4. As in the classical geometrical method in optimal control, the solutions of state equations (8) serve to define a family of *paths in augmented state-space (Fig. 2).*
5. The classical concept of *limiting surface* $\Sigma(C)$, where C is an arbitrary constant, could be profitably used in a more thorough discussion of our problem. Since we have replaced the target which is usually considered in engineering problems by source Θ, the equation of a limiting surface is

$$x_0 = V^o(x,t) + C, \qquad (x,t) \in Y$$

and since $x_0 = s$, it reads

$$s = V^o(x,t) + C, \qquad (x,t) \in Y \; .$$

4. PROPOSITIONS 2 AND 3.

Before we apply Proposition 1 to an example, we need complement it by

Proposition 2. For the problem stated in paragraph 1, if Assumptions 1-5 are satisfied, and if π^* is an optimal path, that is one generated by $u^*(\cdot)$, represented by $x^*(\cdot) : t \to x^*(t)$, $t \in [0,t_1]$, then

(i) $(x^*(t_1),t_1) \in \overset{\circ}{Y} \Rightarrow (x^*(t),t) \in \overset{\circ}{Y}$ for all $t \in (0,t_1]$;

(ii) $(x^*(t_1),t_1) \in \overset{\circ}{Y} \Rightarrow u^*(x^*(t),t) \in \overset{\circ}{K}(x^*(t),t)$
for all $t \in (0,t_1]$.

Further, Proposition 3 is valid under Assumption 1-5 and

Assumption 6.[(8)] $\sum_{j,k} g_{j,k}(x,t)u_j u_k$, $j,k = 1,2,3$, is a quadratic form negative definite, and $g_{44}(x,t) > 0$, for all $(x,t) \in \mathcal{E} \times T^+$.

Proposition 3. For the problem stated in paragraph 1, if Assumptions 1-6 are satisfied, then $\partial V^*(x,t)/\partial t$ tends to infinity as (x,t) tends to a boundary point (x^c,t_c) of Y, with $(x,t) \in \overset{\circ}{Y}$ and $t_c \neq 0$.

The proofs of Proposition 2 and 3 will be given in Appendix 2.

5. EXAMPLE 1.

In Minkowski's space we have

$$ds^2 = c^2dt^2 - dx^2 , \quad dx^2 = dx_1^2 + dx_2^2 + dx_3^2$$

so that

$$G(x,t,u) = (ds^2/dt^2) = c^2 - u^2$$

$$L(x,t,u) = - m_o c^2\sqrt{1 - (u^2/c^2)}$$

From Proposition 2 it follows that, if the end point of an optimal path π^*, generated by $u^*(\cdot)$, represented by $x^*(\cdot): t \to x^*(t)$, $t \in [0,t_1]$, lies in the interior $\overset{\circ}{Y}$ of the reachable set, then all points of π^* except the initial one lie in $\overset{\circ}{Y}$. Moreover, for all $t \in (0,t_1]$ the control $u^*(x^*(t),t)$ lies in the interior $\overset{\circ}{K}(x^*(t),t)$ of the constraint set. Accordingly condition (i) of Proposition 1 implies that the partial derivatives of $H(\lambda(y),\lambda_t(y),y,u)$ with

---

(8) See Appendix 4.

respect to the u_α, $\alpha = 1,2,3$, vanish for all $y = (x^*(t),t)$ on the interval $(0,t_1]$.

We thus deduce from Propositions 1 and 2

$$\frac{\partial V^*}{\partial x_\alpha} = \frac{\partial L}{\partial u_\alpha} = \frac{m_o u_\alpha}{\sqrt{1-(u^2/c^2)}}, \qquad \alpha = 1,2,3 \tag{23}$$

$$-\frac{\partial V^*}{\partial t} = \frac{m_o c^2}{\sqrt{1-(u^2/c^2)}}, \tag{24}$$

with $u = u^*(x^*(t),t)$, $t \in (0,t_1]$.

Moreover since from Assumption 3 each interior point (x,t) of the reachable set is reached by an optimal path and since, from Proposition 2, $u^*(x,t) \in \mathring{K}(x,t)$ at such points, relations (23) and (24), with $u = u^*(x,t)$, hold for all $(x,t) \in \mathring{Y}$.

Though these formulas are well-known, their correct derivation requires Proposition 2.

Since according to (23) and (24)

$$\frac{\partial H(\lambda,\lambda_t,y,u)}{\partial u_\beta} = -\frac{\partial L(x,t,u)}{\partial u_\beta} + \frac{\partial V^*(x,t)}{\partial x_\beta} = 0, \quad \beta = 1,2,3$$

the adjoint equations (2) and (3) read

$$\frac{d}{dt}\left(\frac{\partial V^*}{\partial x_\alpha}\right) = -\frac{\partial H(\lambda,\lambda_t,y,u)}{\partial x_\alpha} = \frac{\partial L(x,t,u)}{\partial x_\alpha} = 0, \qquad \alpha = 1,2,3 \tag{25}$$

$$\frac{d}{dt}\left(\frac{\partial V^*}{\partial t}\right) = -\frac{\partial H(\lambda,\lambda_t,y,u)}{\partial t} = \frac{\partial L(x,t,u)}{\partial t} = 0 \tag{26}$$

where the partial derivatives are computed for $u = u^*(y)$ at state $y \in \mathring{Y}$, and the time derivatives are computed at state y along the optimal path generated by $u^*(\cdot)$ through y.

We conclude that $\partial V^*/\partial x_\alpha$, $\alpha = 1,2,3$, and $\partial V^*/\partial t$ are identical to constants along π^*. Accordingly, from (23) and (24) we deduce

$$u^*(x^*(t),t) \equiv \text{constant} \qquad t \in (0,t_1]$$

Substituting in the state equations (1), and taking account of the initial condition $x^*(0) = 0$, gives

$$x^*(t) = (x^1/t_1)t \qquad t \in [0,t_1]$$

(x^1, t_1) denoting the terminal point of π^*.

It follows that

$$u^*(x,t) = x/t \quad \text{for all} \quad (x,t) \in \overset{\circ}{Y} \tag{27}$$

By substituting in (23) and (24), integrating, and taking account of the condition $V^*(0,0) = 0$, we obtain the optimal value function

$$V^*(x,t) = - m_o c\sqrt{c^2t^2 - x^2} \tag{28}$$

Equivalently we have

$$V^o(x,t) = (-1/m_o c)V^*(x,t) = \sqrt{c^2t^2 - x^2} \tag{29}$$

As we have said in concluding comment 5 of paragraph 3, the family of limiting surfaces $\Sigma(C)$ associated with function $V^o(\cdot)$ is defined by

$$s = V^o(x,t) + C = \sqrt{c^2t^2 - x^2} + C, \qquad (x,t) \in Y$$

where C is an arbitrary constant.

A limiting surface of this family is sketched on Fig.3 .

An important remark is that $\partial V^o(x,t)/\partial t$ (or equivalently $\partial V^*(x,t)/\partial t$), tends to infinity as (x,t) tends to the boundary ∂Y of Y, except for $(x,t) \to (0,0)$ where $u^*(x,t)$ is not defined.

6. EXAMPLE 2.

In the Riemannian space-time produced by a static mass M with spherical symmetry, ds is the line element of Schwarzschild given, in isotropic coordinates (r_1, θ, φ) by

$$ds^2 = - a(dr_1^2 + r_1^2\, d\theta^2 + r_1^2 \sin^2\theta\, d\varphi^2) + bc^2dt^2$$

with

$$a = (1 + (GM/2r_1c^2))^4$$
$$b = (1 - (GM/2r_1c^2))^2/(1 + (GM/2r_1c^2))^2$$

where G is the Newtonian constant of gravitation.

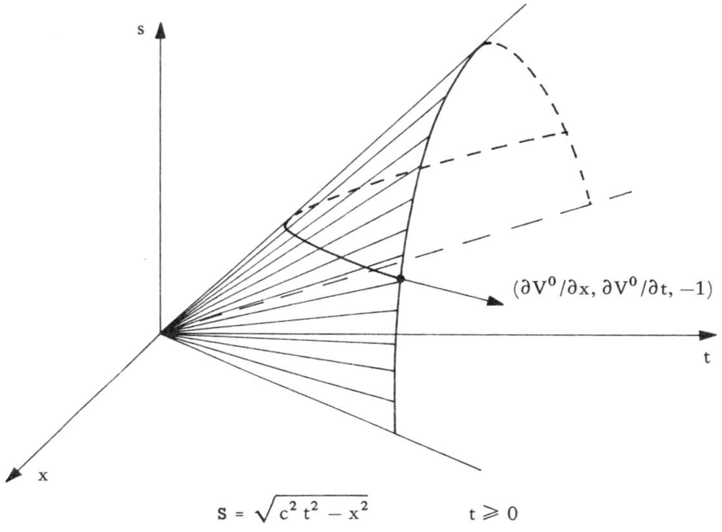

Fig. 3

In Cartesian coordinates x_1, x_2, x_3, t

$$x_1 = r_1 \sin\theta \cos\varphi$$
$$x_2 = r_1 \sin\theta \sin\varphi$$
$$x_3 = r_1 \cos\theta$$

we have

$$ds^2 = bc^2dt^2 - adx^2, \quad dx^2 = dx_1^2 + dx_2^2 + dx_3^2$$

with the provisio that r_1, which occurs in the coefficients a and b, is now expressed as a function of x_1, x_2, x_3, and hence $a = a(x)$ and $b = b(x)$.

In order that $b \neq 0$, our domain $\mathcal{E}$ must not contain points such that

$$r_1 = GM/2c^2$$

In that domain, let us introduce the index $n = n(x)$ defined by

$$n = \sqrt{a/b}$$

and let $\mu_o = \mu_o(x)$ be defined by

$$\mu_o = m_o\sqrt{b}$$

Then we have

$$G(x,t,u) = (ds^2/dt^2) = bc^2 - au^2 = (\mu_o^2 c^2/m_o^2)(1 - n^2(u^2/c^2)) \quad (30)$$

$$L(x,t,u) = - m_o c(ds/dt) = - \mu_o c^2 \sqrt{1 - n^2(u^2/c^2)} \quad (31)$$

$\mu_o = \mu_o(x)$ can be considered as a *modified proper mass*, that is a proper mass modified in such a way that it is a function of the position x of the mass-point in $\mathcal{E}$. The reason of this modification becomes clear if we consider the quite usual case where the term $GM/2r_1c^2$ is small with respect to 1. Then μ_o is approximated by the expression

$$\mu_o \sim m_o(1 - (GM/r_1c^2))$$

In that case, μ_o is *the sum of the proper mass* m_o *in the absence of a gravitational field and of the equivalent mass* $- GMm_o/r_1c^2$ *of the classical potential energy* $- GMm_o/r_1$. We shall return to that remark later.

In the general case we shall let

$$\mu_o = m_o\sqrt{b} = m_o - (\varphi_o(x)/c^2)$$

thus defining function $\varphi_o(\cdot)$ on $\mathcal{E}$.

This function will play, in the General Relativity, a similar role as the classical potential energy in non-relativistic dynamics, except for the fact that it is viewed in a frame of reference where the mass-point is at rest.

Concerning index $n = n(x)$, in Part II it will be shown to be an *optical index of refraction.*

As in Example 1, from Propositions 1 and 2 it follows that, if the end point of an optimal path π^* lies in $\overset{\circ}{Y}$, we have

$$\frac{\partial V^*}{\partial x_\alpha} = \frac{\partial L}{\partial u_\alpha} = \frac{n^2 \mu_o u_\alpha}{\sqrt{1-n^2(u^2/c^2)}} , \qquad \alpha = 1,2,3, \tag{32}$$

$$-\frac{\partial V^*}{\partial t} = \mu_o c^2 \sqrt{1-n^2(u^2/c^2)} + \frac{n^2 \mu_o u^2}{\sqrt{1-n^2(u^2/c^2)}} = \frac{\mu_o c^2}{\sqrt{1-n^2(u^2/c^2)}} \tag{33}$$

with $u = u^*(x^*(t),t)$, $t \in (0,t_1]$.

Note that expressions (32) and (33) still hold if μ_o and n are C^1 functions of x and t .

Moreover since, from Assumption 3, each interior point (x,t) of the reachable set is reached by an optimal path and since, from Proposition 2, $u^*(x,t) \in \overset{\circ}{K}(x,t)$ at such points, relations (32) and (33) with $u = u^*(x,t)$ hold for all $(x,t) \in \overset{\circ}{Y}$.

By an extension to the case of General Relativity of definitions given and discussed by Louis de Broglie [3] we shall define the *total energy* $E = E(x,t)$, the *internal heat* $Q = Q(x,t)$ and the *energy of translation* $E_T = E_T(x,t)$ of the mass-point, at $(x,t) \in \overset{\circ}{Y}$, by

$$E = -\partial V^*/\partial t$$

$$Q = \mu_o c^2 \sqrt{1-n^2(u^2/c^2)}$$

$$E_T = n^2 \mu_o u^2 / \sqrt{1-n^2(u^2/c^2)}$$

with $u = u^*(x,t)$.

Then (33) is written

$$E = Q + E_T$$

By letting

$$\mu = \mu_o/\sqrt{1-n^2(u^2/c^2)}$$

$$m = m_0/\sqrt{1-n^2(u^2/c^2)}$$

$$\varphi = \varphi_o/\sqrt{1-n^2(u^2/c^2)}$$

relation (33) gives

$$E = \mu c^2 = (m - (\varphi/c^2))c^2$$

As we have seen before, when $GM/2r_1c^2$ is small with respect to 1, φ_o is approximated by

$$\varphi_o \sim GMm_o/r_1$$

and accordingly, in that case

$$\varphi \sim GMm/r_1$$

By an extension of a result which holds in the theory of Special Relativity, we shall consider φ as *the potential energy of the mass-point in the frame of reference of the laboratory.*

Then, in that case, *the total energy of the mass-point is the sum of the mass-energy* mc^2 *and of the potential energy* $-\varphi \sim -GMm/r_1$.

As in Example 1, the adjoint equations (2) and (3) read

$$\frac{d}{dt}\left(\frac{\partial V^*}{\partial x_\alpha}\right) = -\frac{\partial H(\lambda,\lambda_t,y,u)}{\partial x_\alpha} = \frac{\partial L(x,t,u)}{\partial x_\alpha}, \quad \alpha = 1,2,3 \tag{34}$$

$$\frac{d}{dt}\left(\frac{\partial V^*}{\partial t}\right) = \frac{\partial H(\lambda,\lambda_t,y,u)}{\partial t} = \frac{\partial L(x,t,u)}{\partial t} = 0 \tag{35}$$

where the partial derivatives are computed for $u = u^*(y)$ at state $y \in \overset{o}{Y}$, and the time derivatives are computed at state y along the optimal path generated by $u^*(\cdot)$ through y .

Equation (35) expresses the fact that *the total energy is constant along an optimal path all of whose points lie in* $\overset{o}{Y}$, *except the initial one.*

Equation (34) can also be written

$$\frac{d}{dt}\left(\frac{\partial V^*}{\partial x_\alpha}\right) = -\frac{\partial Q}{\partial x_\alpha}, \qquad \alpha = 1,2,3 \tag{36}$$

If we identify $\partial V^*/\partial x = (\partial V^*/\partial x_1, \partial V^*/\partial x_2, \partial V^*/\partial x_3)$ with the *momentum* of the mass-point, *equation* (36) *is an extension of the second Newton's law to the case of General Relativity for the problem stated in paragraph 6* .

7. MECHANICAL INDEX OF REFRACTION.

Suppose that $\partial V^*(x,t)/\partial x$ is non-zero on an open subset $\mathcal{Y}$ of $\mathring{Y}$, and let there be associated with function $V^*(\cdot)$ function $v^*(\cdot)$ defined on $\mathcal{Y}$ with range in R^3, such that, for all $(x,t) \in \mathcal{Y}$

$$v^*(x,t) = \alpha\, \partial V^*(x,t)/\partial x \tag{37}$$

with

$$\alpha = -(\partial V^*(x,t)/\partial t)/\, \| \partial V^*(x,t)/\partial x \|^2 \tag{38}$$

Since, from conditions (i) and (ii) of Proposition 1 and from part of Assumption 4, we have

$$\partial V^*(x,t)/\partial t \leqslant L(x,t,0) = -m_o c\sqrt{G(x,t,0)} < 0$$

α is strictly positive.

From (37) and (38) we deduce

$$(\partial V^*/\partial t) + (\partial V^*/\partial x)\cdot v^*(x,t) = 0 \tag{39}$$

where $V^* = V^*(x,t), (x,t) \in \mathcal{Y}$.

$v^*(x,t)$ is termed *the phase-velocity relative to* $V^*(\cdot)$ *at point* $(x,t) \in \mathcal{Y}$.

Then the *mechanical index of refraction* $\tilde{n}(x,t)$ at point

$(x,t) \in Y$ is defined by

$$\tilde{n}(x,t) = c/\|v^*(x,t)\| = c\,\|\partial V^*(x,t)/\partial x\|/(-\partial V^*(x,t)/\partial t) \qquad (40)$$

In the special case where $u^*(x,t) \in \mathring{K}(x,t)$, we deduce from conditions (i) and (ii) of Proposition 1

$$\partial V^*/\partial x_\alpha = \partial L/\partial u_\alpha\,, \qquad \alpha = 1,2,3$$

$$\partial V^*/\partial t = L - (\partial L/\partial u)\cdot u$$

where $\partial L/\partial u = (\partial L/\partial u_1, \partial L/\partial u_2, \partial L/\partial u_3)$, $L = L(x,t,u)$, $u = u^*(x,t)$; and hence

$$\tilde{n} = c\,\|\partial L/\partial u\|/((\partial L/\partial u)\cdot u - L)$$

At last, in the case of Example 2 we find

$$\tilde{n} = n\sqrt{1-(\mu_o c^2/E)^2}$$

It follows that, in the vacuum and in the absence of gravitational field, in which case $n = 1$ and $\mu_o = m_0$, the mechanical index of refraction is

$$\tilde{n} = \sqrt{1-(m_0 c^2/E)^2}$$

It is lesser than 1.

8. THE PRINCIPLE OF PARALLEL TRANSPORT.

In the literature, the problem of finding geodesics in the Riemannian space-time of General Relativity is related to the so-called *principle of parallel transport*. Since we have stated this problem as an optimal control problem, we need discuss the connection between our treatment and this principle.

Let us consider an optimal path π^*, generated by $u^*(\cdot)$, represented by $x^*(\cdot) : t \to x^*(t)$, $t \in [0,t_1]$, whose terminal state $(x^*(t_1),t_1)$ belongs to the interior $\mathring{Y}$ of the reachable set. Then, according to Proposition 2, all states of this path except the initial one lie in $\overset{\circ}{Y}$, and further $u^*(x^*(t),t) \in \overset{\circ}{K}(x^*(t),t)$ for all $t \in (0,t_1]$.

Hence, for all $(x,t) \in \pi^*$, except at the initial point, we are in Case 1 of paragraph 3.

In connection with the Lagrange multipliers rule, which we have met in the special case treated in paragraph 3; that is, in addition to the stationarity condition (14), one can prove by standard methods that we have the adjoint equations

$$\begin{aligned} &\frac{d}{ds}(\partial V^o/\partial x_\alpha) = k(\partial G_s/\partial x_\alpha), \qquad \alpha = 1,2,3 \\ &\frac{d}{ds}(\partial V^o/\partial t) = k(\partial G_s/\partial t) \end{aligned} \tag{41}$$

where the partial derivatives are computed at state $(x,t) \in \pi^*$, $(x,t) \neq (0,0)$, and for $\tilde{u} = \tilde{u}^*(x,t)$ and $\tilde{u}_4 = \tilde{u}_4^*(x,t)$.[9]

By using (12), the set of adjoint equations (41) is written

$$\begin{aligned} &\frac{d}{ds}(\partial G_s/\partial \tilde{u}_\alpha) = \partial G_s/\partial x_\alpha, \qquad \alpha = 1,2,3 \\ &\frac{d}{ds}(\partial G_s/\partial \tilde{u}_4) = \partial G_s/\partial x_4 \end{aligned} \tag{42}$$

Now recall that $G_s(x,t,\tilde{u},\tilde{u}_4)$ is given by (10). Accordingly we have

$$\partial G_s/\partial \tilde{u}_i = 2\sum_{j=1}^{4} g_{ij}\,\tilde{u}_j, \qquad i = 1,2,3,4$$

$$\partial G_s/\partial x_i = \sum_{jk} (\partial g_{jk}/\partial x_i)\tilde{u}_j\tilde{u}_k, \qquad j,k = 1,2,3,4$$

By substituting in (42) and taking account of the state equations (8) we obtain

(9) These equations are also direct consequences of adjoint equations (2) and (3), when taking account of the relation between ds and dt, namely $dt/ds = 1/\sqrt{G(x,t,u)}$.

$$\frac{d}{ds}\left(\sum_{j=1}^{4} g_{ij}\,\tilde{u}_j\right) - \frac{1}{2}\sum_{j,k}(\partial g_{jk}/\partial x_i)\tilde{u}_j\tilde{u}_k =$$
$$= \sum_{j=1}^{4} g_{ij}(d\tilde{u}_j/ds) + \sum_{j,k}\left(\frac{\partial g_{ij}}{\partial x_k} - \frac{1}{2}\frac{\partial g_{jk}}{\partial x_i}\right)\tilde{u}_j\tilde{u}_k = 0 \tag{43}$$
$$i = 1,2,3,4 \; .$$

Since the determinant g of the g_{ij} is non-zero, one can solve equations (43) for the $d\tilde{u}_\alpha/ds$, $\alpha = 1,2,3,4$. One obtains with standard notation

$$d\tilde{u}_\alpha/ds + \sum_{j,k}\Gamma^{\alpha}_{jk}\,\tilde{u}_j\tilde{u}_k = 0, \qquad \alpha = 1,2,3,4 \tag{44}$$

where the coefficients Γ^{α}_{jk} are given by the symbols of Christoffel relative to the metric defined by the g_{ij} .

The system (44) is the *condition of parallel transport*. Clearly it is equivalent to the system (42), which itself is equivalent to the system of adjoint equations (41).

In the literature devoted to the problem of finding geodesics in the Riemannian space-time of General Relativity, the condition of parallel transport (44) is sometimes used as a principle for defining the geodesics.

We insist on the fact that the above derivation holds only for optimal paths all of whose points lie in the interior $\overset{\circ}{Y}$ of the reachable set. We shall discuss this question again in Part II in the case of boundary paths.

9. A CONNECTION WITH WAVE MECHANICS.

At that point it will be essential for further developments to stress two important features of the calculus above:

- first we have been led to increasing the dimension of the space by adding the new coordinate s to the earlier ones x_1, x_2, x_3, t; that is, *we have needed a 5-dimensional space;* and

- secondly, the use of this 5-dimensional space and the Lagrange multipliers rule have led us to replacing the Lagrangian (5), namely $\sqrt{G(x,t,u)}$, by *the new Lagrangian*

$$(1/2) + (1/2)G_s(x,t,\tilde{u},\tilde{u}_4)$$

deduced from (14), with k = 1/2.

It follows that, for the initial problem stated in paragraph 1, the Lagrangian $-m_0 c\sqrt{G(x,t,u)}$ is replaced by the new one(10)

$$\mathcal{L}(x,t,\tilde{u},\tilde{u}_4) = -\ (m_0 c/2) - (m_0 c/2)G_s(x,t,\tilde{u},\tilde{u}_4) \qquad (45)$$

as far as necessary conditions are concerned.

In the theory of Special Relativity, we have

$$G_s(x,t,\tilde{u},\tilde{u}_4) = c^2\tilde{u}_4^2 - \tilde{u}^2$$

then

$$\mathcal{L}(x,t,\tilde{u},\tilde{u}_4) = (m_0 c/2)(\tilde{u}^2 - c^2\tilde{u}_4^2) - (m_0 c/2) \qquad (46)$$

At first sight it is clear that the Lagrangian (46) is simply related to the Lagrangian which we had deduced from completely different arguments in [4] (p.43), in our approach to wave mechanics.

Now we know that the fifth coordinate which proved necessary in that approach is the cost-coordinate of the classical theory of optimal control, when applied to our problem; and furthermore we have here a simple explanation of the unusual form of the Lagrangian in [4].

These two points will be more completely developed elsewhere.

---

(10) This is so since $V^o(x,t) = (-1/m_0 c)V^*(x,t)$, so that (14) is written

$$\frac{\partial}{\partial\tilde{u}_\alpha}(m_0 ck(G_s(x,t,\tilde{u},\tilde{u}_4)-1)+ m_0 c + (\partial V^*/\partial t)\tilde{u}_4 + (\partial V^*/\partial x)\cdot\tilde{u}) = 0 \qquad (14')$$

$$\alpha = 1,2,3,4 \ .$$

PART II

A CONTROLLABILITY PROBLEM IN RELATIVISTIC DYNAMICS OF A MASS-POINT.

Part II is concerned with finding *geodesics of null-length* in the Riemannian space-time of General Relativity. Such geodesics play an important role in particular in connection with the hypothetic motion of photons in a gravitational field. We draw the attention on the fact that this problem is not in general a variational problem, but a problem of (closed-loop) controllability, or equivalently a (one-player) qualitative game[11]. This is so because in the theory of General Relativity the constraint set is closed, and, as will be shown in this chapter under the assumptions of Part I, geodesics of null-length are *boundary paths*; that is, paths all of whose points lie in the boundary ∂Y of the reachable set Y. It follows that this problem must be treated as the so-called *abnormal case* of optimal control theory, which holds for such paths. Considering the large amount of work already published in the area of controllability and qualitative differential games, this chapter cannot be the place for a thorough discussion. Instead we shall illustrate our statement by simple examples.

10. BASIC PROPOSITIONS.

In order to prepare further discussions we shall state here a few basic propositions. Their proofs along similar lines as some developed in [2] will be given in Appendix 1. *They hold for the problem stated in paragraph 1 of Part I and under the corresponding Assumptions 1-4.*

---

(11) According to the terminology of Ref.[2]. Such games are termes *games of kind* by R. Isaacs [5].

Proposition 4. At each $y = (x,t) \in \partial Y$, $(x,t) \notin \Theta$, x is non-zero so that, according Assumption 2, grad $W(x,t)$ and $\partial W(x,t)/\partial t$ are defined and non-zero, and

(i) $$\max_{u \in K(y)} \mathcal{H}(\lambda(y),\lambda_t(y),u) = \mathcal{H}(\lambda(y),\lambda_t(y),u^*(y)),$$

(ii) $$\mathcal{H}(\lambda(y),\lambda_t(y),u^*(y)) = 0,$$

where

$$\mathcal{H}(\lambda,\lambda_t,u) = \lambda_t + \lambda \cdot u, \tag{47}$$

and

$$\lambda(y) = \text{grad } W(x,t)^{(12)}, \quad \lambda_t(y) = \partial W(x,t)/\partial t . \tag{48}$$

Proposition 5. If π^* is an optimal path, that is one generated by $u^*(\cdot)$, represented by $x^*(\cdot)\colon t \to x^*(t), t \in [0,t_1]$, then

(i) $$(x^*(t_1),t_1) \in \partial Y \Rightarrow (x^*(t),t) \in \partial Y \quad \text{for all } t \in [0,t_1] ;$$

(ii) $$(x^*(t_1),t_1) \in \partial Y \Rightarrow u^*(x^*(t),t) \in \partial K(x^*(t),t) \quad \text{for all } t \in (0,t_1] ;$$

According to part of Assumption 4, a direct consequence of Proposition 5 is that

$$(x^*(t_1),t_1) \in \partial Y \Rightarrow G(x^*(t),t,u^*(x^*(t),t)) = 0$$
$$\Rightarrow L(x^*(t),t,u^*(x^*(t),t)) = 0 \qquad \text{for all } t \in (0,t_1]$$

and accordingly the cost of π^* is zero; in other words π^* is an optimal path (i.e., a geodesic) of null-length.

Also we have

Proposition 6. Let π^o be an optimal path of null-length, that is one generated by $u^*(\cdot)$ for which the cost vanishes, represented by $x^*(\cdot)\colon t \to x^*(t)$, $t \in [0,t_1]$, then

---

(12) grad $W(x,t) = \partial W(x,t)/\partial x = (\partial W(x,t)/\partial x_1, \partial W(x,t)/\partial x_2, \partial W(x,t)/\partial x_3)$.

$$u^*(x^*(t),t) \in \partial K(x^*(t),t) \qquad \text{for all } t \in (0,t_1] .$$

and

Proposition 7. All points of an optimal path of null-length lie in ∂Y.

From these propositions, it is clear that geodesics (i.e., optimal paths) of null-length cannot be defined by a stationary condition, and accordingly cannot be obtained through the Euler equations.

In brief, *the problem of finding geodesics of null-length is a problem of (closed-loop) controllability, or equivalently a (one-player) differential qualitative game.*

11. EXAMPLE 1.

In Minkowski's space consider an optimal path π^*, generated by $u^*(\cdot)$, represented by $x^*(\cdot)$: $t \to x^*(t)$, $t \in [0,t_1]$. Let $(x^*(t_1),t_1) \in \partial Y$ so that, according to Proposition 5 and one of its consequences, π^* is a geodesic of null-length, all of whose points lie in the boundary of the reachable set.

Moreover, from Proposition 5 we know that

$$u^*(x^*(t),t) \in \partial K(x^*(t),t) \qquad \text{for all } t \in (0,t_1] .$$

where $\partial K(x,t)$ is given by

$$G(x,t,u) = c^2 - u^2 = 0$$

In view of this constraint, we deduce from conditions (i) and (ii) of Proposition 4

$$u^*(x,t) = (c \text{ grad } W)/(\|\text{grad } W\|) \tag{49}$$

where $W = W(x,t)$, $x = x^*(t)$, $t \in (0,t_1]$; and

$$(1/c^2)(\partial W/\partial t)^2 - \|\text{grad } W\|^2 = 0 \tag{50}$$

which is the so-called *equation of eikonale* in geometrical optics.

Since the constraint set is independent of (x,t) one can prove by simple arguments that the reachable set Y is given by

$$W(x,t) = - mc^2t + mc\sqrt{x_1^2 + x_2^2 + x_3^2} \leqslant 0$$

where $m > 0$ is an arbitrary constant. The boundary ∂Y of Y is the light-cone $W(x,t) = 0$.

In particular, by this example we see that, contrary to ideas currently accepted

1. *in a problem of relativistic dynamics of a mass-point, with proper mass* m_0 *strictly positive, there can exist cases in which the speed of the point is equal to the speed of light c in the vacuum; and*
2. *the proper mass of a photon, when the photon is modelized as a mass-point whose trajectory is a geodesic of null-length along the light-cone, needs not be zero.*

12. EXAMPLE 2.

In the Riemannian space-time produced by a static mass M with spherical symmetry, ds is the line element of Schwarzschild. With the notation of paragraph 6 of Part I, $\partial K(x,t)$ is given by

$$G(x,t,u) = bc^2 - au^2 = 0, \qquad a = a(x), \quad b = b(x) .$$

As in Chapter I, we shall suppose that $b \neq 0$.

At each point (x,t) of a geodesic of null-length except at the initial point, we deduce from conditions (i) and (ii) of Proposition 4

$$u^*(x,t) = c\sqrt{b/a}\ (\text{grad } W)/\ \| \text{grad } W \|) \tag{51}$$

where $W = W(x,t)$; and

$$(n^2/c^2)(\partial W/\partial t)^2 - \| \text{grad } W \|^2 = 0 \tag{52}$$

where $n = \sqrt{a/b}$ in an *optical index of refraction.*

Equation (52) is the equation of eikonale in a medium where the optical index of refraction is a function of x .

A more general definition of an optical index of refraction will be given later.

13. A FAMILY OF SEMI-PERMEABLE SURFACES.

Now let us introduce

Assumption 7. There exists a strategy $\bar{u}(\cdot) \in U$ of class C^1 on $(\mathcal{E} - \{0\}) \times T^+$, such that

$$\max_{u \in K(x,t)} ((\partial W/\partial t) + (\partial W/\partial x)\cdot u) = (\partial W/\partial t) + (\partial W/\partial x)\cdot \bar{u}(x,t) = 0$$

for all $(x,t) \in (\mathcal{E} - \{0\}) \times T^+$; where $W = W(x,t)$ and $\partial W/\partial x = (\partial W/\partial x_1, \partial W/\partial x_2, \partial W/\partial x_3)$.

By referring to the terminology of qualitative games, a non-empty set $\mathcal{C}(C)$ defined by

$$\mathcal{C}(C) = \{(x,t): \ (x,t) \in Y, \ W(x,t) = C\}, \quad C \in R \tag{53}$$

will be called a *semi-permeable surface*.

By varying C one generates a family $\{\mathcal{C}(C)\}$ of such surfaces, and since $\mathcal{C}(0) = \partial Y$, ∂Y is embedded in this family (Fig.4) .

Note that our definition differs from the one given in Ref.[2] by the fact we include in it points of the half-ray $L^+ = \{(x,t): \ x = 0, \ t \geqslant 0\}$ where $\partial W(x,t)/\partial x$ needs not be defined.

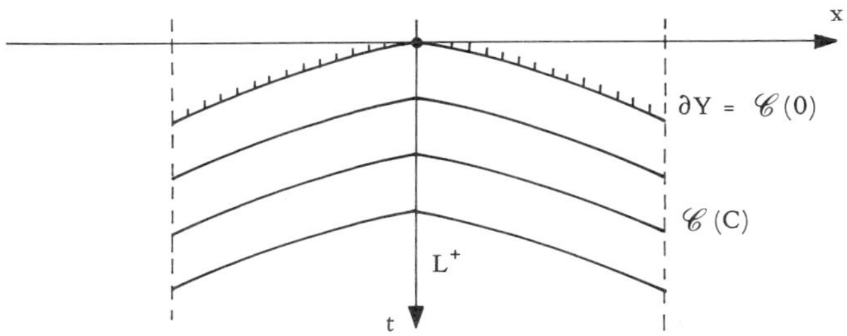

Fig. 4

14. TRAJECTORIES IN SEMI-PERMEABLE SURFACES.

Further one can prove

Proposition 8. $G(x,t,\bar{u}(x,t)) = 0$ for all $(x,t) \in (\mathcal{E} - \{0\}) \times T^+$.

Proof. Let $(x,t) \in (\mathcal{E} - \{0\}) \times T^+$. If $\bar{u}(x,t) \in \mathring{K}(x,t)$, it follows from Assumption 7 that

$$\frac{\partial}{\partial u_\alpha}((\partial W/\partial t) + (\partial W/\partial x) \cdot u) = 0, \qquad \alpha = 1,2,3$$

where the partial derivatives $\partial/\partial u_\alpha$ are computed for $u = \bar{u}(x,t)$ and $W = W(x,t)$. It follows that

$$\partial W/\partial x = 0$$

which contradicts Assumption 2.

Hence $\bar{u}(x,t) \in \partial K(x,t)$. It follows from part of Assumption 4 that

$$G(x,t,\bar{u}(x,t) = 0 \tag{54}$$

Hence Proposition 8 is established.

Then, since from Assumption 4

$$G(x,t,\bar{u}(x,t)) = 0 \Leftrightarrow \bar{u}(x,t) \in \partial K(x,t)$$

from Assumption 7 and Proposition 8 we deduce

Proposition 9. At each $(x,t) \in (\mathcal{E} - \{0\}) \times T^+$

$$\max_{u \in \partial K(x,t)} ((\partial W/\partial t) + (\partial W/\partial x) \cdot u) = (\partial W/\partial t) + (\partial W/\partial x) \cdot \bar{u}(x,t) = 0$$

where $W = W(x,t)$.

Now consider a solution $\bar{x}(\cdot)$ of the state equation (1) generated by $\bar{u}(\cdot)$ on some interval of time $[t_0, t_1]$, *satisfying the condition* $\bar{x}(t) \neq 0$ *for all* $t > t_0$ and the corresponding ordered set of states $\bar{\pi}$

$$\bar{\pi} = \{(x,t): \ x = \bar{x}(t), \qquad t \in [t_0, t_1]\}$$

From Proposition 8 and since $L(x,t,u) = -m_0 c\sqrt{G(x,t,u)}$ it follows that the cost of $\bar{\pi}$ defined by

$$\int_{t_0}^{t_1} L(\bar{x}(t),t,\bar{u}(\bar{x}(t),t))dt$$

vanishes. We shall say that $\bar{\pi}$ has a null-length.

In order to distinguish $\bar{\pi}$ from paths previously defined, we shall call $\bar{\pi}$ a *trajectory*.

The result obtained is inserted in

Proposition 10. A trajectory $\bar{\pi}$, generated by $\bar{u}(\cdot)$, represented by $\bar{x}(\cdot)$: $t \to \bar{x}(t)$, $t \in [t_0,t_1]$, has a null-length; and

$$\bar{u}(\bar{x}(t),t) \in \partial K(\bar{x}(t),t) \qquad \text{for all } t \in (t_0,t_1] \ .$$

At last we shall prove in Appendix 3 the following

Proposition 11. A trajectory $\bar{\pi}$, generated by $\bar{u}(\cdot)$, represented by $\bar{x}(\cdot)$: $t \to \bar{x}(t)$, $t \in [t_0,t_1]$, whose initial point $y^o = (x^o,t_0) = (\bar{x}(t_0),t_0)$ belongs to the open half-ray $\mathcal{L}^+ = \{(x,t)\colon x = 0, t > 0\}$, has all of its points in the semi-permeable surface through y^o; that is $(\bar{x}(t),t) \in Y$ and

$$W(\bar{x}(t),t) \equiv C \qquad t \in [t_0,t_1]$$

where

$$C = W(x^o,t_0) = W(0,t_0)$$

15. OPTICAL INDEX OF REFRACTION.

Let there be associated with function $W(\cdot)$ function $w^*(\cdot)$ defined on $(\mathcal{E} - \{0\}) \times T^+$ with range in R^3, such that, for all $(x,t) \in (\mathcal{E} - \{0\}) \times T^+$

$$w^*(x,t) = \alpha\, \partial W(x,t)/\partial x \tag{55}$$

with

$$\alpha = -\,(\partial W(x,t)/\partial t)/\, \| \partial W(x,t)/\partial x \|^2 \tag{56}$$

Since, from part of Assumption 2, $\partial W(x,t)/\partial t$ is strictly negative, α is strictly positive.

From (55) and (56) we deduce

$$(\partial W/\partial t) + (\partial W/\partial x) \cdot w^*(x,t) = 0 \qquad (57)$$

where $W = W(x,t)$, $(x,t) \in (\mathcal{E} - \{0\}) \times T^+$.

$w^*(x,t)$ is termed *the phase-velocity relative to* $W(\cdot)$ *at point* (x,t).

Then the *optical index of refraction* $n(x,t)$ at point $(x,t) \in (\mathcal{E} - \{0\}) \times T^+$ is defined by

$$n(x,t) = c/\| w^*(x,t) \| = c \| \partial W(x,t)/\partial x \| / (-\partial W(x,t)/\partial t) \qquad (58)$$

Of special interest is the case where there exists a function $\gamma(\cdot): \mathcal{E} \times T^+ \to R$ such that

$$G(x,t,u) = 0 \Leftrightarrow \| u \| = \gamma(x,t) \qquad (59)$$

for all $(x,t) \in \mathcal{E} \times T^+$.

Then, from Proposition 8, Proposition 9 and (59), it follows that

$$\bar{u}(x,t) = \gamma(x,t) \ (\partial W(x,t)/\partial x)/(\| \partial W(x,t)/\partial x\|) \qquad (60)$$

for all $(x,t) \in (\mathcal{E} - \{0\}) \times T^+$; and

$$(\partial W/\partial t) + (\partial W/\partial x) \cdot \bar{u}(x,t) = 0 \qquad (61)$$

where $W = W(x,t)$.

By comparing (57) and (61), and taking account of (55), (56) and (60), we reach the conclusion that

$$w^*(x,t) = \bar{u}(x,t)$$

for all $(x,t) \in (\mathcal{E} - \{0\}) \times T^+$, and hence

$$n(x,t) = c/\| \bar{u}(x,t) \| \qquad (62)$$

For instance, in Example 2, $\| \bar{u}(x,t) \|$ is deduced from Proposition 8 which reads

$$bc^2 - a\bar{u}^2 = 0 \ .$$

We have

$$\| \bar{u} \| = c\sqrt{b/a}$$

and hence, provided that $b \neq 0$

$$n = \sqrt{a/b} = (1 + (GM/2r_1c^2))^3/(1 - (GM/2r_1c^2)) \tag{63}$$

Note that, when $GM/2r_1c^2$ is small with respect to 1, formula (63) is approximated by

$$n \sim 1 + 2(GM/r_1c^2) \tag{64}$$

In the vacuum and in the absence of a gravitational field, the optical index of refraction is

$$n = 1$$

16. EQUATION OF EIKONALE.

As in paragraph 3 of Part I, let us introduce a parameter τ such that $dt/d\tau > 0$, and let

$$\begin{aligned} \tilde{u}_4 &= dt/d\tau > 0 \\ \tilde{u} &= dx/d\tau = (dx/dt)(dt/d\tau) = u\,\tilde{u}_4 \end{aligned} \tag{65}$$

Let $(x,t) \in (\mathcal{E} - \{0\}) \times T^+$ and $u \in \partial K(x,t)$, so that $(\tilde{u},\tilde{u}_4)$ belongs to the *constraint set* $K_\tau(x,t)$ defined by (65) and

$$G_\tau(x,t,\tilde{u},\tilde{u}_4) = 0 \tag{66}$$

where

$$\begin{aligned} G_\tau(x,t,\tilde{u},\tilde{u}_4) &= (ds^2/d\tau^2) = (ds^2/dt^2)(dt^2/d\tau^2) \\ &= G(x,t,u)\tilde{u}_4^2 = \sum_{i,j} g_{ij}(x,t)\tilde{u}_i\tilde{u}_j \\ & \qquad i,j, = 1,2,3,4 \end{aligned} \tag{67}$$

Indeed, it is clear that $u \in \partial K(x,t) \Leftrightarrow G(x,t,u) = 0$
$\Leftrightarrow G_\tau(x,t,\tilde{u},\tilde{u}_4) = 0$.

From Proposition 9 one can readily deduce

$$\max_{(\tilde{u},\tilde{u}_4)\in K_\tau(x,t)} ((\partial W/\partial t)\tilde{u}_4 + (\partial W/\partial x)\cdot\tilde{u})$$
$$= (\partial W/\partial t)\bar{u}_4 + (\partial W/\partial x)\cdot\bar{u} = 0 \tag{68}$$

where $\bar{u}_4 = 1$ and $\bar{u} = \bar{u}(x,t)$.

By similar arguments as in paragraph 3 of Part I; that is, by using the Lagrange multipliers rule, one finds

$$\frac{\partial W(x,t)}{\partial x_\alpha} = k\left.\frac{\partial G_\tau(x,t,\tilde{u},\tilde{u}_4)}{\partial \tilde{u}_\alpha}\right|_{\substack{\tilde{u}=\bar{u}\\ \tilde{u}_4=1}}$$
$$\frac{\partial W(x,t)}{\partial t} = k\left.\frac{\partial G_\tau(x,t,\tilde{u},\tilde{u}_4)}{\partial \tilde{u}_4}\right|_{\substack{\tilde{u}=\bar{u}\\ \tilde{u}_4=1}} \tag{69}$$

where k is a non-zero real number, which depends on (x,t) in general.

Since from (67)

$$\partial G_\tau/\partial\tilde{u}_i = 2\sum_{j=1}^{4} g_{ij}\,\tilde{u}_j, \qquad i = 1,2,3,4 \tag{70}$$

we have

$$\tilde{u}_j = (1/2)\sum_{i=1}^{4} g^{ij}(\partial G_\tau/\partial\tilde{u}_i), \quad j = 1,2,3,4 \tag{71}$$

and from (69) we deduce

$$\bar{u}_j = (1/2k)\sum_{i=1}^{4} g^{ij}(\partial W/\partial x_i), \quad j = 1,2,3,4 \tag{72}$$

Substituting in (68) we obtain

$$\sum_{i,j} g^{ij}(\partial W/\partial x_i)(\partial W/\partial x_j) = 0 \qquad i,j = 1,2,3,4 \tag{73}$$

which is the equation of Hamilton-Jacobi associated with (68). We recognize the equation of eikonale of geometrical optics, which reduces to (50) and (52) in the special cases of Example 1 and Example 2, respectively.

17. AGREEMENT OF $\overline{u}(\cdot)$ AND $u^*(\cdot)$ ON THE BOUNDARY OF THE REACHABLE SET.

If, instead of relying on Proposition 9 we refer to conditions (i) and (ii) of Proposition 4, and we make use of the fact that

$$\left.\begin{array}{l}(x,t) \in \partial Y \\ (x,t) \notin \Theta\end{array}\right\} \Rightarrow u^*(x,t) \in \partial K(x,t)$$

deduced from Proposition 5, together with the fact that Y is closed, we obtain

$$\begin{aligned}&\max_{(\tilde{u},\tilde{u}_4)\in K_\tau(x,t)} ((\partial W/\partial t)\tilde{u}_4 + (\partial W/\partial x)\cdot\tilde{u}) \\ &\qquad = (\partial W/\partial t)u_4^* + (\partial W/\partial x)\cdot u^* = 0\end{aligned} \tag{74}$$

where $u_4^* = 1$, $u^* = u^*(x,t)$ and $(x,t) \in \partial Y$, $(x,t) \notin \Theta$.

By the Lagrange multipliers rule, as in paragraph 16 above, we deduce from (74) together with (66)

$$u_j^* = (1/2k')\sum_{i=1}^{4} g^{ij}(\partial W/\partial x_i), \qquad j = 1,2,3,4 \tag{75}$$

where k' is a non-zero real number.

By comparing (72) and (75) we see that

$$\overline{u}_j = (k'/k)u_j^* , \qquad j = 1,2,3,4 \tag{76}$$

for $\bar{u}_j = \bar{u}_j(x,t)$, $u_j^* = u_j^*(x,t)$, and $(x,t) \in \partial Y$, $(x,t) \notin \Theta$

And since

$$\bar{u}_4 = u_4^* = 1$$

we have $k' = k$

so that $\bar{u}_j = u_j^*, \qquad j = 1,2,3,4.$

This result proves

Proposition 12. For all $(x,t) \in \partial Y$, except for $(x,t) = (0,0)$, we have

$$\bar{u}(x,t) = u^*(x,t)$$

As we have noted in paragraph 13, the boundary ∂Y of the reachable set is embedded in a family of semi-permeable surfaces. Proposition 12 enables us to complement this statement: *optimal boundary paths, that is those generated by* $u^*(\cdot)$ *and all of whose points lie in* ∂Y, *that is geodesics of null-length, are trajectories in the semi-permeable surface* $C(0) = \partial Y$. *They are thus embedded in the family of trajectories in semi-permeable surfaces.*

18. THE PRINCIPLE OF PARALLEL TRANSPORT FOR TRAJECTORIES IN SEMI-PERMEABLE SURFACES.

When dealing with the hypothetic motion of photons in a gravitational field, the literature concerned with this problem admits in general that a photon can be modelized as a point which moves along a geodesic of null-length of the Riemannian space-time of General Relativity. The problem of finding such geodesics is related to the principle of parallel transport.

However as there has been, up to the present paper we believe, no unified treatment of the dynamics of a mass-point in Mechanics and of the motion of photons in geometrical optics, in the case of photons the principle of parallel transport is presented in a different form as in the case of regular mass-points, as though the two cases would be of completely different nature.

In this paragraph we will show how the condition of parallel transport applies to trajectories in semi-permeable surfaces.

Let us consider a trajectory $\overline{\pi}$, generated by $\overline{u}(\cdot)$, represented by $\overline{x}(\cdot): t \to \overline{x}(t)$, $t \in [t_o, t_1]$ with $\overline{x}(t_o) = 0$, $t_o \geqslant 0$. As we have seen before this trajectory entirely belongs to a semi-permeable surface $C(C)$[13]

In connection with the Lagrange multipliers rule, one can prove by standard methods that we have, together with (69), the adjoint equations

$$\frac{d}{dt}(\partial W/\partial x_\alpha) = k(\partial G/\partial x_\alpha) \qquad \alpha = 1,2,3$$
$$\frac{d}{dt}(\partial W/\partial t) = k(\partial G/\partial t) \tag{77}$$

where the partial derivatives are computed at state $(x,t) = (\overline{x}(t),t)$, $t \in (t_o, t_1]$, and for $u = \overline{u}(x,t)$. By letting $dt = |k|d\tau$, that is $\tilde{u}_4 = |k|$ and $\tilde{u} = |k|\overline{u}$, and by using (69), the set of adjoint equations (77) is written

$$\frac{d}{d\tau}(\partial G_\tau/\partial \tilde{u}_\alpha) = \partial G_\tau/\partial x_\alpha , \qquad \alpha = 1,2,3$$
$$\frac{d}{d\tau}(\partial G_\tau/\partial \tilde{u}_4) = \partial G_\tau/\partial x_4 , \qquad x_4 \equiv t \tag{78}$$

Now recall that $G_\tau(x,t,\tilde{u},\tilde{u}_4)$ is given by (67).

---

(13) If $t_o > 0$, we rely on Proposition 11. If $t_o = 0$, then we let $\overline{\pi}$ be an optimal boundary path, that is one generated by $u^*(\cdot)$ and all of whose points lie in ∂Y. As we have seen in paragraph 17, according to Proposition 12, it is a trajectory in the semi-permeable surface $C(0) = \partial Y$.

Accordingly we have

$$\partial G_\tau/\partial \tilde{u}_i = 2 \sum_{j=1}^{4} g_{ij} \tilde{u}_j, \qquad i = 1,2,3,4$$

$$\partial G_\tau/\partial x_i = \sum_{j,k} (\partial g_{jk}/\partial x_i)\tilde{u}_j\tilde{u}_k, \qquad j,k = 1,2,3,4.$$

By substituting in (78) we obtain

$$\frac{d}{d\tau}\left(\sum_{j=1}^{4} g_{ij}\tilde{u}_j\right) - \frac{1}{2}\sum_{j,k} (\partial g_{jk}/\partial x_i)\tilde{u}_j\tilde{u}_k =$$
$$= \sum_{j=1}^{4} g_{ij}(d\tilde{u}_j/d\tau) + \sum_{j,k}\left(\frac{\partial g_{ij}}{\partial x_k} - \frac{1}{2}\frac{\partial g_{jk}}{\partial x_i}\right)\tilde{u}_j\tilde{u}_k = 0 \tag{79}$$

Since the determinant g of the g_{ij} is non-zero, one can solve equations (79) for the $d\tilde{u}_\alpha/d\tau$, $\alpha = 1,2,3,4$. One obtains with standard notation

$$d\tilde{u}_\alpha/d\tau + \sum_{j,k} \Gamma^\alpha_{jk} \tilde{u}_j \tilde{u}_k = 0, \qquad \alpha = 1,2,3,4 \tag{80}$$

where the coefficients Γ^α_{jk} are given by the symbols of Christoffel relative to the metric defined by the g_{ij}.

The system (80) is the *condition of parallel transport along trajectories in semi-permeable surfaces*. It is equivalent to the system (78) which itself is equivalent to the system of adjoint equations (77).

19. CONNECTION WITH WAVE MECHANICS.

By the Lagrange multipliers rule, one deduces from (68) together with the constraint condition (66)

$$\frac{\partial}{\partial \tilde{u}_\alpha}\left(- kG_\tau(x,t,\tilde{u},\tilde{u}_4) + (\partial W/\partial t)\tilde{u}_4 + (\partial W/\partial x)\cdot\tilde{u}\right) = 0 \tag{81}$$

$$\alpha = 1,2,3,4$$

the partial derivatives with respect to the $\tilde{u}_\alpha$, $\alpha = 1,2,3,4$, being computed for $\tilde{u} = \bar{u}(x,t)$ and $\tilde{u}_4 = 1$.

Since $G_\tau(x,t,\bar{u}(x,t),1) = G(x,t,\bar{u}(x,t)) = 0$, (81) can also be written

$$\underset{\tilde{u}}{\text{stat}} \left(- kG_\tau(x,t,\tilde{u},\tilde{u}_4) + (\partial W/\partial t)\tilde{u}_4 + (\partial W/\partial x) \cdot \tilde{u}\right) = (\partial W/\partial t) + (\partial W/\partial x) \cdot \bar{u}(x,t) = 0 \tag{82}$$

where "stat" means *stationary value of*, and the stationarity condition is written for $\tilde{u} = \bar{u}(x,t)$ and $\tilde{u}_4 = 1$.

This amounts to introducing the *Lagrangian*

$$\mathcal{L}(x,t,\tilde{u},\tilde{u}_4) = kG_\tau(x,t,\tilde{u},\tilde{u}_4) \tag{83}$$

where k is a non-zero real number.

In the theory of Special Relativity, we have

$$G_\tau(x,t,\tilde{u},\tilde{u}_4) = c^2 \tilde{u}_4^2 - \tilde{u}^2$$

then

$$\mathcal{L}(x,t,\tilde{u},\tilde{u}_4) = k \left(c^2 \tilde{u}_4^2 - \tilde{u}^2 \right) \tag{84}$$

The use of this Lagrangian will enable us to further develop an extension of our approach to Wave Mechanics [4] [6], namely the one which corresponds to the classical case of a mass-point moving along a geodesic of null-length in the Riemannian space time of General Relativity.

APPENDIX 1. PROOF OF PROPOSITIONS 4 - 7

From Assumption 2, Y is the closure of a domain $D \subset \mathcal{E} \times T^+$. Let $(x^1, t_1) \in D$ and

$$L = \{(x,t): x = x^1, \quad t \in T\} \subset \mathcal{E} \times T,$$

and consider Case 1 and Case 2, namely

Case 1. $x^1 \neq 0$.

In that case, we have $(x^1, 0) \notin Y^+$ since $0 \notin T^+$, and $(x^1, 0) \notin \Theta$. It follows that

$$(x^1, 0) \notin Y$$

since $Y = \Theta \cup Y^+$.

Accordingly we have

$$(x^1, 0) \in L \cap \text{comp } Y \text{ and } (x^1, t_1) \in L \cap Y \tag{1.1}$$

where "comp Y" means the complement of Y in $\mathcal{E} \times T$.

From (1.1) one can easily deduce by dichotomy that there exists (x^1, t_c)

$$(x^1, t_c) \in L \cap \partial Y \quad \text{with} \quad t_c \leqslant t_1,$$

and accordingly

$$W(x^1, t_c) = 0.$$

Case 2. $x^1 = 0$.

Then

$$(x^1, t_c) = (0,0) \in L \cap \partial Y$$

and

$$W(x^1, t_c) = W(0,0) = 0$$

In Case 1, since $(x^1, t_c) \in \partial Y$ and $(x^1, t_1) \in D$, we have

$t_c \neq t_1$, and since $t_c \leqslant t_1$ it follows that

$$t_c < t_1 \tag{1.2}$$

In Case 2, $t_c = 0$ so that again (1.2) is satisfied since $0 < t_1$, which follows from $(x^1, t_1) \in D \subset \mathcal{E} \times T^+$

Since according to part of Assumption 2

$$\partial W(x,t)/\partial t < 0 \quad \text{for all } (x,t) \in \mathcal{E} \times T$$

we have

$$t_c < t_1 \Rightarrow W(x^1, t_1) < W(x^1, t_c)$$

that is

$$W(x^1, t_1) < 0 \tag{1.3}$$

We have thus proved

Proposition a. $(x^1, t_1) \in D \Rightarrow W(x^1, t_1) < 0$.

Let us also prove

Proposition b. At each $y = (x,t) \in \partial Y$, $(x,t) \notin \Theta$, x is non-zero.

Let $y = (x,t) \in \partial Y$, $(x,t) \notin \Theta$, and suppose that $x = 0$.

Indeed $(0,t) \notin \Theta \Rightarrow t \neq 0$, and since $(0,t) \in \overline{D}$ where $D \subset \mathcal{E} \times T^+$ we have $t > 0$.

On the other hand

$$(0,t) \in \partial Y \Rightarrow W(0,t) = 0$$

whereas on the other hand

$$t > 0 \Rightarrow W(0,t) < W(0,0) = 0$$

since

$$\partial W(0,t)/\partial t < 0 \quad \text{for all } t \in T$$

according to part of Assumption 2.

The contradiction concludes the proof.

Now let us prove

Proposition c. If π^* is an optimal path generated by $u^*(\cdot)$, represented by $x^*(\cdot): t \to x^*(t),\ t \in [0,t_1]$, then

$$(x^*(t_1),t_1) \in \partial Y \Rightarrow (x^*(t),t) \in \partial Y \quad \text{for all } t \in [0,t_1] .$$

Suppose that $t \in (0,t_1)$ and that $(x^*(t),t) \in \mathring{Y}$. Then there exists an open ball $B_\varepsilon(x^*(t),t)$ with center $(x^*(t),t)$ and radius ε such that

$$B_\varepsilon(x^*(t),t) \subset \mathring{Y}$$

Let us denote by $B_\alpha(x^*(t_1),t_1)$ an open ball with center $(x^*(t_1),t_1)$ and radius α, such that

$$B_\alpha(x^*(t_1),t_1) \subset \mathcal{E} \times T^+$$

such an open ball exists since $(x^*(t_1),t_1) \subset \mathcal{E} \times T^+$ and since $\mathcal{E} \times T^+$ is an open set.

Since, according to Assumption 3, $u^*(\cdot)$ is of class C^1 on $\mathcal{E} \times T^+$, and since $(x^*(t),t) \in \mathcal{E} \times T^+$ for all $t \in (0,t_1]$, we can rely on the continuity of solutions of ordinary differential equations with respect to terminal conditions as follows:

for all $\varepsilon > 0$ sufficiently small, there exists $\alpha > 0$ such that for all

$$(x^j,t_j) \in B_\alpha(x^*(t_1),t_1)$$

there exists a solution $x''(\cdot): t \to x''(t),\ t \in [t_i,t_j]$ of equation

$$\dot{x} = u^*(x,t) \tag{1.4}$$

such that

$$(x''(t_i),t_i) \in B_\varepsilon(x^*(t),t) \text{ and } x''(t_j) = x^j .$$

Since $(x^*(t_1),t_1) \in \partial Y$ and since by definition of ∂Y we have $\partial Y = \overline{Y} \cap \overline{\text{comp } Y} = Y \cap \text{comp } \mathring{Y}$, one can prove easily by classical arguments that

$$B_\alpha(x^*(t_1),t_1) \cap \text{comp } Y \neq \phi \tag{1.5}$$

According to (1.5) let us choose (x^j, t_j) such that

$$(x^j, t_j) \in B_\alpha(x^*(t_1), t_1) \cap \text{comp } Y . \tag{1.6}$$

From Assumption 3 and since $(x''(t_i), t_i) \in \overset{\circ}{Y}$, there exists an optimal path which reaches $(x''(t_i), t_i)$; that is, there exists a solution of equation (1.4), namely $x'(\cdot): t \to x'(t),\ t \in [0, t_i]$, such that $x'(0) = 0$ and $x'(t_i) = x''(t_i)$.

At last consider the function $x'''(\cdot): t \to x'''(t), t \in [0, t_j]$, such that

$$x'''(t) = x'(t) \quad \text{for all} \quad t \in [0, t_i]$$

$$x'''(t) = x''(t) \quad \text{for all} \quad t \in [t_i, t_j]$$

It is a solution of equation (1.4) which satisfies $x'''(0) = 0$ and $x'''(t_j) = x^j$.

This contradicts the fact that $(x^j, t_j) \notin Y$ according to (6), and hence for $t \in (0, t_1]$ we have $(x^*(t), t) \in \partial Y$.

For $t = 0$, $(x^*(0), 0) = (0, 0) \in \partial Y$.

Hence Proposition c is established.

PROOF OF PROPOSITION 4.

Let $(x^c, t_c) \in \partial Y$, $(x^c, t_c) \notin \Theta$.

From Proposition b, x^c is non-zero so that, according to Assumption 2, grad $W(x,t)$ and $\partial W(x,t)/\partial t$ are defined and non-zero at point (x^c, t_c).

Since, according to Assumption 2, Y is closed, (x^c, t_c) can be reached from Θ along a path π^c, which implies that $t_c > 0$ and hence that $(x^c, t_c) \in \mathcal{E} \times T^+$. Since $\mathcal{E} \times T^+$ is an open set, there exists an open ball $B_\varepsilon(x^c, t_c)$ with center (x^c, t_c) and radius ε, such that

$$B_\varepsilon(x^c, t_c) \subset \mathcal{E} \times T^+ .$$

Let $u \in K(x^c,t_c)$. According to Assumption 1, there exists a strategy $u(\cdot) \in U$ which is continuous and continuously differentiable with respect to x on some open ball $B_\varepsilon(x^c,t_c) \subset \mathcal{E} \times T^+$, with center (x^c,t_c) and radius ε, and such that

$$u(x^c,t_c) = u$$

Since $x^c \neq 0$, for ε sufficiently small

$$(x,t) \in B_\varepsilon(x^c,t_c) \Rightarrow x \neq 0$$

Then there exists a solution $x(\cdot): t \to x(t)$, $t \in [t_c, t_c + \tau_c)$, $\tau_c > 0$, of equation (1.7)

$$\dot{x} = u(x,t) \tag{1.7}$$

such that $x(t_c) = x^c$ and $(x(t),t) \in B_\varepsilon(x^c,t_c)$ for all $t \in [t_c, t_c + \tau_c)$.

By introducing the transfer time $\tau = t - t_c$, we shall write

$$x(t) = x(t_c + \tau) = x^c + u\tau + o(\tau)$$

where $u = u(x^c,t_c)$ and $\text{Lim}\,\| o(\tau) \| / \tau = 0$ as $\tau \to 0^+$.

Let $\bar{\pi}^\varepsilon$ be the ordered set of states

$$\bar{\pi}^\varepsilon = \{(x,t): x = x(t),\ t \in [t_c, t_c + \tau_c)\}$$

One can prove easily that $\pi^c \cup \bar{\pi}^\varepsilon$ is a path, so that $(x(t),t) \in Y$ for all $t \in [t_c, t_c + \tau_c)$.

From Proposition a together with the fact that $(x(t),t) \in Y$ for all $t \in [t_c, t_c + \tau_c)$ we deduce that

$$W(x^c + u\tau + o(\tau),\ t_c + \tau) \leqslant 0 \quad \text{for all } \tau \in [0,\tau_c) \tag{1.8}$$

where $u = u(x^c,t_c)$;
and since $(x^c + u\tau + o(\tau),\ t_c + \tau) \in B_\varepsilon(x^c,t_c)$ for all $\tau \in [0,\tau_c)$, for ε sufficiently small we have

$$x^c + u\tau + o(\tau) \neq 0 \quad \text{for all } \tau \in [0, \tau_c)$$

so that grad $W(x,t)$ and $\partial W(x,t)/\partial t$ are defined and non-zero for all $x = x(t)$ and for all $t \in [t_c, t_c + \tau_c)$.

From (1.8) we deduce

$$W(x^c, t_c) + (\partial W/\partial t)\tau + \text{grad } W \cdot u\tau + o(\tau) \leqslant 0 \tag{1.9}$$

where $W = W(x,t)$ and the partial derivatives are computed at (x^c, t_c).

Moreover, since $(x^c, t_c) \in \partial Y$ we have $W(x^c, t_c) = 0$.

For $\tau > 0$, dividing both sides of (1.9) by τ, then letting $\tau \to 0^+$ gives

$$(\partial W/\partial t) + \text{grad } W \cdot u \leqslant 0 \tag{1.10}$$

and (1.10) holds for all $u \in K(x^c, t_c)$, and for all $(x^c, t_c) \in \partial Y$ provided that $(x^c, t_c) \notin \Theta$.

In the case where $u = u^*(x^c, t_c)$, by relying on Proposition c and on the fact that $u^*(\cdot)$ is of class C^1 on $\mathcal{E} \times T^+$, by similar arguments as above one obtains

$$(\partial W/\partial t) + \text{grad } W \cdot u^*(x^c, t_c) = 0 \tag{1.11}$$

Again, (1.11) holds for all $(x^c, t_c) \in \partial Y$, $(x^c, t_c) \notin \Theta$.

Hence Proposition 4 is established.

PROOF OF PROPOSITION 5.

Now we are ready to prove

Proposition d. If π^* is an optimal path, generated by $u^*(\cdot)$ represented by $x^*(\cdot): t \to x^*(t)$, $t \in [0, t_1]$, then

$$(x^*(t), t) \in \partial Y \Rightarrow u^*(x^*(t), t) \in \partial K(x^*(t), t)$$

provided that $t \in (0, t_1]$.

Let $t \in (0, t_1]$, $(x^*(t), t) \in \partial Y$ and suppose that

$u^*(x^*(t),t) \in \overset{\circ}{K}(x^*(t),t)$. Indeed, from Proposition b since $(x^*(t),t) \notin \Theta$ we have $x^*(t) \neq 0$.

From Proposition 4 we have in that case

$$\frac{\partial}{\partial u_\alpha} ((\partial W/\partial t) + \text{grad } W \cdot u) = 0, \quad \alpha = 1,2,3 \tag{1.12}$$

where $W = W(x,t)$, $x = x^*(t)$, and the partial derivatives with respect to the u_α, $\alpha = 1,2,3$, are computed for $u = u^*(x^*(t),t)$.

We deduce from (1.12) grad $W = 0$ which is in contradiction with part of Assumption 2 together with the fact that $x^*(t) \neq 0$.

This concludes the proof of Proposition d.

Proposition 5 is a direct consequence of Propositions c and d.

PROOF OF PROPOSITION 6.

Suppose that there exists $t_k \in (0,t_1)$ such that

$$G(x^*(t_k),t_k, u^*(x^*(t_k),t_k)) > 0$$

Since

1. $(x^*(t_k),t_k) \in \mathcal{E} \times T^+$;
2. function $G(\cdot)$ is continuous on $\mathcal{E} \times T \times R^3$;
3. $u^*(\cdot)$ is of class C^1 on $\mathcal{E} \times T^+$;
4. $x^*(\cdot)$ is continuous on $[0,t_1]$;

there exists an interval $(t_k - \varepsilon,\ t_k + \varepsilon)$, $\varepsilon > 0$, on which

$$G(x^*(t),t\ ,\ u^*(x^*(t),t)) > 0$$

According to the definition of the cost of a path π, this is in contradiction with the fact that the cost of π^o is zero[14], and hence

(14) Remember that $G(x^*(t),t,u^*(x^*(t),t))$, is non-negative.

$$G(x^*(t_k),t_k,u^*(x^*(t_k),t_k)) = 0 \qquad (1.13)$$

for all $t_k \in (0,t_1)$.

By continuity (1.13) also holds for $t_k = t_1$.

Then, according to part of Assumption 4, it follows that

$$u^*(x^*(t_k),t_k) \in \partial K(x^*(t_k),t_k) \qquad \text{for all } t_k \in (0,t_1]$$

which concludes the proof of Proposition 6.

PROOF OF PROPOSITION 7.

Let π^o be an optimal path of null-length generated by $u^*(\cdot)$, represented by $x^*(\cdot)$: $t \to x^*(t)$, $t \in [0,t_1]$. Suppose that there exists $t_k \in (0,t_1]$ such that

$$(x^*(t_k),t_k) \in D$$

Let $x^k = x^*(t_k)$, and

$$L^k = \{(x,t): \; x = x^k, \; t \in T\} \subset \mathcal{E} \times T .$$

As we have proved before in Appendix 1, there exists (x^k,t_c)

$$(x^k,t_c) \in L^k \cap \partial Y \qquad \text{with} \qquad t_c < t_k .$$

From Assumption 3 and from the fact that Y is closed, there exists an optimal path - say π^k - generated by $u^*(\cdot)$, represented by $x^k(\cdot)$: $t \to x^k(t)$, $t \in [0,t_c]$, which reaches (x^k,t_c) from Θ.

From Proposition c all points of π^k lie in ∂Y, and from Proposition d

$$u^*(x^k(t),t) \in \partial K(x^k(t),t) \qquad \text{for all} \quad t \in (0,t_c] .$$

It follows that

$$G(x^k(t),t,u^*(x^k(t),t)) = 0 \qquad \text{for all} \quad t \in (0,t_c] \qquad (1.14)$$

so that the cost of π^k is zero.

Now consider the solution $x^o(\cdot):t \to x^o(t)$, $t \in [t_c, t_k]$ of equation

$$\dot{x} = 0$$

such that $x^o(t_c) = x^k$, and let $\bar{\pi}^k$ be the ordered set of states

$$\bar{\pi}^k = \{(x,t): x = x^k, \quad t \in [t_c, t_k]\}.$$

According to part of Assumption 4

$$G(x^o(t),t,0) > 0 \quad \text{for all} \quad t \in [t_c, t_k] \tag{1.15}$$

It follows that the cost of $\bar{\pi}^k$ is strictly negative, since this cost is

$$\int_{t_c}^{t_k} (-m_o c\sqrt{G(x^o(t),t,0)})dt$$

One can prove easily that $\pi^k \cup \bar{\pi}^k$ is a path and that, as a consequence of (1.14) and (1.15), its cost is strictly negative.

This contradicts the fact that π^o is optimal and hence Proposition 7 is proved.

APPENDIX 2. PROOF OF PROPOSITIONS 2 AND 3

First we shall prove

Proposition 2a. $(x,t) \in \overset{\circ}{Y} \Rightarrow u^*(x,t) \in \overset{\circ}{K}(x,t)$.

Let $(x^1, t_1) \in \overset{\circ}{Y}$. According to Assumption 3, there exists an optimal path π^*, generated by $u^*(\cdot)$, represented by $x^*(\cdot): t \to x^*(t)$, $t \in [0, t_1]$ such that $x^*(t_1) = x^1$.

Since π^* and ∂Y are closed sets, $\pi^* \cap \partial Y$ is a closed set, and since π^* is bounded, $\pi^* \cap \partial Y$ is compact. Since $x^*(\cdot)$ is continuous, the set S

$$S = \{t: (x^*(t),t) \in \pi^* \cap \partial Y\}$$

is compact, and hence there exists a time $t_M \in S$ such that

$$t \in S \Rightarrow t \leqslant t_M$$

Since $t_M \leqslant t_1$ and $(x^*(t_1),t_1) \in \overset{\circ}{Y}$, we have $t_M < t_1$; and since

$$(x^*(t),t) \notin \partial Y \quad \text{for all} \quad t \in (t_M,t_1]$$

we have

$$(x^*(t),t) \in \overset{\circ}{Y} \quad \text{for all} \quad t \in (t_M,t_1].$$

Moreover, since $(x^*(t_M),t_M) \in \partial Y$, from Proposition 5 proved in Appendix I, we have

$$(x^*(t),t) \in \partial Y \quad \text{for all} \quad t \in [0,t_M] \tag{a}$$

and if $t_M \neq 0$,

$$u^*(x^*(t),t) \in \partial K(x^*(t),t) \quad \text{for all} \quad t \in (0,t_M] \tag{b}$$

Consider the partition (Δ_o,Δ_1) of $(t_M,t_1]$, where

$$\Delta_o = \{t: \ t \in (t_M,t_1], \quad u^*(x^*(t),t) \in \partial K(x^*(t),t)\}$$

$$\Delta_1 = \{t: \ t \in (t_M,t_1], \quad u^*(x^*(t),t) \in \overset{\circ}{K}(x^*(t),t)\}$$

Indeed $(t_M,t_1] = \Delta_o \cup \Delta_1$; and according to (15) and (22), since $(x^*(t),t) \in \overset{\circ}{Y}$ for all $t \in (t_M,t_1]$, we have

$$t \in \Delta_o \Rightarrow \sum_{i,j} g^{ij}(\partial V^o/\partial x_i)(\partial V^o/\partial x_j) = 0 \tag{2.1}$$

$$t \in \Delta_1 \Rightarrow \sum_{i,j} g^{ij}(\partial V^o/\partial x_i)(\partial V^o/\partial x_j) = 1 \tag{2.2}$$

where $V^o = V^o(x,t)$, and $x = x^*(t)$.

Since $V^o = (-1/m_oc)V^*$ and since, according to the adjoint equations (2) and (3), the functions

$$\left.\frac{\partial V^*(x,t)}{\partial x_\alpha}\right|_{x = x^*(t)} \qquad \alpha = 1,2,3,4$$

are continuous on $(t_M, t_1]$, and since the g^{ij} are of class C^1 on $\mathcal{E} \times T$, the function $\sigma(\cdot): t \to \sigma(t)$, $t \in (t_M, t_1]$ defined by

$$\sigma(t) = \sum_{i,j} g^{ij} (\partial V^o/\partial x_i)(\partial V^o/\partial x_j)\Big|_{x = x^*(t)}$$

is continuous on $(t_M, t_1]$.

Suppose that Δ_o and Δ_1 are both non-empty sets, and let

$$t' \in \Delta_o \quad \text{and} \quad t'' \in \Delta_1$$

Then, according to (2.1) and (2.2)

$$\sigma(t') = 0 \quad \text{and} \quad \sigma(t'') = 1$$

and since $\sigma(\cdot)$ is continuous on $[t', t''] \subset (t_M, t_1]$, $\sigma(\cdot)$ must take every value in the range $[0,1]$.

Since it is not so, Δ_o and Δ_1 cannot be both non-empty; that is

$$\Delta_o \neq \phi \Rightarrow \Delta_1 = \phi \text{ ; and}$$

$$\Delta_1 \neq \phi \Rightarrow \Delta_o = \phi .$$

Suppose that $u^*(x^*(t_1), t_1) \in \partial K(x^*(t_1), t_1)$. Then $t_1 \in \Delta_o$ and hence $\Delta_1 = \phi$. It follows that

$$u^*(x^*(t), t) \in \partial K(x^*(t), t) \quad \text{for all} \quad t \in (t_M, t_1]$$

and since, from condition (b) above, if $t_M \neq 0$

$$u^*(x^*(t), t) \in \partial K(x^*(t), t) \quad \text{for all} \quad t \in (0, t_M]$$

we have

$$u^*(x^*(t), t) \in \partial K(x^*(t), t) \quad \text{for all} \quad t \in (0, t_1].$$

Accordingly

$$G(x^*(t),t,u^*(x^*(t),t)) = 0 \quad \text{for all} \quad t \in (0,t_1]$$

so that the cost of π^* is zero.

We shall exhibit a path π^{**} generated by a strategy $u^{**}(\cdot)$, represented by $x^{**}(\cdot): t \to x^{**}(t), \quad t \in [0,t_1]$, such that

$$x^{**}(0) = 0, \quad x^{**}(t_1) = x^*(t_1) \text{ ; and}$$

$$\int_0^{t_1} (- m_o c\sqrt{G(x^{**}(t),t,u^{**}(x^{**}(t),t))}\,)dt < 0$$

which will contradict the fact that π^* is optimal and consequently will prove that $u^*(x^*(t_1),t_1) \in \mathring{K}(x^*(t_1),t_1)$.

Consider the ray L_1

$$L_1 = \{(x,t): \quad x = x^*(t_1), \quad t \in T\}$$

and point $(x^*(t_1),0) \in L_1$.

From Assumption 2 we have

$$(x^*(t_1),t_1) \in D \quad \text{and} \quad (x^*(t_1),0) \notin D$$

since $\mathring{Y} = D \subset \mathcal{E} \times T^+$, and since $0 \notin T^+$.

One can easily prove that

$$L_1 \cap \partial Y \neq \phi$$

Indeed, if $x^*(t_1) = 0$ this result is trivial, and if $x^*(t_1) \neq 0$ it is proved by dichotomy.

Let

$$(x^c,t_c) \in L_1 \cap \partial Y$$

Since, from part of Assumption 2, Y is closed, if $t_c \neq 0$ there exists a path π^c, generated by $u^*(\cdot)$, represented by $x^c(\cdot): t \to x^c(t), \quad t \in [0,t_c]$ such that

$$x^c(t_c) = x^c$$

Then, from Proposition 5, since $(x^c(t),t_c) \in \partial Y$, we have

$$(x^c(t),t) \in \partial Y \quad \text{for all} \quad t \in [0,t_c]; \text{ and} \qquad (a')$$

$$u^*(x^c(t),t) \in \partial K(x^c(t),t) \quad \text{for all} \quad t \in (0,t_c] \qquad (b')$$

so that

$$G(x^c(t),t,u^*(x^c(t),t)) = 0 \text{ for all } \quad t \in (0,t_c]$$

Accordingly, the cost of π^c is zero.

Now consider the solution $x^o(\cdot): t \to x^o(t), \quad t \in [t_c,t_1]$ of state equations

$$\dot{x} = 0$$

such that

$$x^o(t_c) = x^c(t_c) = x^c$$

Since

$$(x^o(t),t) \in L_1 \qquad \text{for all} \quad t \in [t_c,t_1]$$

we have

$$x^o(t_1) = x^*(t_1)$$

Let us denote by π^o the ordered set of states

$$\pi^o = \{(x,t): \quad x = x^o(t), \qquad t \in [t_c,t_1]\}$$

From part of Assumption 4 we have

$$G(x^o(t),t,0) > 0 \qquad \text{for all} \quad t \in [t_c,t_1]$$

so that

$$\int_{t_c}^{t_1} (-m_o c\sqrt{G(x^o(t),t,0)})dt < 0 \tag{2.3}$$

At last let us define function $u^{**}(\cdot)$ as follows :

If $t_c = 0$ we let

$$u^{**}(x,t) = 0 \qquad \text{for all} \qquad (x,t) \in \mathcal{E} \times T^+ .$$

If $t_c \neq 0$, we let

$$u^{**}(x,t) = u^{*}(x,t) \quad \text{for all } (x,t) \in \mathcal{E} \times (0,t_c] \text{ ; and}$$

$$u^{**}(x,t) = 0 \quad \text{for all } (x,t) \in \mathcal{E} \times (t_c, +\infty) \text{ .}$$

In both cases the function thus defined satisfies the constraint

$$G(x,t,u^{**}(x,t)) \geqslant 0 \text{ for all } (x,t) \in \mathcal{E} \times T^+ \text{ .}$$

It is of class C^1 on $\mathcal{E}$ and piece-wise continuous on T^+, so that *it is a strategy*. This strategy generates path π^{**} on the interval $[0,t_1]$, namely

if $t_c = 0$, $\pi^{**} = \pi^o$; and

if $t_c \neq 0$, $\pi^{**} = \pi^c \cup \pi^o$

the cost of π^{**} is given by (2.3), which contradicts the fact that π^* is optimal, so that as we have said before

$$u^*(x^1,t_1) \in \mathring{K}(x^1,t_1)$$

and since this result holds for all $(x^1,t_1) \in \mathring{Y}$, Proposition 2a is established.

Now we will prove

Proposition 2b.[15] $t_M = 0$.

Suppose that $t_M \neq 0$. Then

$$(x^M,t_M) \in (\mathcal{E} \times T^+) \cap \partial Y$$

where $x^M = x^*(t_M)$.

Accordingly there exists an open ball $B_r(x^M,t_M)$ with center

(15) This part of the proof of Proposition 2 is based on weaker assumptions than the one given in Ref.(7). It is due to Mlle M. Pauchard, Laboratoire d'Automatique Théorique, Université de Paris 7.

(x^M, t_M) and radius $r > 0$ such that

$$B_r(x^M, t_M) \subset \mathcal{E} \times T^+$$

and for all $\rho > 0$, $\rho \leqslant r$

$$B_\rho(x^M, t_M) \cap \text{comp } Y \neq \phi$$

where "comp Y" denotes the complement of Y in $\mathcal{E} \times T^+$.

Since $(x^1, t_1) \in \mathring{Y}$, there exists an open ball $B_\varepsilon(x^1, t_1)$ with center (x^1, t_1) and radius $\varepsilon > 0$ such that

$$B_\varepsilon(x^1, t_1) \subset \mathring{Y} \tag{2.4}$$

Since $u^*(\cdot)$ is of class C^1 on $\mathcal{E} \times T^+$, for ε sufficiently small there exist $\rho \leqslant r$ *and* a solution $x'(\cdot): t \to x'(t)$, $t \in [t_i, t_j]$, of state equation

$$\dot{x} = u^*(x, t) \tag{2.5}$$

such that

$$(x'(t_i), t_i) \in B_\rho(x^M, t_M) \cap \text{comp } Y \tag{2.6}$$

and

$$(x'(t_j), t_j) \in B_\varepsilon(x^1, t_1) \tag{2.7}$$

Since from (2.4) and (2.7) $(x'(t_j), t_j) \in \mathring{Y}$, and since $u^*(\cdot)$ is optimal on Y, there exists a solution $x''(\cdot): t \to x''(t)$, $t \in [0, t_j]$ of state equation (2.5) such that

$$(x''(t), t) \in Y \quad \text{for all} \quad t \in [0, t_j]$$

and

$$(x''(t_j), t_j) = (x'(t_j), t_j)$$

But in view of (2.6) this is in contradiction with the uniqueness of the solutions of (2.5) on $\mathcal{E} \times T^+$, and hence $t_M = 0$.

Proposition 2 is a direct consequence of Propositions 2a and 2b .

Now we shall prove Proposition 3.

Remember that $V^o(x,t) = (-1/m_o c)V^*(x,t)$. Then for $(x,t) \in \overset{\circ}{Y}$, it follows from Proposition 2a and (4) that

$$\begin{aligned}\partial V^o/\partial t &= \sqrt{G(x,t,u)} - \sum_{\alpha=1}^{3} (1/2\sqrt{G(x,t,u)})(\partial G(x,t,u)/\partial u_\alpha)u_\alpha \\ &= (2G(x,t,u) - \sum_{\alpha=1}^{3} (\partial G(x,t,u)/\partial u_\alpha)u_\alpha)/(2\sqrt{G(x,t,u)})\end{aligned} \tag{2.8}$$

where $V^o = V^o(x,t), u = u^*(x,t)$.

Now since

$$G(x,t,u) = g_{44} + 2\sum_{i=1}^{3} g_{4i}\, u_i + \sum_{j,k} g_{jk}\, u_j\, u_k, \qquad j,k = 1,2,3$$

and accordingly

$$\partial G(x,t,u)/\partial u_\alpha = 2g_{4\alpha} + 2\sum_{j=1}^{3} g_{j\alpha}\, u_j, \qquad \alpha = 1,2,3$$

we obtain

$$2G(x,t,u) - \sum_{\alpha=1}^{3} (\partial G(x,t,u)/\partial u_\alpha)u_\alpha = 2g_{44} + 2\sum_{i=1}^{3} g_{4i}\, u_i$$

For $u = u^*(x^c,t_c)$, $(x^c,t_c) \in \partial Y$, $t_c \neq 0$, we have

$$G(x^c,t_c,u) = 0 \tag{2.9}$$

This is so because, from Proposition d of Appendix 1, we have

$$u^*(x^c,t_c) \in \partial K(x^c,t_c)$$

Hence

$$g_{44} + 2\sum_{i=1}^{3} g_{4i}\, u_i = -\sum_{j,k} g_{jk}\, u_j\, u_k, \qquad j,k = 1,2,3$$

where indeed

$$g_{\alpha\beta} = g_{\alpha\beta}(x^c, t_c), \qquad \alpha,\beta = 1,2,3,4 \ .$$

It follows that, as $(x,t) \to (x^c, t_c)$

$$\left(2G(x,t,u) - \sum_{\alpha=1}^{3} (\partial G(x,t,u)/\partial u_\alpha) u_\alpha\right)\Bigg|_{u = u^*(x,t)}$$

which is continuous in (x,t) tends to

$$\left(g_{44}(x^c,t_c) - \sum_{j,k} g_{jk}(x^c,t_c) u_j \, u_k\right)\Bigg|_{u^*(x^c,t_c)} \qquad j,k = 1,2,3 \tag{2.10}$$

Since, from Assumption 6, $g_{44}(x,t) > 0$ and $\sum_{j,k} g_{jk}(x,t)u_j \, u_k$ $j,k = 1,2,3$, is negative definite, this limit is *strictly positive*.

Then since $G(x,t,u^*(x,t))$ in the denominator of (2.8), which is continuous in (x,t), tends to zero as $(x,t) \to (x^c,t_c)$ according to (2.9), $\partial V^o/\partial t$ tends to infinity, which concludes the proof of Proposition 3.

APPENDIX 3. PROOF OF PROPOSITION 11

Consider a trajectory $\bar{\pi}$, generated by $\bar{u}(\cdot)$, represented by $\bar{x}(\cdot)\colon t \to \bar{x}(t)$, $t \in [t_0, t_1]$, whose initial point $y^o = (x^o, t_0) = (\bar{x}(t_0), t_0)$ belongs to the open half-ray $\mathcal{L}^+ = \{(x,t)\colon x = 0,\ t > 0\}$.

From the definition of a trajectory

$$\bar{x}(t) \neq 0 \quad \text{for all} \quad t > t_0 \ .$$

Since, from Assumption 7, $\bar{u}(\cdot)$ is of class C^1 on $(\mathcal{E} - \{0\}) \times T^+$, and since, from Assumption 2, grad $W(x,t)$ and $\partial W(x,t)/\partial t$ are defined and non-zero on $(\mathcal{E} - \{0\}) \times T^+$, and accordingly for all $x = \bar{x}(t)$, $t \in (t_o, t_1]$, we deduce from Assumption 7

$$\left.\frac{dW(x,t)}{dt}\right|_{x = \bar{x}(t)} = 0 \quad \text{for all} \quad t \in (t_o, t_1]$$

Hence there exists a constant C such that

$$W(\bar{x}(t), t) = C \quad \text{for all} \quad t \in (t_o, t_1] \ .$$

Since, from Assumption 2, $W(\cdot)$ is continuous on $\mathcal{E} \times T$, it follows that

$$W(\bar{x}(t_o), t_o) = W(x^o, t_o) = W(0, t_o) = C \ .$$

The fact that $(\bar{x}(t), t) \in Y$ for all $t \in [t_o, t_1]$ comes from the property that point $(0, t_o)$ can be reached from source Θ along path $\pi = \{(x,t): x = 0, \ t \in [0, t_o]\}$ and that $(\bar{x}(t), t)$ can be reached from $(0, t_o)$ along trajectory $\bar{\pi}$. Then $(\bar{x}(t), t)$ can be reached from Θ along path $\pi \cup \bar{\pi}$.

Hence Proposition 11 is established.

APPENDIX 4. SOME GEOMETRICAL REMARKS RELATED TO ASSUMPTION 6

According to an oral remark of Prof. A. Lichnerowicz, one can replace Assumption 6 by a condition which is relativistically invariant. We shall report this condition.

Here, we shall make use of the following convention, namely, in the summations Greek indices α, β will take the values 1,2,3,4, and Roman indices i,j,k, will take the values 1,2,3.

First remember that

$$G(x,t,u) = g_{44}(x,t) + 2 \sum_i g_{4i}(x,t) u_i + \sum_{j,k} g_{jk}(x,t) u_j \, u_k$$

and note that the condition

$$G(x,t,0) = g_{44}(x,t) > 0$$

has been already introduced as part of Assumption 4.

It is equivalent to stating that the vector

$$e_4 = (0,0,0,1)$$

lies in the interior of the cone $\tilde{K}(x,t)$

$$\tilde{K}(x,t) = \{u = (u,u_4) = (u_1,u_2,u_3,u_4): \sum_{\alpha,\beta} g_{\alpha\beta}(x,t)u_\alpha u_\beta \geqslant 0\}$$

On the other hand, the part of Assumption 6 according to which the quadratic form

$$\sum_{j,k} g_{jk}(x,t)u_j\, u_k$$

is negative definite, is equivalent to stating that no vector $(u,0) \neq (0,0)$ belongs to $\tilde{K}(x,t)$. (Fig.5).

Further, our proof of Proposition 3, in Appendix 3, makes use of the expression

$$Q = G(x,t,u) - (1/2) \sum_i (\partial G(x,t,u)/\partial u_i)u_i$$

$$= g_{44}(x,t) + \sum_i g_{4i}(x,t)u_i$$

Q has a simple geometrical meaning; it is the scalar product(16) relative to the given fundamental tensor $(g_{\alpha\beta})$, of the two vectors

$$e = (u,1) \quad \text{and} \quad e_4$$

Let us denote this scalar product by

$$e \cdot e_4$$

Indeed, we have

$$e_4 \cdot e_4 = g_{44}(x,t)$$

Now let

$$S(x,t) = \{(u,1): G(x,t,u) \geqslant 0\} \subset \tilde{K}(x,t)$$

---

(16) In the extended sense of Relativity.

One can easily prove that

$$e \in S(x,t) \Rightarrow e \cdot e_4 > 0$$

Indeed $e \in S(x,t)$ implies that

$$g_{44}(x,t) + 2\sum_i g_{4i}(x,t)u_i + \sum_{j,k} g_{jk}(x,t)u_j u_k \geq 0$$

which in turn implies that

$$2(g_{44}(x,t) + \sum_i g_{4i}(x,t)u_i) \geq g_{44}(x,t) - \sum_{j,k} g_{jk}(x,t)u_j u_k$$

Since the right-hand side of this inequality is strictly positive, the property is established.

In particular, the property $e \cdot e_4 > 0$ holds for those vectors e which are isotropic; that is, which belong to the boundary of $\tilde{K}(x,t)$. This is one of the properties on which the proof of Proposition 2 is based.

What we have proved, starting from the assumption that the matrix (g_{ij}) is negative definite, is equivalent to the statement that the set $S(x,t)$ lies in one of the open half-spaces defined by the plane Π

$$\Pi : \sum_\alpha g_{4\alpha}(x,t)u_\alpha = 0$$

namely, the open half-space which contains e_4.(17)

The same conclusion can be reached starting from another assumption. Note that

$$\sum_{\alpha,\beta} g_{\alpha\beta}u_\alpha u_\beta = \frac{1}{g_{44}}(\sum g_{4\alpha}u_\alpha)^2 + \sum \gamma_{ij}u_i u_j$$

$$\gamma_{ij} = g_{ij} - \frac{g_{4i}\, g_{4j}}{g_{44}}$$

Then if, instead of assuming that the matrix (g_{ij}) is negative

(17) Indeed $e_4 \in S(x,t)$

definite, we assume with A. Lichnerowicz that the matrix (γ_{ij}) is negative definite,[18] it is clear that

$$\Pi \cap S(x,t) = \phi$$

The conclusion follows readily.

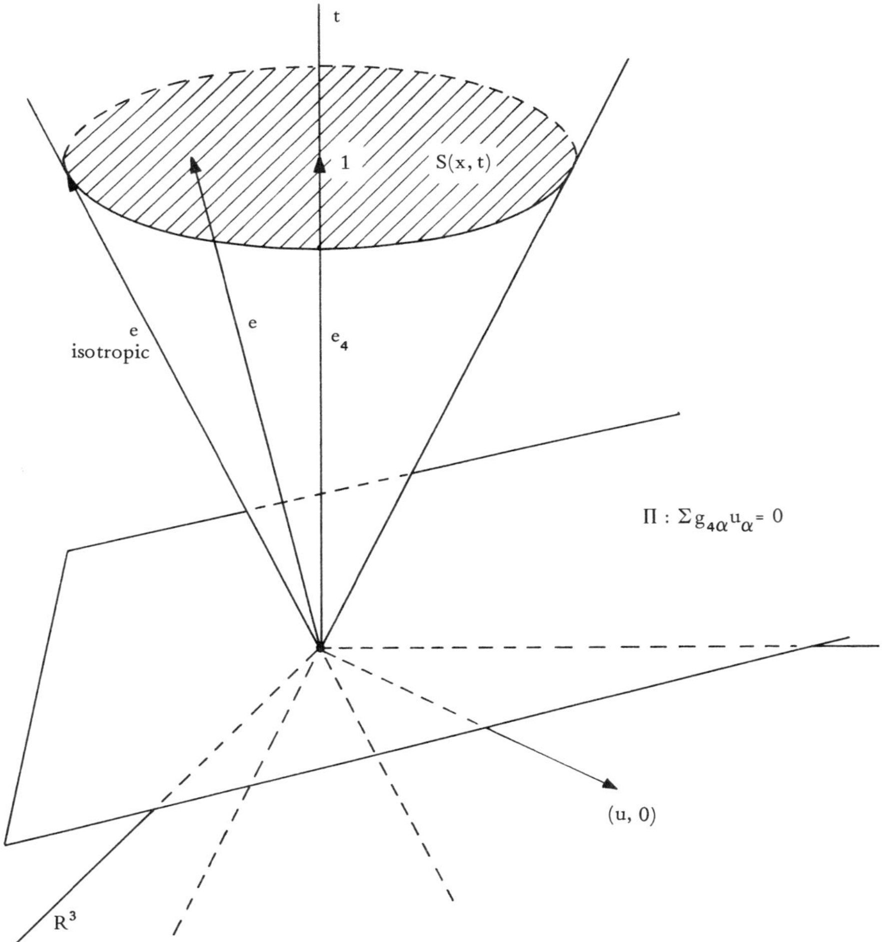

Fig. 5

(18) That is, the signature of the metric is + ---

REFERENCES

[1] Gelfand, M., Fomin, S.V., *Calculus of Variations*, Prentice-Hall, Englewood Cliffs, New Jersey, 1963.

[2] Blaquière, A., Gérard, F., Leitmann, G., *Quantitative and Qualitative Games*, Academic Press, New York, New York, 1969.

[3] De Broglie, L., Thermodynamique relativiste et mécanique ondulatoire, Ann. Inst. Henri Poincaré, Vol. IX, n.2, 1968, pp. 89-108. Section A : Physique théorique.

[4] Blaquière, A., Wave Mechanics as a Two-Player Game, in *Dynamical Systems and Microphysics*, A. Blaquière, F. Fer, A. Marzollo, Editors, Springer-Verlag, Wien-New York, 1980.

[5] Isaacs, R., *Differential Games*, Wiley, New York, New York, 1965.

[6] Blaquière, A., A new approach to the theory of the spin, Int. J. Non-Linear Mechanics, Vol. 15, n.4/5, pp.275-285, 1980.

[7] Blaquière, A., Optimization and controllability in problems of relativistic dynamics and of geometrical optics, Research Report No. 41, C.I.S.M., Udine, 1981.

BIBLIOGRAPHY

Pham Mau Quan, Inductions electromagnétiques en relativité générale et principe de Fermat, Archive for Rational Mechanics and Analysis, Vol.1, n.1, 1957, pp.54-80.

Pham Mau Quan, Sur la dynamique analytique du point en relativité, Acta Physica Austriaca, Vol.14, n.2, 1961, pp.232-238.

Lichnérowicz, A., Eléments de calcul tensoriel, Armand Colin 1951.

Blaquière, A., A controllability problem in relativistic dynamics of a mass-point. Systems and Control Letters, Vol.1, n.1, 1981.

SOME RELATIONS BETWEEN OPTIMAL CONTROL THEORY AND CLASSICAL MECHANICS

A. Blaquière
Laboratoire d'Automatique Théorique
Université Paris VII.

A. Marzollo
Faculté des Sciences,
Université d'Udine
et
Laboratoire d'Automatique Théorique
Université Paris VII.

INTRODUCTION.

The aim of this paper is to discuss some relations between optimal control theory and some aspects of the calculus of variations, in their applications to classical mechanics.

Special attention will be devoted to the possibility of expressing Hamilton's principle in the case of non-holonomic constraints in the framework of optimization theory.

1. OPTIMAL CLOSED-LOOP CONTROL AND CALCULUS OF VARIATIONS.

In this paragraph we shall consider some relations between the approach to the problem of optimal closed-loop control contained in (1), and in the first part of (2), and some conditions of the calculus of variations.

Let us consider a problem of finding an optimal closed-loop control similar to the one exposed in detail in (1). We shall follow the same pattern as in (2), using when possible the same notation. Our assumptions will be of the same kind as in (1) and (2).

Let the system under consideration be characterized by

(a) a set $E \times T$ of *states* (x,t), where E is a domain of the n-dimensional Euclidean space R^n, containing the *initial state*

DYNAMICAL SYSTEMS
AND MICROPHYSICS

ISBN 0-12-068720-8

(x^o, t^o), and $T = (-\infty, +\infty)$; let $T^+ = (t^o, +\infty)$;

(b) a set U of *closed-loop*, or feedback *controls* $u(\cdot)$, defined on $E \times T^+$, of class C^1 on E, piecewise continuous on T^+, and such that, for each $(x,t) \in E \times T^+$,

$$u(x,t) \in K(x,t) \subset R^n$$

with $K(x,t)$ defined by

$$\varphi_\alpha(x,t,u) \geqslant 0, \qquad \alpha = 1,2,\ldots m\ ; \qquad m < n\ . \tag{1.1}$$

The given *constraint* functions $\varphi_\alpha(\cdot)$ are of class C^2 on $E \times T^+ \times R^n$, and are such that, for each $(x,t) \in E \times T^+$, the matrix

$$\left(\frac{\partial \varphi_\alpha}{\partial u_k}(x,t,u)\right), \qquad \alpha = 1,2,\ldots m, \qquad k = 1,2,\ldots n$$

has rank m for all $u \in R^n$ satisfying (1.1).

Let $\mathring{K}(x,t)$ be the interior of $K(x,t)$.

(c) The class of *state equations*

$$\dot{x} = u(x,t), \qquad u(\cdot) \in U \tag{1.2}$$

(d) The *cost function* $f(\cdot)$, which is a prescribed C^2 function on $E \times T^+ \times R^n$.

Given a fixed (x^o, t_o), a closed-loop control $u(\cdot) \in U$ is *admissible at* $(x^1, t^1) \in E \times T^+$ if it generates a solution $x(\cdot)$ of (1.2) on the interval $[t^o, t^1]$ such that $x(t^o) = x^o$ and $x(t^1) = x^1$.

Let a *path* $\pi = \{(x,t): x = x(t),\ \ t \in [t^o, t^1]\}$ be a set of states ordered by t, and $I(x,t)$ be the set of closed-loop controls admissible at (x,t). Let the *cost* of path π generated by $u(\cdot)$ be defined and to be given by

$$V(x^1, t^1, u(\cdot)) = \int_{t^o}^{t^1} f(x(\tau), \tau, u(x(\tau), \tau))\, d\tau \tag{1.3}$$

Let the *reachable set* be denoted by Y, where
$Y = \{(x^o, t^o)\} \cup Y^+$, and

$$Y^+ = \{(x,t): \ (x,t) \in E \times T^+, \ \ \exists u(\cdot) \in I(x,t)\}\ .$$

Let $\mathring{Y}$ be the interior of Y. We shall deal with paths contained in $\mathring{Y}$ only.

Rephrasing in this set-up Assumptions 1, 3, 5 of [2], we make the following Assumptions :

(i) for all $(x,t) \in Y^+$, and for all $u \in K(x,t)$, there exists a closed-loop control $u(\cdot) \in U$, which is continuous, and continuously differentiable with respect to x, and such that $u(x,t) = u$;

(ii) there exists an *optimal closed-loop control* $u^*(\cdot) \in U$, that is, such that $u^*(\cdot) \in I(x,t)$ and

$$V(x,t,u^*(\cdot)) \leqslant V(x,t,u(\cdot))$$

for all $u(\cdot) \in I(x,t)$ and for all $(x,t) \in \mathring{Y}$.

$u^*(\cdot)$ is of class C^1 on $\mathring{Y}$.

The solution of (1.2) generated by $u^*(\cdot)$ will be denoted by $x^*(\cdot)$.

Let the *optimal value function* $V^*(\cdot)$ be defined by $V^*(x,t) = V(x,t,u^*(\cdot))$ for $(x,t) \in \mathring{Y}$, and $V^*(x^o,t^o) = 0$.

(iii) The function $V^*(\cdot)$ is of class C^2 on $\mathring{Y}$.

Under the stated assumptions, an optimal closed-loop control $u^*(\cdot)$ on $\mathring{Y}$ obeys the necessary conditions contained in

PROPOSITION 1.1' .

For each $(x,t) \in \mathring{Y}$,

(i)
$$\max_{u \in K(x,t)} \left(- f(x,t,u) + \sum_{i=1}^{n} \frac{\partial V^*}{\partial y_i}(y,t)\Big|_{y=x} u_i \right)$$
$$= - f(x,t,u^*(x,t)) + \sum_{i=1}^{n} \frac{\partial V^*}{\partial y_i}(y,t)\Big|_{y=x} u_i^*(x,t)$$

(ii)
$$- f(x,t,u^*(x,t)) + \sum_{i=1}^{n} \frac{\partial V^*}{\partial y_i}(y,t)\Big|_{y=x} u_i^*(x,t) + \frac{\partial V^*}{\partial t}(x,t) = 0$$

with $V^*(x^o,t^o) = 0$.

Sketch of the proof

Let (x,t) be arbitrarily given in $\mathring{Y}$, and $u \in R^n$ arbitrarily chosen in $\mathring{K}(x,t)$. Let $x(\cdot)$ be such that $\dot{x} = u$, $x(t) = x$. The value of $x(\cdot)$ at $t-\varepsilon$ is therefore $x - u\varepsilon$; for ε small enough, $(x-u\varepsilon\, ,\, t-\varepsilon) \in \mathring{Y}$. In correspondence to the path π_ε generated by $u^*(\cdot) \in I(x-u\varepsilon\, ,\, t-\varepsilon)$ on $[t_o,\ t-\varepsilon]$ and by the constant function u on $(t-\varepsilon, t]$ we have

$$T_u(\varepsilon) = V^*(t-\varepsilon,\ x-u\varepsilon) + \int_{t-\varepsilon}^{t} f(x(\tau),\tau,u)d\tau \geqslant V^*(x,t) = T_u(0)$$

and therefore, for every $u \in \mathring{K}(x,t)$

$$0 \leqslant \lim_{\eta \to 0^+} \left.\frac{dT_u(\varepsilon)}{d\varepsilon}\right|_{\varepsilon=\eta} = - \left.\frac{\partial V^*}{\partial \tau}(x,\tau)\right|_{\tau=t} - \sum_{i=1}^{n} \left.\frac{\partial V^*}{\partial y_i}(y,t)\right|_{y=x} u_i + f(x,t,u) .$$

By continuity, the above inequality holds for $u \in K(x,t)$.

On the other hand

$$T_{u^*}(\varepsilon) = V^*(t-\varepsilon,\ x^*(t-\varepsilon)) + \int_{t-\varepsilon}^{t} f(x^*(\tau),\tau,u^*(x^*(\tau),\tau))d\tau$$

and therefore $T_{u^*}(\varepsilon) = V^*(x,t) = T_{u^*}(0)$, and

$$0 = \lim_{\eta \to 0^+} \left.\frac{dT_{u^*}(\varepsilon)}{d\varepsilon}\right|_{\varepsilon=\eta} = - \left.\frac{\partial V^*}{\partial \tau}(x,\tau)\right|_{\tau=t} - \sum_{i=1}^{n} \left.\frac{\partial V^*}{\partial y_i}(y,t)\right|_{y=x} u_i^*(x,t) + f(x,t,u^*(x,t))$$

which concludes the proof.

Let us define on $R^n \times E \times R^1 \times R^n$ the "prehalmitonian" function $H(\cdot)$, namely

$$H(\lambda,x,t,u) = \sum_{i=1}^{n} \lambda_i u_i - f(x,t,u)$$

and on $[t^o,t^1]$ the vectors $\lambda(t) = (\lambda_1(t),\lambda_2(t),\ldots\lambda_n(t)) \in R^n$ and $\lambda_t(t) \in R$, defined by

$$\lambda_i(t) = \frac{\partial V^*}{\partial x_i}(x,t)\Big|_{x=x^*(t)}, \quad i = 1,2,\ldots n .$$

$$\lambda_t(t) = \frac{\partial V^*}{\partial t}(x,t)\Big|_{x=x^*(t)} .$$

Then we have the following

PROPOSITION 1.1".

The following equations

$$(1.3) \qquad \dot{\lambda}_k(t) = -\frac{\partial H}{\partial x_k}(\lambda,x,t,u) - \sum_{i=1}^{n} \frac{\partial H}{\partial u_i}(\lambda,x,t,u)\frac{\partial u_i^*}{\partial x_k}(x,t), \quad k = 1,2,\ldots n$$

$$(1.3') \qquad \dot{\lambda}_t(t) = -\frac{\partial H}{\partial t}(\lambda,x,t,u) - \sum_{i=1}^{n} \frac{\partial H}{\partial u_i}(\lambda,x,t,u)\frac{\partial u_i^*}{\partial x_k}(x,t),$$

hold along the optimal path generated by $u^*(\cdot)$, that is, for $x = x^*(t)$, $u = u^*(x^*(t),t)$.

Proof.

Consider the line integral

$$(1.4) \qquad \int_{x^o,t^o}^{x^1,t^1} \left(\sum_{i=1}^{n} \frac{\partial V^*}{\partial x_i}(x,\tau)dx_i + \frac{\partial V^*}{\partial t}(x,\tau)d\tau \right)$$

evaluated along a curve ξ,
$\xi = \{(x,t) : x = x(t),\ x(t^o) = x^o,\ x(t^1) = x^1\} \subset \mathring{Y}$,
with $x(\cdot)$ of class C^1.

Recalling (ii) of Proposition 1.1', and putting $dx_i = u_i(x(\tau),\tau)d\tau$, (1.4) may be rewritten as

$$\int_{t^o}^{t^1} \psi(x(\tau),\tau,u(x(\tau),\tau))d\tau$$

where

$$\psi(x,t,u) = \sum_{i=1}^{n} \frac{\partial V^*}{\partial y_i}(y,t)\Big|_{y=x} u_i$$

$$- H(\text{grad } V^*(x,t),x,t,u^*(x,t)) .$$

Since (1.4) is independent of curve ξ defined above, it follows that the Euler-Lagrange equations

$$\frac{d}{dt}\left(\frac{\partial\psi}{\partial u_k}\right) = \frac{\partial\psi}{\partial x_k} , \qquad k = 1,2,\ldots n$$

$$\frac{d}{dt}\left(\psi - \sum_{i=1}^{n} \frac{\partial\psi}{\partial u_i} u_i\right) = \frac{\partial\psi}{\partial t}$$

are satisfied along ξ .

Computing the above expression for the optimal closed-loop control $u^*(\cdot)$ and the corresponding $x^*(\cdot)$, we have the *adjoint-equations* (1.3) and (1.3').

In the following, we shall show how some classical conditions of the calculus of variations may be established as a consequence of Propositions 1.1' and 1.1", by making direct reference to the explicit form (1.1) of the constraints.

Putting

$$F(\mu,x,t,u) = f(x,t,u) - \sum_{\alpha=1}^{m} \mu_\alpha \varphi_\alpha(x,t,u)$$

by applying Kuhn-Tucker theorem (see for example [3]), from (i) of Proposition 1.1' we have the existence of a function $\mu(\cdot)$, defined and non-negative on $\mathring{Y}$, $\mu(x,t) = 0$ for $u^*(x,t) \in \mathring{K}(x,t)$, such that

(1.5) $$\frac{\partial}{\partial u_k} F(\mu(x,t),x,t,u)\Big|_{u=u^*(x,t)} - \frac{\partial V^*}{\partial x_k}(x,t) = 0,$$

$$k = 1,2,\ldots n .$$

It may be proved that, under the given hypotheses, $\mu(\cdot)$ is of class C^1 on $\overset{\circ}{Y}$.

Because of (1.5) and of (ii) of Proposition 1.1',

$$(1.6)\qquad I(u(\cdot)) = \int_{t^o}^{t^1} \Big(f(x(\tau),\tau,u^*(x(\tau),\tau)) - \sum_{k=1}^{n} \frac{\partial F}{\partial u_k}(\mu(x(\tau),\tau),x(\tau),\tau,u)\Big|_{u=u^*(x(\tau),\tau)} (u_k^*(x(\tau),\tau)-u_k(x(\tau),\tau))\Big) d\tau$$

is independent of the chosen $u(\cdot) \in I(x^1,t^1)$, that is, of the corresponding path between (x^o,t^o) and (x^1,t^1).

In the particular case of absence of constraints, or for paths such that $u^*(x(t),t) \in \overset{\circ}{K}(x(t),t)$, $t \in [t^o,t^1]$, (1.6) is an Hilbert integral and therefore, using the terminology of the calculus of variations, the pair $(\overset{\circ}{Y},u^*(\cdot))$ with $u^*(\cdot)$ of class C^1 on $\overset{\circ}{Y}$ forms a Mayer field (see for example [6]). In general, we may call the pair $(\overset{\circ}{Y},u^*(\cdot))$ with $u^*(\cdot)$ of class C^1 on $\overset{\circ}{Y}$ such that the integral (1.6) is independent of $u(\cdot) \in U$, a *generalized Mayer field*.

Defining

$$E(\mu,x,t,u^*,u) = f(x,t,u) - F(x,t,u^*) - \sum_{k=1}^{n} \frac{\partial F}{\partial u_k}(\mu,x,t,u)\Big|_{u=u^*(x,t)} (u_k - u^*(x,t))$$

from (i) of Proposition 1.1' and (1.5) we have, for $(x,t) \in \overset{\circ}{Y}$

$$(1.7)\qquad E(\mu(x,t),x,t,u^*(x,t),u) \geqslant 0 \qquad u \in K(x,t).$$

In an analogous way as above, we may call (1.7) a *generalized Weierstrass condition*.

By defining in an obvious way the function $F(\cdot)$ such that (1.5) becomes

$$(1.8) \qquad \frac{\partial}{\partial u_k} F(x,t,u)\bigg|_{u=u^*(x,t)} = \frac{\partial V^*}{\partial x_k}(x,t), \qquad k=1,2,\ldots n$$

and by using the inverse function theorem under proper assumptions for this case, (1.8) gives the existence of a function $v(\cdot)$ on a neighborhood of (x,t,p), with $p = \text{grad } V^*(x,t)$, such that, in that neighborhood

$$\frac{\partial}{\partial u_k} F(x,t,u)\bigg|_{u=v(x,t,p)} = p_k, \qquad k=1,2,\ldots n$$

Introducing function $v(\cdot)$ in (ii) of Proposition 1.1' we have

$$\frac{\partial V^*}{\partial t}(x,t) + H\left(\frac{\partial V^*}{\partial x}(x,t),x,t,v(x,t,\frac{\partial V^*}{\partial x}(x,t))\right) = 0$$

Defining the hamiltonian $\mathcal{H}(\cdot)$ in an obvious way, we may write the classical Hamilton-Jacobi equation :

$$(1.9) \qquad \frac{\partial V^*}{\partial t}(x,t) + \mathcal{H}\left(\frac{\partial V^*}{\partial x}(x,t),x,t\right) = 0$$

Taking into account the form (1.1) of the constraints, and applying Kuhn-Tucker theorem, a simple computation (see for example [4]) shows that the adjoint equations (1.3) take the form of generalized Euler-Lagrange equations

$$(1.10) \qquad \frac{d}{dt}\left(\frac{\partial F}{\partial u_k}(\mu(x^*(t),t),x^*(t),t,u)\right)\bigg|_{u=u^*(x^*(t),t)} =$$

$$= \frac{\partial F}{\partial x_k}\left(\mu(x^*(t),t),x,t,u^*(x^*(t),t)\right)\bigg|_{x=x^*(t)}$$

Our optimal closed-loop control problem may be considered as a closed-loop version of the classical "problem of Lagrange" of the calculus of variations (see for example [6]).

Summarizing the above results : Propositions 1.1' and 1.1" imply a generalization to the constrained case of some well known conditions of elementary calculus of variations, namely the existence of a Mayer field, Weierstrass condition, Euler-Lagrange equations and, under proper invertibility conditions, Hamilton-Jacobi equation.

Let us now remark that the two first above mentioned conditions alone, more precisely the existence of a C^1 function $u^*(\cdot)$ on $\mathring{Y}$ such that the integral (1.6) is independent of the path between (x^o,t^0) and (x^1,t^1) in $\mathring{Y}$, and condition (1.7), imply that $u^*(\cdot)$ is an optimal closed-loop control on $\mathring{Y}$, that $V^*(\cdot)$ is of class C^2, and that (ii) of Proposition 1.1' holds.

Taking these conditions as hypotheses, we have

$$\int_{t^0}^{t^1} f(x(\tau),\tau,u(x(\tau),\tau))d\tau =$$

$$= I(u(\cdot)) + \int_{t^0}^{t^1} E(\mu(x(\tau),\tau),x(\tau),\tau,u^*(x(\tau),\tau),u(x(\tau),\tau))d\tau \geqslant$$

$$\geqslant I(u(\cdot)) = I(u^*(\cdot)) = \int_{t^0}^{t^1} f(x^*(\tau),\tau,u^*(x^*(\tau),\tau))d\tau$$

$$u(\cdot) \in \mathcal{I}(x^1,t^1) \ .$$

Moreover, from the above inequality one can see readily that the value of the functional $I(\cdot)$, which only depends on (x^1,t^1) for (x^o,t^0) given, coincides with the value of the optimal value function $V^*(\cdot)$.

Therefore, if we define

$$(1.11)\quad V^*(x^1,t^1) = \int_{x^o,t^0}^{x^1,t^1} \Bigg(f(x,\tau,u^*(x,\tau))d\tau - \sum_{i=1}^{n} \frac{\partial F}{\partial u_i}(\mu(x,\tau),x,\tau,u)\bigg|_{u=u^*(x,t)} (u_i^*(x,\tau)d\tau - dx_i)\Bigg)$$

where $dx_i = u_i(x(t),t)dt$, we have

$$(1.12)\quad \frac{\partial V^*}{\partial y_k}(y,t)\bigg|_{y=x} = -\frac{\partial F}{\partial u_k}(\mu(x,t),x,t,u)\bigg|_{u=u^*(x,t)}$$

(1.13) $$\frac{\partial V^*}{\partial t}(x,t) = f(x,t,u^*(x,t)) - \sum_{i=1}^{n} \frac{\partial F}{\partial u_i}(\mu(x,t),x,t,u)\bigg|_{u=u^*(x,t)} u_i^*(x,t)$$

From the continuity properties of the right hand sides of (1.12) and (1.13), we see that $V^*(\cdot)$ is of class C^2 ; moreover, using (1.12), we see that (1.13) coincides with (ii) of Proposition 1.1'.

Note that the existence of a C^1 function $\tilde{u}(\cdot)$ on $\mathring{Y}$, such that the integral (1.11) (with $u^*(\cdot)$ replaced by $\tilde{u}(\cdot)$) is independent of the path between given (x^o,t^o) and (x,t) in $\mathring{Y}$, enables to define on $\mathring{Y}$ function $V(\cdot)$ as done in (1.11), also if $\tilde{u}(\cdot)$ is not an optimal closed-loop control.

Equalities (1.12) and (1.13) would likewise follow with $u^*(\cdot)$ replaced by $\tilde{u}(\cdot)$, and $V^*(\cdot)$ by $V(\cdot)$, which also would be of class C^2. In other words a generalized Mayer field does not need to correspond to an optimal closed-loop control.

From the above, we may conclude that, in the case of constraints (1.1), the problem of optimal closed-loop control was faced in [1], [2], using assumptions which, in the framework of the calculus of variations, consist in

a) the existence of a (generalized) Mayer field; and

b) the (generalized) Weierstrass condition.

As it is well known in the calculus of variations (see for example [5]), condition a), or the possibility of defining a function $V(\cdot)$ of class C^2 on $\mathring{Y}$ as above, is linked to the non-existence of points conjugate to (x^o,t^o) on path $\tilde{\pi}$ generated by $\tilde{u}(\cdot)$ as above. In view of applications to classical mechanics, whose paths may well include points conjugate to the initial point (x^o,t^o), condition a) will not appear in the conditions for optimality of controls which we shall expose in the next paragraph.

2. HAMILTON'S PRINCIPLE IN THE FRAMEWORK OF OPTIMIZATION THEORY.

We start this paragraph by considering an open-loop (see [1]) extremal (minimal or maximal) control problem, which parallels the closed-loop optimal (minimal) control problem of paragraph 1, since we shall deal with similar state equations and constraints on admissible controls. An important difference will consist in dropping the differentiability assumptions on the optimal value function, thus avoiding the limitations mentioned at the end of paragraph 1. Since this paragraph aims at discussing relations between optimization theory and classical mechanics, we shall restrict the mathematical hypotheses to those usually made in mechanics; in particular we shall deal with equality constraints of a particular form.

Let there be given :

- (x^o,t^o) and (x^1,t^1) in $R^n \times R^1$;
- a cost function $f(\cdot)$ of class C^2 on $R^n \times R^1 \times R^n$;
- a set X of functions $x(\cdot)$ with range in R^n, of class C^1

with respect to their argument t, such that $x(t^o) = x^o$, $x(t^1) = x^1$, and such that for all $t \in [t^o,t^1]$ the constraints (2.1) are satisfied, namely

$$\text{(2.1)} \qquad \sum_{i=1}^{n} a_{\alpha i}(x(t),t)\dot{x}_i(t) + b_\alpha(x(t),t) = 0$$

$$\alpha = 1,2,\ldots m \; ; \qquad m < n$$

where the constraint functions $a_{\alpha i}(\cdot)$, $b_\alpha(\cdot)$, are of class C^2 on $R^n \times R^1$, and rank $(a_{\alpha i}(x(t),t)) = m$, for $x(\cdot) \in X$, $t \in [t^o,t^1]$.

Let us suppose that $\bar{x}(\cdot) \in X$ generating path $\bar{\pi}$ is of class C^2, is normal(†) and such that

$$\int_{t^o}^{t^1} f(\bar{x}(\tau),\tau,\dot{\bar{x}}(\tau))d\tau \leqslant J(x(\cdot)) = \int_{t^o}^{t^1} f(x(\tau),\tau,\dot{x}(\tau))d\tau$$

---

†) for a definition of normality see [7], p.273 .

or

$$\int_{t^o}^{t^1} f(\bar{x}(\tau),\tau,\dot{\bar{x}}(\tau))d\tau \geqslant J(x(\cdot))$$

for all $x(\cdot) \in X$ such that $(x(t),t)$ belongs to some neighborhood of $\bar{\pi}$.

Then, as a particular case of Theorem 3.1, p.257 in [7] or under slightly broader hypotheses, of Theorem 23, p.298 in [8], the following proposition i) holds, namely

i) there exist C^1 functions $\mu(\cdot) = (\mu_1(\cdot),\mu_2(\cdot),\ldots\mu_m(\cdot))$, and $\lambda(\cdot) = (\lambda_1(\cdot),\lambda_2(\cdot),\ldots\lambda_n(\cdot))$, on $[t^o, t^1]$, such that

$$\dot{\lambda}_k = -\frac{\partial H}{\partial x_k}(\mu,\lambda,x,t,u)\Bigg|_{\substack{x=\bar{x}(t)\\ u=\dot{\bar{x}}(t)\\ \mu=\mu(t)\\ \lambda=\lambda(t)}} , \quad k = 1,2,\ldots n . \tag{2.2}$$

where $\bar{x}(\cdot) \in X$ and

$$H(\mu,\lambda,x,t,u) = -f(x,t,u) + \sum_{k=1}^{n} \lambda_k u_k + \sum_{\alpha=1}^{m} \mu_\alpha \varphi_\alpha(x,t,u),$$

$$\varphi_\alpha(x,t,u) = \sum_{i=1}^{n} a_{\alpha i}(x,t)u_i + b_\alpha(x,t),$$

and

$$\frac{\partial H}{\partial u_k}(\mu(t),\lambda(t),\bar{x}(t),t,u)\Bigg|_{u=\dot{\bar{x}}(t)} = 0, \qquad k = 1,2,\ldots n, \tag{2.3}$$

for all $t \in [t^o,t^1]$.

It may be shown (see Appendix) that i) is equivalent to the following proposition ii) namely

ii) there exist C^1 functions $\mu_\alpha(\cdot)$, $\alpha = 1,2,\ldots m$, on $[t^o,t^1]$, such that the following equations are satisfied $(k = 1,2,\ldots n)$.

$$\frac{d}{dt}\left(\frac{\partial f}{\partial u_k}\right) - \frac{\partial f}{\partial x_k} = \sum_{\alpha=1}^{m} \frac{d\mu_\alpha}{dt} a_{\alpha k} + \sum_{j=1}^{n} \sum_{\alpha=1}^{m} \mu_\alpha \left(\frac{\partial a_{\alpha k}}{\partial x_j} - \frac{\partial a_{\alpha j}}{\partial x_k}\right) \dot{x}_j +$$

$$(2.4) \qquad + \sum_{\alpha=1}^{m} \mu_\alpha \left(\frac{\partial a_{\alpha k}}{\partial t} - \frac{\partial b_\alpha}{\partial x_k}\right)$$

where all functions are computed for $x = \bar{x}(t)$, $u = \dot{\bar{x}}(t)$, $\mu_\alpha = \mu_\alpha(t)$, and $\bar{x}(\cdot) \in X$.

In the framework of optimization theory, for given (x^o, t^0), (x^1, t^1), cost function $f(\cdot)$ and constraints (2.1), we shall define as *stationary relative to the varied paths* $x(\cdot) \in X$ those paths $\bar{\pi}$ such that $\bar{x}(\cdot) \in X$ is of class C^2 and such that proposition i), or equivalently proposition ii), above holds.

Note that (2.3) gives

$$(2.5) \qquad - \frac{\partial f}{\partial u_k}(\bar{x}(t), t, u)\Big|_{u=\dot{\bar{x}}(t)} + \lambda_k(t) + \sum_{\alpha=1}^{m} \mu_\alpha(t) a_{\alpha k}(\bar{x}(t), t) = 0$$

$$k = 1, 2, \ldots n$$

and since, by differentiating the constraints (2.1) with respect to $u = \dot{x}(t)$ at $(\bar{x}(t), t)$ one deduces

$$\sum_{k=1}^{n} \sum_{\alpha=1}^{m} \mu_\alpha(t) a_{\alpha k}(\bar{x}(t), t) du_k = 0$$

it follows from (2.5) that

$$\sum_{k=1}^{n} \left(- \frac{\partial f}{\partial u_k} (\bar{x}(t), t, u)\Big|_{u = \dot{\bar{x}}(t)} + \lambda_k(t)\right) du_k = 0$$

that is, if $\bar{\pi}$ is stationary relative to the varied paths $x(\cdot) \in X$, then the function

$$(2.6) \qquad H(\lambda, x, t, u) = - f(x, t, u) + \sum_{k=1}^{n} \lambda_k(t) u_k$$

is stationary with respect to u at $x = \bar{x}(t)$, $u = \dot{\bar{x}}(t)$, under the constraints

$$(2.7) \qquad \sum_{i=1}^{n} a_{\alpha i}(\bar{x}(t),t)u_i + b_\alpha(\bar{x}(t),t) = 0, \qquad \alpha = 1,2,\ldots m .$$

Now in classical mechanics, let us consider a system whose *representative point* $q = (q_1, q_2, \ldots q_n) \in R^n$ evolves with the changing of time. Let its motion be described by a C^2 function of time, namely $\bar{q}(\cdot)$ on $[t^o,t^1]$. We call a *trajectory* or *mechanical path* with initial point $q^o = x^o$ and terminal point $q^1 = x^1$ the ordered set of points

$$\gamma = \{(q,t): \; q = \bar{q}(t), t \in [t^o,t^1] , \; q(t^o) = q^o, \; q(t^1) = q^1 \} .$$

Let *mechanical constraint functions* $\tilde{a}_{\alpha i}(\cdot), \tilde{b}_\alpha(\cdot)$, $\alpha = 1,2,\ldots m$, $i = 1,2,\ldots n$, $m < n$, of class C^2 be given on $R^n \times R^1$, rank $(\tilde{a}_{\alpha i}(\bar{q}(t),t)) = m$, and γ satisfy the constraints

$$(2.8) \qquad \sum_{i=1}^{n} \tilde{a}_{\alpha i}(\bar{q}(t),t)\dot{\bar{q}}_i(t) + \tilde{b}_\alpha(\bar{q}(t),t) = 0, \qquad \alpha = 1,2,\ldots m .$$

Let the *Lagrangean function* $L(\cdot)$ be of class C^2 on $R^n \times R^1 \times R^n$, $L(\cdot) = T(\cdot) - \mathcal{U}(\cdot)$, where at $(\bar{q}(t),t,\dot{\bar{q}}(t))$

$$L(\bar{q}(t),t,\dot{\bar{q}}(t)) = T(\bar{q}(t),t,\dot{\bar{q}}(t)) - \mathcal{U}(\bar{q}(t),t,\dot{\bar{q}}(t))$$

and $\mathcal{U}(\bar{q}(t),t,\dot{\bar{q}}(t))$ is the generalized potential. This potential is supposed to be such that

$$f_i^{(a)}(\bar{q}(t),t,\dot{\bar{q}}(t)) = - \frac{\partial \mathcal{U}}{\partial y_i}(y,t,\dot{\bar{q}}(t))\Bigg|_{y=\bar{q}(t)} + \frac{d}{d\tau}\frac{\partial \mathcal{U}}{\partial \dot{y}_i}(\bar{q}(\tau),\tau,\dot{y})\Bigg|_{\substack{\dot{y}=\dot{\bar{q}}(t)\\ \tau=t}}$$

The $f_i^{(a)}(\bar{q}(t),t,\dot{\bar{q}}(t))$, $i = 1,2,\ldots n$, are the components of the "total active force" acting at $(\bar{q}(t),t,\dot{\bar{q}}(t))$.

The *virtual displacements* $\delta q(t) = (\delta q_1(t), \delta q_2(t), \ldots \delta q_n(t))$, where the $\delta q_i(\cdot)$ are C^1 functions on $[t^o,t^1]$, $\delta q(t^o) = \delta q(t^1) = 0$, are defined by

$$(2.9) \qquad \sum_{i=1}^{n} \tilde{a}_{\alpha i}(\bar{q}(t),t,\delta q_i(t) = 0 \qquad t \in [t^o,t^1], \quad \alpha = 1,2,...m .$$

The "forces of constraint" are supposed not to work in correspondence to virtual displacements.

It is well known in classical mechanics (see for example [9], [10]) that Newton's law and the principle of virtual work imply that, along γ, the following *mechanical equations of motion* are satisfied

$$(2.10) \qquad \frac{d}{d\tau}\left(\frac{\partial L}{\partial \dot{q}_k}(\bar{q}(\tau),\tau,\dot{q})\Big|_{\substack{\dot{q}=\dot{\bar{q}}(\tau)\\ \tau=t}}\right) - \frac{\partial L}{\partial q_k}(q,t,\dot{\bar{q}}(t))\Big|_{q=\bar{q}(t)} =$$

$$= \sum_{\alpha=1}^{m} \frac{d\mu_\alpha}{d\tau}(\tau)\Big|_{\tau=t} \tilde{a}_{\alpha k}(\bar{q}(t),t), \qquad k = 1,2,...n .$$

$$\bar{q}(t^o) = q^o, \qquad \bar{q}(t^1) = q^1,$$

where the $\mu_\alpha(\cdot)$ are C^1 functions.

The above statement holds both in the case of *holonomic* constraint functions, that is $\tilde{a}_{\alpha i}(\cdot)$, $\tilde{b}_\alpha(\cdot)$ such that

$$(2.11) \qquad \left.\begin{aligned} \frac{\partial \tilde{a}_{\alpha i}}{\partial y_j}(y,t)\Big|_{y=q} &= \frac{\partial \tilde{a}_{\alpha j}}{\partial y_i}(y,t)\Big|_{y=q} \\ \frac{\partial \tilde{b}_{\alpha}}{\partial y_j}(y,t)\Big|_{y=q} &\quad \frac{\partial \tilde{a}_{\alpha j}}{\partial \tau}(q,\tau)\Big|_{\tau=t} \end{aligned}\right\} \quad \begin{aligned} \alpha &= 1,2,...m \\ i,j &= 1,2,...n . \end{aligned}$$

for all $(q,t) \in R^n \times R^1$,

and of *non-holonomic* constraint functions $\tilde{a}_{\alpha i}(\cdot)$, $\tilde{b}_\alpha(\cdot)$ for which (2.11) does not hold.

Recalling the definition ii) of stationarity relative to the varied paths $x(\cdot) \in X$, we have the following

PROPOSITION 2.1.

For $f(\cdot) = L(\cdot)$, equation (2.4) in condition ii) of stationarity becomes (2.10) if functions $a_{\alpha i}(\cdot)$, $b_\alpha(\cdot)$, $g^\gamma_\alpha(\cdot)$ ($\alpha = 1,2,\ldots m$; $i = 1,2,\ldots n$) are chosen such that, for all $q(\cdot) \in X$ such that $(q(t),t)$ lies in some neighborhood $N(\gamma)$ of γ, the following holds:

$$(2.12)\quad a_{\alpha i}(q(t),t) = \frac{\partial g^\gamma_\alpha}{\partial q_i}(q,t)\Big|_{q=q(t)} \; ; \; b_\alpha(q(t),t) = \frac{\partial g^\gamma_\alpha}{\partial \tau}(q(t),\tau)\Big|_{\tau=t}$$

$$(2.13)\quad a_{\alpha i}(\bar{q}(t),t) = \tilde{a}_{\alpha i}(\bar{q}(t),t) \; ; \; b_\alpha(\bar{q}(t),t) = \tilde{b}_\alpha(\bar{q}(t),t)$$

$$\alpha = 1,2,\ldots m; \qquad i = 1,2,\ldots n; \qquad t \in [t^o,t^1] .$$

Proposition 2.1 is an obvious consequence of the fact that functions $g^\gamma_\alpha(\cdot)$ are of class C^2 on $N(\gamma)$, and that therefore

$$(2.14)\quad \left.\begin{aligned} \frac{\partial a_{\alpha i}}{\partial y_j}(y,t)\Big|_{y=q} &= \frac{\partial a_{\alpha j}}{\partial y_i}(y,t)\Big|_{y=q} \\ \frac{\partial b_\alpha}{\partial y_j}(y,t)\Big|_{y=q} &= \frac{\partial a_{\alpha j}}{\partial \tau}(q,\tau)\Big|_{\tau=t} \end{aligned}\right\} \quad \begin{aligned} \alpha &= 1,2,\ldots m \\ i,j &= 1,2,\ldots n \\ &\text{for } (q,t) \in N(\gamma) . \end{aligned}$$

Constraint functions $a_{\alpha i}(\cdot)$, $b_\alpha(\cdot)$ satisfying (2.14) may be called *locally holonomic on* $N(\gamma)$.

In other words, Proposition 2.1 states that the mechanical equations of motion satisfied along a trajectory γ coincide with the equations satisfied by a stationary path $\bar{\pi}$ of optimization theory, with $\gamma = \bar{\pi}$, provided that the constraints of optimization theory are chosen to be locally holonomic on $N(\gamma)$. This is true also for non-holonomic mechanical constraint functions.

Obviously, the above locally holonomic constraint functions depend on the given trajectory γ and are not unique.

The above clarifies the often debated question (see for example [11], [12], [13], [14], [15], [16], [17]) whether or not stationarity of a path as defined in stating Hamilton's principle coincides with stationarity of the same path as already defined in i) or ii), also for non-holonomic mechanical constraint functions.

The literature on this subject is rather confusing, also because different definitions are used by different authors concerning Hamilton's principle for this case (see for example [17] as opposed to [9], [10]).

We shall state Hamilton's principle following the standard definition given for example in [10]. In order to do that, let us define, for each ε, $|\varepsilon| < \varepsilon_0$, each $\bar{q}(\cdot)$ satisfying (2.8), and each function $\delta q(\cdot)$ corresponding to virtual displacements, the functional $I(\cdot)$:

$$I(\bar{q}(\cdot)+\varepsilon\delta q(\cdot)) \triangleq \int_{t^0}^{t^1} (L(\bar{q}(t)+\varepsilon\delta q(t),t,\dot{\bar{q}}(t)+\varepsilon\delta\dot{q}(t))-L(\bar{q}(t),t,\dot{\bar{q}}(t)))dt$$

Then $\bar{q}(\cdot)$ of class C^2 on $[t^0,t^1]$ is *stationary relative to virtual displacements* if

$$\frac{\partial}{\partial\varepsilon} I(\bar{q}(\cdot) + \varepsilon\delta q(\cdot)) \Big|_{\varepsilon=0} = 0$$

for every $\delta q(\cdot)$ satisfying (2.9) and $\delta q(t^0) = \delta q(t^1) = 0$.

Hamilton's principle states that trajectories coincide with paths which are stationary relative to virtual displacements.

In other words, Hamilton's principle states that stationary paths as defined above coincide with those which satisfy equations (2.8) and (2.10). Recalling Proposition 2.1 and using, in a neighborhood $N(\gamma)$ of γ, locally holonomic constraints satisfying (2.12) and (2.13), we see that the definition of stationarity i), or ii), is equivalent to the definition of stationarity given above in stating Hamilton's principle.

Let us recall that the locally holonomic constraint functions

$a_{\alpha i}(\cdot)$, $b_\alpha(\cdot)$ may be chosen arbitrarily, provided that they satisfy (2.12) and (2.13) on $N(\gamma)$.

It is interesting to note that the corresponding "varied paths", that is those generated by functions $\bar{q}(\cdot) + \varepsilon\Delta q(\cdot)$ such that

$$(2.15)\quad \sum_{i=1}^{n} a_{\alpha i}(\bar{q}(t)+\varepsilon\Delta q(t),t)(\dot{\bar{q}}_i(t)+\varepsilon\Delta\dot{q}_i(t)) + b_\alpha(\bar{q}(t)+\varepsilon\Delta q(t),t = 0$$

do not coincide in general with the "varied paths" appearing in the last definition of stationarity, that is paths generated by functions $\bar{q}(\cdot) + \varepsilon\delta q(\cdot)$ such that $\delta q(\cdot)$ satisfies (2.9).

It may be easily proved (for example by a slight modification of the proof of the Lemma at p.356 in [17]) that, for $\delta q(\cdot) = \Delta q(\cdot)$, the difference of the first members of (2.15) and (2.9) tends to zero with the same rate as ε^2, as ε tends to zero.

For a particular choice of functions $a_{\alpha i}(\cdot)$, $b_\alpha(\cdot)$, which corresponds to putting

$$g_\alpha^\gamma(q(t),t) = \sum_{i=1}^{n} \tilde{a}_{\alpha i}(\bar{q}(t),t)(q_i(t) - \bar{q}_i(t))$$

for $q(\cdot) \in X$, $(q(t),t) \in N(\gamma)$, the two kinds of varied paths mentioned above coincide.

In fact, we have

$$\frac{d}{d\tau} g_\alpha^\gamma(q(\tau),\tau)\Big|_{\tau=t} = \sum_{i=1}^{n} \tilde{a}_{\alpha i}(\bar{q}(t),t)\dot{q}_i(t) - \sum_{i=1}^{n} \tilde{a}_{\alpha i}(\bar{q}(t),t)\dot{\bar{q}}_i(t) + $$
$$+ \sum_{i=1}^{n} \frac{d\tilde{a}_{\alpha i}}{d\tau}(\bar{q}(\tau),\tau)\Big|_{\tau=t} (q_i(t) - \bar{q}_i(t)) .$$

We may therefore put on $N(\gamma)$:

$$a_{\alpha i}(q(t),t) = \tilde{a}_{\alpha i}(\bar{q}(t),t)$$

$$b_\alpha(q(t),t) = -\sum_{i=1}^{n} \tilde{a}_{\alpha i}(\bar{q}(t),t)\dot{\bar{q}}_i(t) + \sum_{i=1}^{n} \frac{d\tilde{a}_{\alpha i}}{d\tau}(\bar{q}(\tau),\tau)\Big|_{\tau=t} (q_i(t)-\bar{q}_i(t))$$

It is clear that, on $N(\gamma)$

$$a_{\alpha i}(\overline{q}(t),t) = \tilde{a}_{\alpha i}(\overline{q}(t),t)$$

and that, since (2.8) holds along γ

$$b_\alpha(\overline{q}(t),t) = -\sum_{i=1}^{n} \tilde{a}_{\alpha i}(\overline{q}(t),t)\dot{\overline{q}}_i(t) = \tilde{b}_\alpha(\overline{q}(t),t) \ .$$

If we put $\varepsilon\delta q(\cdot) = \varepsilon\Delta q(\cdot)$ with $\Delta q(t) = q(t) - \overline{q}(t)$, $\delta q(t^o) = \delta q(t^1) = 0$, then

$$g^\gamma_\alpha(q(t),t) = 0 \qquad \text{for all} \qquad t \in [t^o,t^1], \quad \alpha = 1,2,\ldots m,$$

is equivalent to (2.15), and also to (2.9) which concludes the proof.

APPENDIX.

Let us start from Proposition i) in the definition of stationarity relative to the varied paths $x(\cdot) \in X$.

With simplified notation, all functions below being computed for $x = \overline{x}(t)$, $u = \dot{\overline{x}}(t)$, $\mu_\alpha = \mu_\alpha(t)$ and $\overline{x}(\cdot) \in X$, we deduce from (2.3)

$$\text{(a)} \qquad \lambda_k = \frac{\partial f}{\partial u_k} - \sum_{\alpha=1}^{m} \mu_\alpha a_{\alpha k}$$

and since

$$\frac{\partial H}{\partial x_k} = -\frac{\partial f}{\partial x_k} + \sum_{j=1}^{n} \sum_{\alpha=1}^{m} \mu_\alpha\left(\frac{\partial a_{\alpha j}}{\partial x_k} \dot{x}_j + \frac{\partial b_\alpha}{\partial x_k}\right)$$

it follows from (a) above and (2.2), by direct computation, that (2.4) of Proposition ii) is satisfied. Hence Proposition i) implies Proposition ii).

Now , let us start from Proposition ii). Let

$$\text{(b)} \qquad H(\mu,\lambda,x,t,u) \triangleq -f(x,t,u) + \sum_{k=1}^{n} \lambda_k u_k + \sum_{\alpha=1}^{m} \mu_\alpha \varphi_\alpha(x,t,u)$$

$$\text{(c)} \qquad \tilde{\lambda}_k(\mu,x,t,u) \triangleq \frac{\partial f}{\partial u_k}(x,t,u) - \frac{\partial}{\partial u_k}\left(\sum_{\alpha=1}^{m} \mu_\alpha\varphi_\alpha(x,t,u)\right)$$

$$\text{(d)} \qquad \lambda_k(t) \triangleq \tilde{\lambda}_k(\mu(t),\overline{x}(t),t,\dot{\overline{x}}(t)) \ .$$

Then one readily verifies by direct computation that (2.4), together with (b), (c) and (d), implies (2.2). Furthermore (b), (c) and (d) obviously imply (2.3). Hence Proposition ii) implies Proposition i).

This concludes the proof of the equivalence of Propositions i) and ii) in the definition of stationarity relative to the varied paths $x(\cdot) \in X$.

An additional remark will be in order. Let

$$\text{(e)} \qquad \lambda_t = \lambda_t(t) \triangleq -H(\mu(t),\lambda(t),\overline{x}(t),t,\dot{\overline{x}}(t));$$

we have

$$\dot{\lambda}_t = -\frac{\partial H}{\partial \tau}(\mu(\tau),\lambda(t),\overline{x}(t),\tau,\dot{\overline{x}}(t))\Big|_{\tau=t} -$$

$$-\sum_{k=1}^{n} \frac{\partial H}{\partial \lambda_k}(\mu(t),\lambda,\overline{x}(t),t,\dot{\overline{x}}(t))\Big|_{\lambda=\lambda(t)} \dot{\lambda}_k(t) -$$

$$-\sum_{k=1}^{n} \frac{\partial H}{\partial x_k}(\mu(t),\lambda(t),x,t,\dot{\overline{x}}(t))\Big|_{x=\overline{x}(t)} \dot{\overline{x}}_k(t) -$$

$$-\sum_{k=1}^{n} \frac{\partial H}{\partial u_k}(\mu(t),\lambda(t),\overline{x}(t),t,u)\Big|_{u=\dot{\overline{x}}(t)} \frac{d\dot{\overline{x}}_k}{dt}(t) .$$

With simplified notation, the following partial derivatives being computed for $\lambda = \lambda(t)$, $x = \overline{x}(t)$, $u = \dot{\overline{x}}(t)$, $\mu = \mu(t)$, we have

$$\frac{\partial H}{\partial \lambda_k} = \dot{\overline{x}}(t), \qquad \frac{\partial H}{\partial x_k} = -\dot{\lambda}_k(t), \qquad \frac{\partial H}{\partial u_k} = 0,$$

$$k = 1,2,\ldots n ,$$

it follows that

(f) $$\dot{\lambda}_t = -\frac{\partial H}{\partial \tau}(\mu(\tau),\lambda,x,\tau,u)\Big|_{\substack{\tau = t\\ \lambda = \lambda(t)\\ x = \overline{x}(t)\\ u = \dot{\overline{x}}(t)}}$$

Formula (f) is a consequence of Proposition i). It may be convenient, in some applications, to use it in addition to relations (2.2).

At last let us briefly discuss a third definition of stationarity relative to the varied paths $x(\cdot) \in X$, which is equivalent to definitions i) and ii) given previously.

Let

$$\Delta J(\overline{x}(\cdot);\xi(\cdot)) =$$

$$\int_{t^o}^{t^1} f(\overline{x}(\tau) + \xi(\tau),\tau,\dot{\overline{x}}(\tau) + \dot{\xi}(\tau))d\tau - \int_{t^o}^{t^1} f(\overline{x}(\tau),\tau,\overline{x}(\tau))d\tau,$$

and let us denote by $\delta J(\overline{x}(\cdot);\cdot)$ the principal linear part of the functional $\Delta J(\overline{x}(\cdot);\cdot)$, defined on X with the usual norm; that is

$$\delta J(\overline{x}(\cdot);\xi(\cdot)) =$$

$$= \int_{t^o}^{t^1} \sum_{k=1}^{n} \left(\frac{\partial f}{\partial x_k}(x,t,\dot{\overline{x}}(t))\Big|_{x=\overline{x}(t)} - \frac{d}{dt}\left(\frac{\partial f}{\partial \dot{x}_k}(\overline{x}(t),t,\dot{x})\Big|_{\dot{x}=\dot{\overline{x}}(t)}\right)\right) \xi_k(t)dt$$

Also, for $\xi(\cdot) \in X$, let

$$\delta K_\alpha(\overline{x}(t);\xi(t)) =$$

$$\sum_{k=1}^{n} \frac{\partial \varphi_\alpha}{\partial x_k}(x,t,\dot{\overline{x}}(t))\Big|_{x=\overline{x}(t)} \xi_k(t) + \sum_{k=1}^{n} \frac{\partial \varphi_\alpha}{\partial \dot{x}_k}(\overline{x}(t),t,\dot{x})\Big|_{\dot{x}=\dot{\overline{x}}(t)} \dot{\xi}_k(t)$$

$$\alpha = 1,2,\ldots m, \qquad t \in [\, t^o,t^1 \,]$$

where
$$\varphi_\alpha(x,t,u) = \sum_{i=1}^{n} a_{\alpha i}(x,t)u_i + b_\alpha(x,t)$$

Then it may be shown that i), or ii), is equivalent to the following proposition iii), namely

iii) there exist C^1 functions $\nu_\alpha(\cdot)$, $\alpha = 1,2,\ldots,m$, on $[t^o,t^1]$ such that

$$\left.\begin{array}{l}\displaystyle\int_{t^o}^{t^1} \nu_\alpha(t)\,\delta K_\alpha(\overline{x}(t)\,;\,\xi(t))\,dt = 0 \\ \qquad\qquad \alpha = 1,2,\ldots,m\end{array}\right\} \Rightarrow \delta J(\overline{x}(\cdot)\,;\,\xi(\cdot)) = 0$$

One can readily verify that

$$\int_{t^o}^{t^1} \nu_\alpha(t)\,\delta K_\alpha(\overline{x}(t)\,;\,\xi(t))\,dt =$$

$$\int_{t^o}^{t^1} \sum_{k=1}^{n}\left(\frac{\partial(\nu_\alpha\varphi_\alpha)}{\partial x_k}(x,t,\dot{\overline{x}}(t))\Bigg|_{x=\overline{x}(t)} - \frac{d}{dt}\left(\frac{\partial(\nu_\alpha\varphi_\alpha)}{\partial \dot{x}_k}(\overline{x}(t),t,\dot{x})\Bigg|_{\dot{x}=\dot{\overline{x}}(t)}\right)\right)\xi_k(t)\,dt$$

Now if proposition iii) is satisfied, there exist constant multipliers μ'_α, $\alpha = 1,2,\ldots,m$, such that

$$\delta J(\overline{x}(\cdot)\,;\,\xi(\cdot)) + \sum_{\alpha=1}^{m} \mu'_\alpha \int_{t^o}^{t^1} \nu_\alpha(t)\,\delta K_\alpha(\overline{x}(t)\,;\,\xi(t))\,dt = 0$$

holds for all $\xi(t) \in X$.

Condition (2.4) of proposition ii) follows by letting $\mu_\alpha(t) = \mu'_\alpha \nu_\alpha(t)$, $\alpha = 1,2,\ldots,m$.

Conversely if proposition ii) is satisfied, proposition iii) follows by letting $\nu_\alpha(t) = \mu_\alpha(t)$, $\alpha = 1,2,\ldots,m$.

REFERENCES.

(1) Leitmann, G. : "Optimality and Reachability with Feedback Control", this volume.

(2) Blaquière, A. : "Optimization and Controllability in Problems of Relativistic Dynamics and of Geometrical Optics", this volume.

(3) Mangasarian, O. :*Non-Linear Programming*, McGraw Hill, 1969.

(4) Blaquière, A., Gérard, F. and Leitmann, G. : *Quantitative and Qualitative Games*, Academic Press, New York 1969.

(5) Gelfand, I.M., Fomin, S.V., *Calculus of Variations*, Prentice Hall, 1963.

(6) Bliss, G.A., *Lectures on the Calculus of Variations*, University of Chicago Press, Chicago 1946.

(7) Hestenes, M.R., *Calculus of Variations and Optimal Control Theory*, R.E. Krieger, Hungtington, New York 1980.

(8) Pontryaghin, L.S. et al.,*The Mathematical Theory of Optimal Processes*, Interscience 1962.

(9) Wittaker, E.T.,*A Treatise on Analytical Dynamics*, Dover 1944.

(10) Goldstein, H., *Classical Mechanics*, Second edition, Addison Wesley 1980.

(11) Pars, L.A., Variational Principles in Dynamics, Quat. Journ. Mech. and Applied Math., vol VII, Pt 3 (1954).

(12) Hertz, H., *The Principles of Mechanics*, 1894, Chapter 5.

(13) Holder, O., Ueber die Principien von Hamilton und Maupertuis, Gött. Nachrichten, 1896, p.122-157.

(14) Voss, A., Ueber die Principe von Hamilton und Maupertuis, Gött. Nachrichten, 1900, p.322-327.

(15) Capon, R., Hamilton's Principle in Relation to Non-holonomic Mechanical Systems, Quat. Journ. Mech. and Applied Math., vol V, Pt 4 (1952)

(16) Jeffrey, H., "What is Hamilton's Principle ? ", the same journal, vol VII, Pt 3 (1954)

(17) Rund, H., *The Hamilton-Jacobi Theory in the Calculus of Variations*, Van Nostrand, 1966, Ch. 15 .

MODELING OF DYNAMICAL SYSTEMS USING EXTERNAL AND INTERNAL VARIABLES WITH APPLICATIONS TO HAMILTON SYSTEMS

J.C. Willems and A.J. van der Schaft

Mathematics Institute
P.O. Box 800
9700 AV Groningen
The Netherlands

INTRODUCTION

If one looks at the (deterministic) theories of dynamics as one usually encounters these in physics then, as a system theorist, one is struck by the fact that it is basically a set of first order differential equations which is heralded as the typical example of a mathematical model for physical phenomena. The paradigm of this situation is Hamiltonian mechanics but one can also recognize this point of view in the axiomatization of statistical mechanics (taking into account stochasticity), in quantum mechanics, etc. Since physics, together with chemistry, biology, parts of economics, etc. can be regarded as the prototype of the descriptive sciences, this vantage point may be quite natural. Indeed, the ultimate purpose here is to explain observations which presumably have been obtained against a fixed environmental background and it is therefore natural to attempt to catalogize the laws in terms

ISBN 0-12-068720-8

of the evolution of the state of the system. If the state variables have been choosen appropiately and under suitable smoothness assumptions, this evolution law can then be described by a differential equation which is of first order in the state variables.*

In the field of systems theory however, the approach has been radically different in that a dynamical system is above all viewed as a map which transforms input signals into output signals, the classical example of such a map being a convolution operator, or, in terms of Laplace transforms, a transfer function. Since mathematical systems theory finds its roots in control theory, this set-up is quite natural for example if one identifies inputs with control variables and outputs with observations. Problem formulations of this type are very natural in the prescriptive sciences (as engineering, operations research, and economics) where the problems are laid out in terms of the decision variables (inputs) which should be choosen on the basis of certain observations (outputs) and where the mathematical model involves principally the interrelation between these variables.

This dichotomy between the autonomous differential equation approach of physics and the input/output approach of control and systems theory is in fact much less natural than it may appear at

---

* We like to draw attention to the paper by Brockett [1] which has an interesting discussion relevant to our introduction and which has been a source of inspiration to us also on the technical level.

first sight. *Firstly*, eventhough it may be true that control theory is primarily concerned with the input/output behaviour of a system it has recognized the notion of state as a central idea in mathematical modeling and it is clear that the growth of mathematical systems theory in the last 25 years has for a significant part been due to the fact that the field has learned to think in terms of an input/state/output pattern, both for mathematical modeling and for problems of systems synthesis. This state point of view has been put forward consistently in the work of Kalman [2]. *Secondly*, eventhough it may be true that the axiomatics of certain areas of physics, as Hamiltonian mechanics, do not involve inputs and ouputs in an explicit way, this situation may not be tenable in the long run. Hamiltonian mechanics for example cannot even deal successfully with Newton's second law, and hence much is to be said for involving external variables as forces in the basic mathematical structure. We will return to this in a later section of the paper. Another example is (deterministic) thermodynamics which is a field where a mathematical model which does not involve something like external inputs and outputs is totally inadequate.

INPUTS, OUTPUTS, AND EXTERNAL VARIABLES

As we have already mentioned, the input/output setting is natural in the context of control theory. However, in the modeling of physical systems this cause/effect (input/output) relation is often not clear a priori. In [3] we have given a number of examples of physical systems in which it is not possible to start with such an input/output structure. There are other objections which may be made against the assumption that there is an intrinsically defined input space. In [1] it is argued convincingly that in a differential geometric setting the assumption that the input space is independent of the state is awkward from a mathematical point of view (see also [4]). All this motivates working with an axiomatic framework in which the external variables are not a priori divided into cause and effect variables.

In the present paper we will do the following:

1. Give a mathematical framework in which a dynamical system is defined in terms of external variables.
2. Introduce the concept of state in this context.
3. Indicate the relationship between models given in external and in state space form.
4. Outline how all this may be used to provide a conceptual basis for the basic laws of (deterministic) thermodynamics.
5. Introduce differential geometric structure in this set-up.
6. Show how this allows one to define Hamiltonian systems which incorporate external forces and observations.
7. Generalize Noether's theorem on symmetries and conservation laws to the situation at hand.

The main ideas of this paper are based on our previous work [2,5, 6] and we refer to these references for proofs and more details.

BASIC DEFINITIONS [2]

For simplicity we will only consider continuous time systems with time axis $\mathbb{R}$.

Definition 1: A *(external) dynamical system* is a subset $\Sigma_e \subset W^{\mathbb{R}}$, i.e., a family of trajectories $w: \mathbb{R} \to W$, where W is called the *(external) signal space*.

Important qualitative properties of systems are:

time-invariance: $\forall\ T \in \mathbb{R}$ we must have $S_T\Sigma_e = \Sigma_e$ with

$$(S_T w)(t) := w(t - T)$$

time-reversibility: $R\Sigma_e = \Sigma_e$ with $(R_e w)(t) := w(-t)$

linearity: W is a vector space and Σ_e is linear subspace of $W^{\mathbb{R}}$

autonomous systems: $\exists$ a bijection $f: W^{(-\infty,0]} \to W^{[0,\infty)}$ such that

$$\{w \in \Sigma_e\} \Leftrightarrow \{w|_{[0,\infty)} = f(w|_{(-\infty,0]})\}$$

Example 1: Let $F: \underbrace{\mathbb{R}^q \times \ldots \times \mathbb{R}^q}_{n\text{-times}} \times \mathbb{R}^q \times \mathbb{R} \to \mathbb{R}^r$

and consider the differential equation $F(w(t), \dot{w}(t), \ldots, w^{(n)}(t), t) = 0$. This obviously defines a system in above sense, namely $\Sigma_e := \{w: \mathbb{R} \to \mathbb{R}^q \mid w$ is n-times differentiable and satisfies $F(w(t), \dot{w}(t), \ldots, w^{(n)}(t), t) = 0\ \forall t\}$. It is time-invariant if $F(\sigma_0, \sigma_1, \ldots, \sigma_n, t)$ is independent of t. If F is independent of σ_{2i+1} $i = 0, 1, 2, \ldots$, and of t, then Σ_e is time-reversible. If F is linear, then so is Σ_e. If $F(\sigma_0, \sigma_1, \ldots, \sigma_n, t)$ is of the form $\sigma_n + F'(\sigma_0, \ldots, \sigma_{n-1}, t)$ with F' smooth, then Σ_e will be autonomous.

In control theory, in physics, in computer science, in stochastic process theory, etc. a very important role is played by the notion of state. In our context the appropriate definition is:

Definition 2: A *dynamical system in state space form* is defined as a subset $\Sigma_i \subset (X \times W)^{\mathbb{R}}$ (i.e. a system over X × W) where X is called the *state space*; Σ_i satisfies the following axiom:

$$\{(x_1,w_1), (x_2,w_2) \in \Sigma_i, \text{ and } x_1(t) = x_2(t)\}$$
$$\Rightarrow \{(x,w) \in \Sigma_i\}, \text{ where } (x,w)(\tau) := \begin{cases} (x_1,w_1)(\tau) & \text{for } \tau < t \\ (x_2,w_2)(\tau) & \text{for } \tau \geq t \end{cases}$$

The idea expressed in this axiom is that of 'state': it says that the present state is all what one needs to know about the past trajectory in order to be able to specify all possible future trajectories. One can also state this axiom with a (set theoretic) Markov flavor: the past and the future are conditionally independent given the present state.

Example 2: Let $f: \mathbb{R}^n \times \mathbb{R}^m \to \mathbb{R}^n$ and $g: \mathbb{R}^n \times \mathbb{R}^m \to \mathbb{R}^q$, and consider the differential equation $\dot{x} = f(x,u)$; $w = g(x,u)$. This defines a time-invariant system in state space form, as follows:
$\Sigma_i := \{(x,w): \mathbb{R} \to \mathbb{R}^n \times \mathbb{R}^q \mid x$ is absolutely continious and $\exists u: \mathbb{R} \to \mathbb{R}^m$ such that $\dot{x}(t) = f(x(t), u(t))$ for almost all t and $w(t) = g(x(t),u(t))\}$.

THE RELATIONSHIP BETWEEN EXTERNAL AND STATE SPACE SYSTEMS [2]

One of the main topics considered in mathematical systems theory (and in theoretical computer sciece) is what is called 'realization theory'. The problem there is the following: starting from an input/output map one tries to represent it efficiently in terms of a model which is in state space form (this question also has a stochastic analogue). It is possible to generalize many of these ideas to the present context.

Definition 3: Let Σ_i be a state space dynamical system over $X \times W$. Then $\Sigma_e := \{w \mid \exists x \ni (x,w) \in \Sigma_i\}$ is called its *external behaviour*. We denote this by $\Sigma_i \Rightarrow \Sigma_e$. If Σ_e is given and if Σ_i is such that $\Sigma_i \Rightarrow \Sigma_e$, then Σ_i is called a *(state space) realization* (or a *state space representation*) of Σ_e.

The problem of finding a state space representation for Σ_e involves coming up with the state space X. The basic ideas which go in to a realization of time-invariant systems will now be explained. Let $\cdot$ denote concatenation product. Thus if $w^- : (-\infty,0) \to W$ and $w^+ : [0,\infty) \to W$ then $w^- \cdot w^+$ is defined by

$$(w^- \cdot w^+)(t) := \begin{cases} w^-(t) & \text{for } t < 0 \\ w^+(t) & \text{for } t \geq 0 \end{cases} .$$

Let Σ_e be a time-invariant system. Then Σ_e^- and Σ_e^+ denote respectively the restrictions of Σ_e to $(-\infty,0)$ and $[0,\infty)$. Formally $\Sigma_e^- := \{w^- : (-\infty,0) \to W \mid \exists w \in \Sigma_e \ni w^- = w|_{(-\infty,0)}\}$ and Σ_e^+ is similarly defined. For a state space system Σ_i we define $\Sigma_i(a)$ as those trajectories which have $x(0) = a$, $\Sigma_i(a) := \{(x,w) \in \Sigma_i \mid x(0) = a\}$. Let $\Sigma_i^-(a)$ and $\Sigma_i^+(a)$ be defined in the obvious way. Note that the axiom of state says precisely that $\Sigma_i(a) = \Sigma_i^-(a) \cdot \Sigma_i^+(a)$. Define analogously for the external behavior Σ_e of Σ_i, the sets $\Sigma_e(a)$, $\Sigma_e^-(a)$, and $\Sigma_e^+(a)$ as the external trajectories with internal state $x(0) = a$. Obviously we also have $\Sigma_e(a) = \Sigma_e^-(a) \cdot \Sigma_e^+(a)$. Since $\Sigma_e = \bigcup_{a \in X} \Sigma_e^-(a) \cdot \Sigma_e^+(a)$ it is clear that a realization allows one to write Σ_e, viewed as a subset of the product set $W^{(-\infty,0)} \cdot W^{[0,\infty)}$, into a union (with elements parametrized by elements of X) of rectangular subsets. It is easily seen that the converse also holds, i.e., that any such representation of a time-invariant system Σ_e also yields a realization.

It is not difficult to use these ideas in order to arrive at 'specific' realizations. The procedure is based on a generalization of the ideas of Nerode equivalence from input/output theory. Thus for a given time-invariant system Σ_e one defines an equivalence relation R^+ on Σ_e by taking

$\{w_1 R^+ w_2\} :\Leftrightarrow \{w_1|_{(-\infty,0)} \cdot w^+ \in \Sigma^e \text{ iff } w_2|_{(-\infty,0)} \cdot w^+ \in \Sigma^e\}$.

Now take $X^+ := \Sigma_e (\text{mod } R^+)$ and define the state at 0 corresponding to a $w \in \Sigma_e$ to be the equivalence class to which w belongs. Shift invariance then allows one to define the whole state trajectory corresponding to this w. It is not difficult to verify that this defines a realization, denoted by Σ_i^+. Analogously one may define R^- by $\{w_1 R^- w_2\} :\Leftrightarrow \{w^- \cdot w_1|_{[0,\infty)} \in \Sigma_e \text{ iff } w^- \cdot w_2|_{[0,\infty)} \in \Sigma_e\}$. Denote the resulting realization, which is completely analously defined, with state space $X^- := \Sigma_e (\text{mod } R^-)$, by Σ_i^-.

These realizations have the pleasing property of minimality. A realization $\Sigma_i \Rightarrow \Sigma_e$ is *minimal* if $\bigcup_{a \in X' \subset X} \Sigma_e(a)$ is not a rectangle whenever X' consists of more than one point. Now two realizations Σ_i^1, Σ_i^2 of Σ_e with respective state spaces X_1 and X_2 are called *equivalent* if there exists a bijection $S: X_1 \to X_2$ such that $\{(x_1,w) \in \Sigma_i^1\} \Leftrightarrow \{(Sx_1,w) \in \Sigma_i^2\}$ with $(Sx_1)(t) := Sx_1(t)$, naturally. For input/output systems it may by shown (on a set theoretic level) that all minimal realizations are equivalent. This, however, breaks down in our case:

Theorem 1: *Let* Σ_e *be a time-invariant system. Then* Σ_i^+ *and* Σ_i^- *define minimal realizations of* Σ_e. *Furthermore all minimal realizations of* Σ_e *are equivalent iff* $R^+ = R^-$.

The condition $R^+ = R^-$ may be shown to hold in two important cases:

1. (trivially) for autonomous systems
2. (see [2]) for linear time-invariant finite dimensional systems.

It would be of interest to prove a general theorem which yields for which cases this uniqueness modulo equivalence of minimal realizations holds. It is easy to see that this comes down to proving that for all $a' \neq a''$ there should hold $\Sigma_e^-(a') \cap \Sigma_e^-(a'') = \emptyset$ and $\Sigma_e^+(a') \cap \Sigma_e^+(a'') = \emptyset$.

We summarize what we view to be the relevance of these ideas: in physics it is usually implicitly assumed that one knows a priori which variables constitute the state of the system and in systems theory it is usually assumed that one knows a priori which variables constitute the inputs (causes) and outputs (effects). In our set-up we postulate only that we know the laws governing the system, yielding Σ_e as a family of trajectories. The state space which goes along with this may be deduced from Σ_e via realization theory. Often this state space is unique modulo minimality and equivalence. Also, in some situations (e.g. linear time-invariant systems) it is possible to deduce input/output representations as well (see [2]).

Of course, in order to obtain non trivial properties of our systems we will have to introduce more structure. Those interested in mechanics think hereby on differential geometric structure, while a control theorist would undoubtedly go into the direction of linearity. However, we will hold off on this for a while. Indeed, one area of physics where it does no seem natural to introduce much additional structure in order to express its basic laws is thermodynamics. We will briefly outline in the next section how one may state the first and second law of thermodynamics in our language.

DISSIPATIVE SYSTEMS AND THERMODYNAMICS [2]

The first and second law of thermodynamics are respectively a conservation law and a dissipation law. In terms of our abstract state space framework these properties may be formalized as follows:

Definition 4: Let Σ_i be a time-invariant state space systems, $\Sigma_i \subset (X \times W)^{\mathbb{R}}$, with external behavior Σ_e. Let ρ: $W \to \mathbb{R}$ be such that $\forall w \in \Sigma_e$: $\rho(w(\cdot)) \in L_1^{loc}$ and let $\Lambda : X \to \mathbb{R}$. The triple $\{\Sigma_i, \rho, \Lambda\}$ is said to define a *dissipative system* if

$$\Lambda(x(t_0)) + \int_{t_0}^{t_1} \rho(w(\tau))d\tau \geq \Lambda(x(t_1)) \qquad \text{(DIE)}$$

holds for all $(x,w) \in \Sigma_i$ and $t_1 \geq t_0$. It is said to define a *conservative system* if (DIE) holds with equality.

The defining inequality (DIE) is easily interpreted if we identify ρ with a *supply rate* and Λ with a *storage function*.

Consider now a thermodynamic body at uniform temperature T which is heated from its environment at a rate Q and which does work on its environment at a rate W. The physical laws then define a family of possible time functions T(•), Q(•), W(•) which together define an (assumed time-invariant) external dynamical systems $\Sigma_e \subset (\mathbb{R}^+ \times \mathbb{R} \times \mathbb{R})^{\mathbb{R}}$. In order to express the constraints imposed by the first and the second law we assume that we have a state space model Σ_i with external behavior Σ_e (alternatively, we could assume that Σ_i has been deduced from Σ_e from realization theory). Now, the first and second law postulate the existence of an internal energy function $E: X \to \mathbb{R}$ and and entropy function $S: X \to \mathbb{R}$ which are such that

(i) (<u>first law</u>:) $\{\Sigma_i, Q - W, E\}$ is conservative

(ii) (<u>second law</u>:) $\{\Sigma_i, -\frac{Q}{T}, -S\}$ is dissipative

It may not seem to be very natural to have to work with an internal state space model of a thermodynamic systems in order to express the constraints imposed by the first and second law since in the long run these should be constraining the external behaviour. We will now show how one can state these constraints purely in terms of this external behaviour. The idea of dissipativeness of a dynamical system in external form is:

<u>Definition 5</u>: Let Σ_e be a time-invariant system, $\Sigma_e \subset W^{\mathbb{R}}$, and let $\rho: W \to \mathbb{R}$ be such that $\rho(w(\cdot)) \in L_1^{loc}$ $\forall w \in \Sigma_e$. Then $\{\Sigma_e, \rho\}$ is said to define an *(externally) dissipative system* if for all periodic $w \in \Sigma_e$ there holds

$$\frac{1}{T} \int_0^T \rho(w(\tau))d\tau \geq 0 \qquad \text{(DIE)'}$$

where $T > 0$ is the period of w. If (DIE)' holds with equality then we will call the system *(externally) conservative.*

It is obvious that there will be a close relation between dissipativeness of Σ_i and of its external behaviour. We will call Σ_i *regular w.r.t. periodic motions* if $\{(x,w) \in \Sigma_i$, w periodic with period $T > 0\}$ implies $\{x$ periodic with period $T > 0\}$. It is easily verified that both the minimal realizations Σ_i^+ and Σ_i^- have this property (and consequently so do all minimal realizations if $R^+ = R^-$).

Theorem 2: *Let Σ_i be time-invariant and regular w.r.t. periodic motions, and let Σ_e be its external behaviour. Then $\{\Sigma_e, \rho\}$ is dissipative (resp. conservative) iff there exists Λ such that $\{\Sigma_e, \rho, \Lambda\}$ is dissipative (resp. conservative).*

The implication of Theorem 2 to thermodynamic systems is clear: it allows one to conclude the existence of E and S based solely on restrictions imposed on the external behavior of the system. This yields the equivalence of some seemingly distinct statements of the second law. A different issue is in how far the functions E and S are uniquely defined. Of course, starting from the external behavior Σ_e, there is non-uniqueness introduced by the fact that for a given Σ_e satisfying (DIE)' there will exist different (: non-equivalent) realizations. However, if we concentrate on a given realization $\Sigma_i \Rightarrow \Sigma_e$ then under mild assumptions on Σ_i (e.g. that its state space is *connected* which means that

$\forall x_0, x_1 \in X\ \exists (x,w) \in \Sigma_i$ and $t_1 \geq t_0$ such that $x(t_0) = x_0$ and $x(t_1) = x_1$), there will exist an energy function E which is unique up to an additive constant. The situation is quite differrent however for the entropy S and other than assuming that the second law holds with equality (i.e., that $\{\Sigma_i, -\frac{Q}{T}, -S\}$ is conservative) there seem to be no natural assumptions which allow one to conclude the existence of a unique S in a minimal realization of a thermodynamic system.

DIFFERENTIAL GEOMETRIC STRUCTURE [1,2,4]

As we have already mentioned, in order to obtain results beyond the purely set theoretic- hence rather unsophisticated - level, one needs to introduce more structure. For differential systems this yields:

Definition 6: A *dynamical system in differential form* is described by:

(i) smooth manifolds X,W

(ii) a smooth fibre bundle $\pi: B \to X$

(iii) a smooth map $f: B \to TX \times W$

such that the diagram

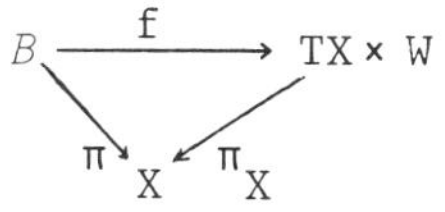

commutes. This defines a state space system $\Sigma_i := \{(x,w) : \mathbb{R} \to X \times W \mid x$ is abs. cont. and $(\dot{x}(t), w(t)) \in f(\pi^{-1}(x(t)))$ a.e$\}$. We will denote this system by $\Sigma(X,W,B,f)$.

In the above structure we think of w as the external variables and of $\pi^{-1}(x)$ as the input space when the system is in state x. In this sense the inputs (the fibres of B) serve as the variables which 'explain' the motions observed in Σ_e, and we donot need to consider them in general to be part of w. In this sense the above picture departs in an essential way from the one in [1] (see also [4]) where the idea of involving the input/output/state bundle $\mathcal{B}$ was first used. The traditional input/output point of view is recovered by assuming that $\mathcal{B} = X \times U$ is trivial, $W = U \times Y$, and $f: (x,u) \to (g(x,u), (u,r(x,u)))$. The autonomous vector field case is obtained by letting $\mathcal{B} = X$.

In local coordinates, $\Sigma(X,W,\mathcal{B},f)$ may be written as

$$\dot{x} = g(x,u)$$
$$w = h(x,u)$$

with x, (x,u), and w the local coordinates of X, $\mathcal{B}$, and W, respectively.

Of special interest for the representation of systems is the situation in which some of the external variables may be identified with input variables. In the above context this comes down to assuming that $\frac{\partial h}{\partial u}(x,u)$ has rank equal to $\dim \mathcal{B} - \dim X$. It is then possible to solve (locally) for u in terms of w in order to obtain, as an input representation,

$$\dot{x} = g(x,w_1)$$
$$w_2 = r(x,w_1) \qquad w = (w_1,w_2)$$

It is in fact possible to prove a (global) result on the

input representation of smooth linear finite dimensional time-invariant systems. This is stated in Theorem 1 of [7]. The theorem says that one may represent such systems minimally ($\Leftrightarrow$ observability of (A,C) in such representations) as:

$$\dot{x} = Ax + Bw_1$$

$$w_2 = Cx + Dw_1$$

$$w = (w_1, w_2)$$

where w_1, w_2 is a *componentwise* partition of w. If one does not insist on having a componentwise partition then it is possible to choose D = 0. A similar result holds for discrete time systems. This shows that for such systems one can recover the usual input/output point of view with the exception however that now controllability is not a property which may be assumed to hold without loss of generality when one starts with systems specified from an external point of view.

HAMILTONIAN SYSTEMS [1,5]

In this paragraph we will give a definition of a Hamiltonian system with external forces which fits very naturally into the framework described before, particularly in the previous section. Before going on to the definitions, we would like to say a few words about the concept of force in classical mechanics. Very roughly speaking one could say that in the older literature on mechanics forces have a basic place in the theory, while in the modern textbooks (see e.g. [8,9]) there has been a growing tendency to ignore the idea of force. Most of the time only *conservative* forces are considered, which by adding an external potential to the internal one, can be incorporated into the system while keeping it autonomous. We now want to show that in some sense it is natural and illuminating to maintain forces as basic variables, and that they can be included in the definition of a mechanical system *ab initio*.

Example 3 *(Newton's second law)*: consider a pointmass m with position q_1, influenced by some external force F_1. Then the relation between q_1 and F_1 as functions of time is given by $m\ddot{q}_1 = F_1$. This really expresses a compatibility relation between the force $F_1(t)$ (seen as a *basic* variable) and the position $q_1(t)$. Note that from a system theoretic point of view it is natural to consider the state space description of this system:

$$\dot{q}_1 = q_2$$
$$\dot{q}_2 = F_1 \ ; \quad y_1 = q_1$$

with input F_1, state $(q_1, \dot{q}_1)$, and output q_1.

Example 4 *(Euler-Lagrange equations)*: Let $L(q_1,..,q_n,\dot{q}_1,..,\dot{q}_n)$ be the Lagrangian of a mechanical system. Then the Euler-Lagrange equations are qiven by

$$\frac{d}{dt}\frac{\partial L}{\partial \dot{q}_i} - \frac{\partial L}{\partial q_i} = F_i \qquad i = 1,...,n$$

where F_i are the external forces. We can interpret this as an input/output system by adding the output equations $y_i = q_i$, $i = 1,..., n$.

In both examples the forces can be interpreted as the inputs, and the positions as the outputs. However this is not always the role of forces and positions as is shown in the next example:

Example 5: Consider a particle of unit mass in a potential field $V(q_2)$. This defines a force F_2 as follows:

$$F_2 = \frac{dV}{dq}(q_2)$$

One can view this again as a mechanical system relating the functions $q_2(t)$ and $F_2(t)$. Now it is natural to view q_2 as the input and F_2 as the output.

Furthermore one can interconnect these systems. For instance, interconnection of the mechanical systems of Examples 3 and 5 by

$$q_1 = q_2$$
$$F_1 = -F_2 \qquad \text{(in fact, \textit{Newton's third law})}$$

yields the system without inputs

$$m\ddot{q} + \frac{dV}{dq}(q) = 0$$

As output we could take the position $q = q_1 = q_2$, or the position together with $F = -\frac{dV}{dq}(q) = F_1 = -F_2$, which is now the internal force acting on the mass m. This shows how one can obtain more complicated mechanical systems composed out of simple building blocks. In the sequel we will show how one can define a "Hamiltonian structure" which underlies the examples mentioned, including the kind of interconnection as above. As may be expected, the natural framework for this is the language of symplectic geometry (see for instance [8,9]).

First we will briefly review the definition of a Hamiltonian vectorfield. Let M^{2n} be a 2n-dimensional manifold (think of the phase space), provided with a 2-form ω which is nondegenerate and closed (i.e. $d\omega = 0$). We will call (M,ω) a *symplectic* manifold with symplectic form ω.

<u>Definition 7</u>: A vector field Z on (M,ω) is called *locally Hamiltonian* if $L_Z\omega = 0$. Here $L_Z\omega$ denotes the Lie derivative of ω w.r.t. Z.

From the properties of L_Z and Poincaré's lemma, it follows that there exists (locally) a function $H: M \to \mathbb{R}$ such that $\omega(Z,-) = dH$. By Darboux's theorem we can find local coordinates $(q_1,\ldots, q_n, p_1,\ldots, p_n)$ for M such that $\omega = \sum_{i=1}^{n} dq_i \wedge dp_i$ (such coordinates are called *canonical*). In these coordinates we then obtain the usual Hamiltonian equations:

$$\begin{aligned} \dot{q}_i &= \frac{\partial H}{\partial p_i} \\ \dot{p}_i &= -\frac{\partial H}{\partial q_i} \end{aligned} \qquad i = 1,\ldots,n$$

Instead of asking $L_Z\omega = 0$, we will find it convenient to use another , but equivalent, definition of a local Hamiltonian vectorfield (from now on we will drop the adjective "local"). First we give

<u>Definition 8</u>: A submanifold N of (M^{2n},ω) is called *lagrangian* if

(i) $\omega(Y,Z) = 0 \qquad \forall\, Y,Z \in TN$

(ii) $\dim N = n$

In canonical coordinates $(q_1,\ldots, q_n, p_1,\ldots, p_n)$ lagrangian submanifolds are characterized by the following property. Locally there exist a function S (the so-called generating function) such that N is given by

$$q_i = \frac{\partial S}{\partial p_i}, \qquad p_j = -\frac{\partial S}{\partial q_j}$$

where the index i runs through a part of $\{1,2,\ldots,n\}$ and the index j runs through the complementary part.

When (M,ω) is a symplectic manifold we can also define a symplectic form, denoted by $\dot{\omega}$ on TM as follows. Since ω is nondegenerate it defines a bundle isomorphism $\overline{\omega}$: $TM \to T^*M$ by setting $\overline{\omega}(Y) := \omega(Y,-)$, $Y \in TM$. Because T^*M is a cotangent bundle, it has a natural symplectic form Ω (see [8,9]). Then one can prove that $\dot{\omega} := \overline{\omega}^*\Omega$ is a symplectic form on TM, which in local coordinates is given by

$$\dot{\omega} = \sum_i dq_i \wedge d\dot{p}_i + d\dot{q}_i \wedge dp_i$$

(with (q,p) canonical coordinates, and $\dot{q}_i$, $\dot{p}_i$ defined by $\dot{q}_i(v) := dq_i(v)$, $\dot{p}_i(v) := dp_i(v)$, for $v \in TM$). Now one can see the following equivalence:

<u>Proposition</u>: A vectorfield $Z: M \to TM$ is *Hamiltonian* iff the graph of Z is a Lagrangian submanifold of $(TM,\dot{\omega})$.

Now we return to the description of a smooth system given in Def. 6 through the commutative diagram

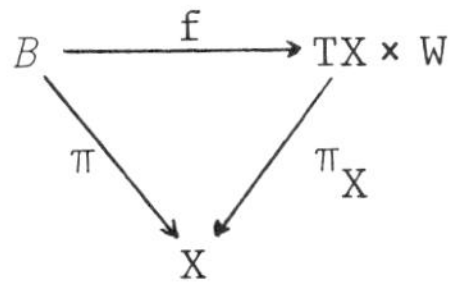

To define a Hamiltonian system with external variables we ask that

i) X (the state space) is a symplectic manifold with symplectic form ω (dim X = 2n). In this case we shall denote the state space by M, to make clear that the state space is a symplectic manifold. We shall denote the system by $\Sigma(M,W,\mathcal{B},f)$.

ii) W (the space of external variables) is a symplectic manifold with symplectic form ω^e (dim W = 2m).

Since ω induces a symplectic form $\dot{\omega}$ on TM, we can define the symplectic form $\Omega := \pi_1{}^*\dot{\omega} - \pi_2{}^*\omega^e$ on TM × W (π_1, resp. π_2, denotes the projection of TM × W on TM, resp. W). Then we have:

Definition 9: a. $\Sigma(M,W,\mathcal{B},f)$ defines a *full Hamiltonian system* if $f(\mathcal{B})$ is a Lagrangian submanifold of $(TM \times W, \Omega)$

b. $\Sigma(M,W,\mathcal{B},f)$ defines a *degenerate Hamiltonian system* if there exists a full Hamiltonian system $\Sigma(M,W,\mathcal{B}',f')$ such that $f(\mathcal{B}) \subset f'(\mathcal{B}')$.

In local coordinates this gives (using the properties of a Lagrangian submanifold and the commutativity of the diagram in Definition 6):

Theorem 3: a. *Let* $\Sigma(M,W,\mathcal{B},f)$ *be a full Hamiltonian system. Let* $(q_1,\ldots,q_n,p_1,\ldots,p_n)$ *be canonical coordinates for* M, *and* $(w_1,\ldots,w_{2m})$ *be canonical coordinates for* W. *Then* $f(\mathcal{B})$ *can locally be parametrized by* $(q_1,\ldots,q_n,p_1,\ldots,p_n)$ *and* m *of the functions* w_j. *Denote these by* $u_1,\ldots,u_m$ *and the remaining by* $y_1,\ldots,y_m$. *Then locally there exists a function* $H(q_1,\ldots,q_n,p_1,\ldots,p_n,u_1,\ldots,u_m)$ *such that* $f(\mathcal{B})$ *is given by*

$$\dot{q}_i = \frac{\partial H}{\partial p_i}(q,p,u) \qquad\qquad i = 1,\ldots,n$$
$$\dot{p}_i = -\frac{\partial H}{\partial q_i}(q,p,u)$$
$$y_j = c_j \frac{\partial H}{\partial u_j}(q,p,u) \qquad j = 1,\ldots,m$$

with $c_j = \pm 1$ *such that* $\omega^e = \sum_{i=1}^{m} c_j \, dy_j \wedge du_j$

b. *In the degenerate Hamiltonian case we obtain the same representation for the full Hamiltonian system* $\Sigma(M,W,\mathcal{B}',f')$ *such that* $f(\mathcal{B}) \subset f'(\mathcal{B}')$. *Because dim* $f(\mathcal{B}) = n+k$, *with* $k \leq m$, *we can take a subset* $u_1,\ldots, u_k$ *of* $u_1,\ldots, u_m$ *such that* $f(\mathcal{B})$ *is parametrized by* $(q_1,\ldots, q_n, p_1,\ldots, p_n, u_1,\ldots, u_k)$.

We can interpret Theorem 3 in the following way. Locally we can split the external variables in a set of inputs $\{u_1,\ldots, u_m\}$ (think of external forces) and a set of outputs $\{y_1,\ldots, y_m\}$ (think of positions). Then the theorem says that for constant inputs, the dynamics are described by a Hamiltonian vectorfield on M. Furthermore the outputs depend in a specified way on the state variables and the inputs, and are in some sense *dual* to the inputs. Finally $c_j = \pm 1$ denotes the fact that we possibly obtain a *hybrid* representation (as it is called in electrical network theory).

We will now give some examples of our general definition:

Example 6: Let $\Sigma(M,W,\mathcal{B},f)$ be a minimal full Hamiltonian system (for a definition of minimality in a differential geometric setting see [5]). Suppose that W is a cotangent bundle, i.e. $W = T^*Y$, for some manifold Y. Also suppose that the map $f: \mathcal{B} \to TM \times T^*Y$ is factored as (g,h) with $g: \mathcal{B} \to TM$, $h: \mathcal{B} \to T^*Y$ such that h is a fiber respecting map, and linear and bijective as a map from fibers (of $\mathcal{B}$) to fibers (of T^*Y). Then one can prove that locally our system is given by

$$\dot{x} = A(x) + \sum_i u_i B_i(x) \quad ; \; y_i = C_i(x) \quad i = 1,\ldots,m$$

with x coordinates for M, $(y_1,\dots,y_m)$ coordinates for Y and $(u_1,\dots,u_m)$ the corresponding natural coordinates for the fibers of T*Y; $C := (C_1,\dots,C_m)$ is a map from M to Y, and A and B_i are vectorfields on M such that $L_A\omega = 0$ and $\omega(B_i,-) = dC_i$, $i = 1,\dots,m$. One can interpret this situation in the following way. A is a Hamiltonian vectorfield on M. Let the function $C = (C_1,\dots,C_m)$ denote the *observations* made on the system. Then it is quite natural to assume that the possibility of influencing the system corresponds to adding a function to the Hamiltonian, and this function depends only on the observations.

Example 7 (*Lagrangian Systems*): Consider again Example 6 and assume moreover that M = T*Q with Q the configuration space, and that $C: T^*Q \to Y$ is given by $(q_1,\dots,q_n,p_1,\dots,p_n) \to (q_1,\dots,q_k)$, $k \leq n$, where (q,p) are canonical coordinates for T*Q. Then locally there exists a function $L: TQ \to \mathbb{R}$ such that the system is described by

$$\frac{d}{dt}\frac{\partial L}{\partial \dot{q}_i} - \frac{\partial L}{\partial q_i} = \begin{cases} u_i & \text{for } i = 1,\dots,k \\ 0 & \text{for } i = k+1,\dots,n \end{cases}$$

$$y_i = q_i \ , \quad i = 1,\dots,k$$

These are the Euler-Lagrange equations with external forces (for a global definition of a Lagrangian system we refer to [4,5]).

Example 8: Consider a linear input/output system

$\dot{x} = Ax + Bu$; $y = Cx$; $w = (u,y)$ A,B,C matrices

(this can be considered as a special case of a general system $\Sigma(X,W,\mathcal{B},f)$, see [2,7]). Then this is a full Hamiltonian system iff $A^TJ + JA = 0$; $B^TJ = C$ where J is a linear symplectic form on the linear space X. When we denote the transfer function of this system by G(s) we can show that {G(s) has a Hamiltonian realization} $\Leftrightarrow$ {$G(s) = G^T(-s)$} $\Leftrightarrow$ {G(s) has a minimal realization which is Hamiltonian}.

Actually this set-up for Hamiltonian systems also gives a strong *variational principle*.

Let $\Sigma(M,W,\mathcal{B},f)$ be a (full) Hamiltonian system, then the state/external signal trajectories generated by $\Sigma(M,W,\mathcal{B},f)$ satisfy

$$\int_{t_1}^{t_2} \omega^e_{w(t)}(\delta_1 w(t), \delta_2 w(t))dt = \omega_{x(t_2)}(\delta_1 x(t_2), \delta_2 x(t_2)) - \omega_{x(t_1)}(\delta_1 x(t_1), \delta_2 x(t_1))$$

where $(\delta_1 x(\cdot), \delta_1 w(\cdot))$ and $(\delta_2 x(\cdot), \delta_2 w(\cdot))$ denote arbitrary variations along a state/external signal trajectory $(x(\cdot), w(\cdot))$. For instance a variation $\delta w(\cdot)$ along $w(\cdot)$ is defined as follows. Take a sequence of external signal trajectories $w_n(\cdot)$ such that $w_n(t) \to w(t)$, for every t. This defines for every t a tangent vector $\delta w(t) \in T_{w(t)}W$. The equation above can be interpreted as a generalization of the classical Hamilton's principle (for references, see [5]). Notice that while most variational principles use only variations over the position varibles, the above equation uses variations over the whole set of w-variables. When we interpret them as the position variables together with the external force variables, we could also use only variations over the external forces. This seemst to correspond to the so-called *principle of complementary virtual work*.

We now like to point out briefly how by defining a natural

type of interconnection the system resulting from interconnecting Hamiltonian systems will again be Hamiltonian.

Definition 10: Let (W_1,ω_1^e) and (W_2,ω_2^e) be two spaces of external variables. An *interconnection* is a submanifold $I \subset W_1 \times W_2$. It is called a. *full Hamiltonian* if I is a Lagrangian submanifold of $(W_1 \times W_2, \pi_1{}^*\omega_1^e + \pi_2{}^*\omega_2^e)$, where π_1 and π_2 denote the projections of $W_1 \times W_2$ onto W_1 and W_2 respectively.

b. *degenerate Hamiltonian* if I is a coisotropic submanifold (coisotropic means $(TI)^{\perp} \subset TI$, where $\perp$ denotes the orthogonal complement w.r.t. the symplectic form).

As is well known (see [8] for references), the idea of a Lagrangian submanifold underlies many of the interconnections considered in physics and biology which have a *reciprocal* character. In fact, the interconnection we have seen earlier:

$$q_1 = q_2 \quad ; \quad F_1 = -F_2$$

is a prototype of a Lagrangian submanifold. It is easy to see that it forms a Lagrangian subspace of

$$\mathbb{R}^4 = \{(q_1,q_2,F_1,F_2) \mid q_i,F_i \in \mathbb{R}\}$$

with the natural symplectic form defined by $\begin{pmatrix} 0 & -I_2 \\ I_2 & 0 \end{pmatrix}$

The degenerate Hamiltonian interconnection allows one to define interconnections where not all variables are linked to each other. In fact, let $I \subset W_1 \times W_2$ be a degenerate Hamiltonian interconnection, then the distribution $I^{\perp}$ $(:= \{ Z \in TI \mid \pi_1{}^*\omega_1^e + \pi_2{}^*\omega_2^e)(Z,Y) = 0, \forall Y \in TI\})$ is involutive since I is coisotropic. Therefore, at least locally, we can factor out I by $I^{\perp}$, and we will obtain again a *symplectic* manifold. This last manifold can then be interpreted as the set of variables which are *not* interconnected and therefore remain free.

Finally we can prove (under regularity assumptions on intersections of manifolds)

Theorem 4: *Let* $\Sigma_i(M_i, W_i, \mathcal{B}_i, f_i)$, $i = 1,2$ *be two Hamiltonian systems (full or degenerate). Let* I *be a Hamiltonian interconnection, then the interconnected system (with state space* $M_1 \times M_2$ *and external signal space* $W_1 \times W_2$*) is again Hamiltonian.*

A GENERALIZATION OF NOETHER'S THEOREM [6]

In this section we want to show how the notion of symmetries and conservation laws for Hamiltonian vectorfields can be extended to Hamiltonian systems with external forces. Using this we will obtain a generalization of Noether's theorem (which loosely speaking states that for Hamiltonian systems a symmetry law generates a conservation law and *vice versa).*
For proofs we refer to [6]. First we want to define what we mean by a *symmetry* law.

Definition 11: Let $\Sigma(X, W, \mathcal{B}, f)$ be dynamical system. An *(infinitesimal) symmetry* is defined by a 3-tuple (R,S,T), with R, S and T vector fields on $\mathcal{B}$, X and W respectively, such that the diagram shown on the next page, commutes for all $t \in \mathbb{R}$ for which R_t, S_t, and T_t, the one parameter groups generated by R,S, and T, exist

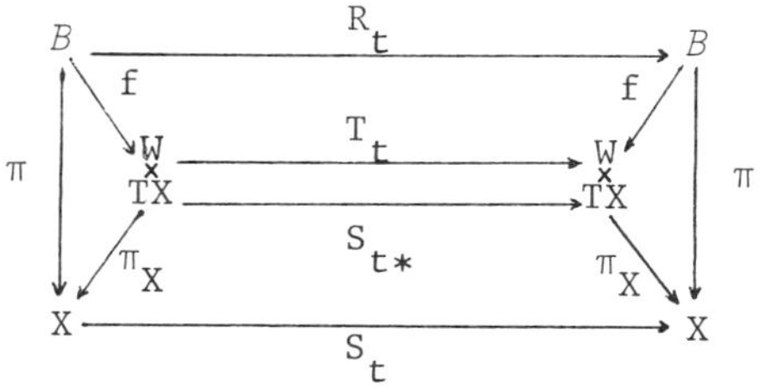

This condition of commutativity states that the one parameter group (S_t,T_t) working on $X\times W$ takes a feasible state/external signal trajectory into another such pair. We obtain the following

Proposition: The commutativity of the above diagram is equivalent to the conditions: (i) $g_*R = \dot{S}$

(ii) $h_*R = T$

where $(g,h) := f:\mathcal{B}\to TX\times W$ and $\dot{S}$ is the vectorfield on TX with one parameter group S_{t*}.

For Hamiltonian systems this definition specializes to :

Definition 12: A symmetry (R,S,T) for a Hamiltonian system $\Sigma(M,W,\mathcal{B},f)$ is called a *Hamiltonian symmetry* if $L_S\omega = 0$ and $L_T\omega^e = 0$.

Let $\Sigma(M,W,\mathcal{B},f)$ be a Hamiltonian system. According to Theorem 3 in local coordinates the dynamics of the system are described by (apart from the output equations):

$$\dot{q}_i = \frac{\partial H}{\partial p_i}(q,p,u) \qquad i = 1,\dots,n$$
$$\dot{p}_i = -\frac{\partial H}{\partial q_i}(q,p,u)$$

Hence for every constant u we obtain a Hamiltonian vectorfield on M which we shall denote by Z^u. Then we can derive:

Theorem 5: *Let* (R,S,T) *be a Hamiltonian symmetry for*

$\Sigma(M,W,\mathcal{B},f)$. *Then locally there exist functions* $F: M \to \mathbb{R}$, *and* $F^e: W \to \mathbb{R}$, *with* $S = Z_F$ *and* $T = Z_{F^e}$ *(i.e.,* F *and* F^e *are Hamiltonian functions corresponding to the vectorfields* S *and* T *), such that for every* $(x,u) \in \mathcal{B}$ *(with* $x \in M$*):* $Z^u(F)(x) = F^e(h(x,u))$ *(where again* $(g,h) := f: \mathcal{B} \to TM \times W$)

Definition 13: A pair (F,F^e) as above is called a *conservation law* for $\Sigma(M,W,\mathcal{B},f)$.

To interpret Theorem 5 recall that a conservation law for a Hamiltonian vectorfield Z on M is a function $F:M \to \mathbb{R}$ such that $Z(F) = 0$. In our framework the theorem gives that the change of F along the trajectories of the system is not zero but is a function *of only the external variables*. Hence from the knowledge of the external variables (the behaviour at the boundary of the system) and the initial conditions we can deduce the behaviour of F as a function of time. In fact, the triple $\{\Sigma(M,W,\mathcal{B},f), F^e, F\}$ defines a conservative system in the sense of Definition 4.

Recall now the set-up for Noether's theorem for Hamiltonian vectorfields (see [8,9]). A vectorfield S on (M,ω) is called a *symmetry* for a Hamiltonian vectorfield Z_H on M if

(i) $L_S\omega = 0$

(ii) $S(H) = 0$

From (i) it follows that S has locally a corresponding Hamilton function $F:M \to \mathbb{R}$. Then (ii) implies that $Z_H(F) = 0$. Also, given a $F:M \to \mathbb{R}$ such that $Z_H(F) = 0$, then $S = Z_F$ satisfies (i) and (ii), so S is a symmetry.

Concluding we can state Noether's theorem as follows: if S is a symmetry for Z_H, then there exists a conservation law F for Z_H (where $S = Z_F$). Conversely, if F is a conservation law for Z_H, then Z_F is a symmetry. In our framework we obtain as a consequence of Theorem 5 the following generalization

Theorem 6 *(Noether's theorem generalized): Let* (R,S,T) *be a Hamiltonian symmetry for* $\Sigma(M,W,\mathcal{B},f)$. *Then (locally) there exists a conservation law* (F,F^e). *Conversely, given a conservation law* (F,F^e), *we can construct a symmetry* (R,S,T) *such that* $S = Z_F$, $T = Z_{F^e}$.

We now like to specialize these results to the subclass of Hamiltonian systems treated in Example 6, which in local coordinates take the form

$$\dot{x} = A(x) + \sum_{i=1}^{m} u_i B_i(x)$$
$$y_i = C_i(x) \qquad\qquad i = 1,\ldots,m$$

with $L_A\omega = 0$, i.e. locally there exists a $H: M \rightarrow \mathbb{R}$ such that $\omega(A,-) = dH$, and $\omega(B_i,-) = dC_i, i = 1,\ldots,m$, and $C: = (C_1,\ldots,C_m) : M \rightarrow Y$. For this class of systems (we will call them *affine* Hamiltonian systems) we obtain:

Theorem 7: *Let* (R,S,T) *be a Hamiltonian symmetry for the above affine Hamiltonian system. Then (locally) there exist functions* K_i, $i = 1,\ldots,m$ *and* V *on* Y, *and a function* F *on* M *such that*

$$\{H,F\} = V \circ C$$
$$\{C_i,F\} = K_i \circ C \qquad\qquad i = 1,\ldots,m$$

where { , } *is the Poisson bracket on* (M,ω) *defined by*

$\{H,G\}:=\omega(Z_H,Z_G)$ *and* $A = Z_H$, $S = Z_F$ *and* $T = Z_G$, *with*
$G(y,u):=\sum_{i=1}^{m} u_iK_i(y) + V(y)$ (y,u) *canonical coordinates for* T^*Y).

We can even sharpen this result:

Theorem 8: *Let* (R,S,T) *be a Hamiltonian symmetry as above. Then* π_*T *is a well-defined vectorfield on* Y *(*π *is the projection of* T^*Y *on* Y*). Suppose* π_*T *is nowhere zero. Then locally we can construct a function* $P: Y \to \mathbb{R}$ *such that*

$$\{H + P\circ C, F\} = 0.$$

This last theorem states that by adding an extra *potential* term P to the system (depending only on the output) one can make $F: M \to \mathbb{R}$ into a conservation law for the autonomous system obtained by setting the external force zero, i.e. F is a conservation law for the Hamiltonian vectorfield with Hamilton function $H + P\circ C$.

Example 9: Consider a particle in $\mathbb{R}^3$ with mass m in a potential field V, and subject to the external force F. In input/state/output form the system is given by

$$m\,\ddot{q}_i = \frac{\partial V}{\partial q_i} + F \qquad i = 1,2,3$$

$$y_i = q_i \qquad \text{(y is the output)}$$

Suppose the equations $m\,\ddot{q}_i = \frac{\partial V}{\partial q_i}$ are invariant under rotation around the e_1-axis (this is equivalent to V or $L = \frac{1}{2}m\,\dot{q}^2 - V$ invariant). Then we know that the rotations around the e_1-axis generate a symmetry S on $T^*\mathbb{R}^3$, the phase space of the system. Moreover for zero external force the angular momentum around the e_1-axis is preserved, i.e.

$$\frac{dI}{dt} = 0, \text{ with } I := \langle \dot{q} \times mq, e_1 \rangle$$

However for a nonzero external force F we obtain:

$$\frac{dI}{dt} = \langle q \times F, e_1 \rangle = \langle y \times F, e_1 \rangle$$

Now $I^e := \langle y \times F, e_1 \rangle$ is a function on the space of inputs and outputs $W := \{(y,F) \mid y \in \mathbb{R}^3, F \in \mathbb{R}^3\}$. Hence (I, I^e) forms a conservation law in the sense of Def. 13. This conservation law induces a Hamiltonian symmetry in the sense of Def. 12 in the following way. With the natural symplectic form $\begin{bmatrix} 0 & -I_3 \\ I_3 & 0 \end{bmatrix}$ on W, the function I^e gives a Hamiltonian vectorfield $T := Z_{I^e}$ on W. In coordinates: $I^e = y_2F_3 - F_2y_3$ and T is given by

$$\frac{d}{dt}\begin{bmatrix} y_1 \\ y_2 \\ y_3 \end{bmatrix} = \begin{bmatrix} 0 & 0 & 0 \\ 0 & 0 & -1 \\ 0 & 1 & 0 \end{bmatrix}\begin{bmatrix} y_1 \\ y_2 \\ y_3 \end{bmatrix}, \quad \frac{d}{dt}\begin{bmatrix} F_1 \\ F_2 \\ F_3 \end{bmatrix} = \begin{bmatrix} 0 & 0 & 0 \\ 0 & 0 & -1 \\ 0 & 1 & 0 \end{bmatrix}\begin{bmatrix} F_1 \\ F_2 \\ F_3 \end{bmatrix}$$

This expresses the fact that the output corresponding to an external force which is rotated around the e_1-axis is obtained by rotating the output in the same way, Finally the Hamiltonian symmetry is given by (R,S,T), where R on $\mathcal{B} = T^*\mathbb{R}^3 \times \mathbb{R}^3$ (phase space and space of external forces) is given by (S,0) (R has a zero component on $\mathbb{R}^3$, the space of external forces).

CONCLUSIONS

In this paper we have sketched the basic ideas which go into a theory of dynamical systems with external variables and internal states and the relationships between these two classes of models. We have also shown how to add differential geometric structure in this set-up. These ideas were finally used to obtain a framework for studying Hamiltonian systems with external forces. We also showed how a generalization of Noether's theorem could be obtained this way. Our hope is that all this might serve as a starting point for a rapprochement between the concepts in mathematical systems theory and the problems in (theoretical) physics.

REFERENCES

[1] R.W. Brockett: *"Control Theory and Analytical Mechanics"*, pp. 1-46 of *Geometric Control Theory* (Editors: C. Martin and R. Hermann), Vol. VII of Lie Groups: History, Frontiers and Applications. Math. Sci. Press, 1977.

[2] J.C. Willems: *"System Theoretic Models for the Analysis of Physical Systems"*, Richerche di Automatica, Special Issue on Systems Theory and Physics, Vol. X, Dec. 1979, No. 2, pp. 71-106.

[3] R.E. Kalman, P.L. Falb and M. Arbib: *"Topics in Mathematical Systems Theory"*, McGraw-Hill, 1969.

[4] F. Takens: *"Variational and Conservative Systems"*, Rapport ZW-7603, Mathematics Institute, Groningen, 1976.

[5] A.J. van der Schaft: *"Hamiltonian Dynamics with External Forces and Observations"*, to appear in Mathematical Systems Theory

[6] A.J. van der Schaft: *"Symmetries and Conservation Laws for*

Hamiltonian Systems with Inputs and Outputs: A Generalization of Noether's Theorem", Systems & Control Letters, Vol. 1, No. 2, pp. 108-115, 1981.

[7] J.C. Willems: "*System Theoretic Foundations for Modelling Physical Systems*" in *Dynamical Systems and Microphysics*, Eds.: A. Blaquière, F. Fer, and A. Marzollo, CISM Lecture Notes No. 261, Springer, 1980, pp. 279-290.

[8] R. Abraham and J.E. Marsden, *Foundations of Mechanics*, Benjamin/Cummings, 1978.

[9] V.I. Arnold: *Mathematical Methods of Classical Mechanics*. Springer, 1978, (translation of the 1974 Russian edition).

PART III

LAGRANGIAN AND HAMILTONIAN FORMULATIONS OF MECHANICS

SOME REFLECTIONS ON HAMILTONIAN FORMALISM

R. Thom

Institut des Hautes Etudes Scientifiques,
Bures-sur-Yvette.

A few years ago André Avez asked me the following question : "What makes Hamiltonian systems so important ?" This paper is simply an attempt to answer this question. The multiplicity of the answers (three) shows that -very likely- none of them is really satisfactory.

1st INTERPRETATION : CALCULUS OF VARIATIONS AND LAGRANGIAN FORMALISM.

Let X be the space $R^n(x_1,\ldots,x_n)$; one considers a function $F(x,\dot{x})$ of class C^∞. We try to find the differentiable paths $s : \underset{(o,1)}{I} \xrightarrow{f} X$, such that $f(o) = x_o$, $f(1) = x_1$ and which extremalize (resp. maximalize) the integral

$$I(f) = \int_0^1 F(x,\dot{x})ds$$

Let us consider the family of functionals

$$J(f) = \int_0^1 \{ p,\dot{x} - F(x,\dot{x}) \} ds$$

Since $\int_0^1 \{p,\dot{x}\}ds = p,\ (x(1)-x(o))$ does not depend on the path f,

ISBN 0-12-068720-8

our problem is equivalent to making J(f) extremal (resp. minimal). Now, in order to make J(f) extremal (resp. minimal), it is convenient to make the integrand $\{p,\dot{x} - F(x,\dot{x})\} = L(x,\dot{x})$ extremal (resp. minimal) with respect to the (multidimensional) parameter p . It, therefore, is possible to forget the boundary conditions $x(o) = x_o$, $x(1) = x_1$. This gives us

$$0 = \frac{\partial L}{\partial \dot{x}} = p - F_{\dot{x}}(x,\dot{x}), \qquad x,p \text{ fixed.} \tag{1}$$

If F is strictly convex with respect to $\dot{x}$ (as is the case for the kinetic energy $T = \frac{1}{2}(\dot{x})^2$), then, for each value of p, there corresponds at most one value $\dot{x} = \xi(x,p)$: the equation (1) has just one solution, smooth in (x,p). (Fig.1). But uniqueness is not generally ensured. Let us choose one particular solution $\xi(x,p)$.

One then puts $H(x,p) = L(x,\xi(x,p))$. H is the "local Hamiltonian", associated with this particular solution. Hence, there are as many local Hamiltonians in (x,p) as local solutions of (1). Each of these solutions gives rise to a field of Euler-Lagrange extremals for the initial problem.

Let us recall the equation for the extremals

$$I(f) = \int_0^1 (F_x(x,\dot{x})\delta x + F_{\dot{x}}(x,\dot{x})\delta\dot{x})ds,$$

Writting $\delta\dot{x} = \frac{d}{ds}(\delta x)$, one has

$$\int_0^1 F_{\dot{x}}(x,\dot{x})\frac{d}{ds}(\delta x) = \left|F_{\dot{x}}(x,\dot{x})\delta x\right|_0^1 - \frac{d}{ds}F_{\dot{x}}(x,\dot{x})\delta x ds \tag{2}$$

from which follows the vanishing-integrand condition (coefficient of δx equal zero) :

$$F_x(x,\dot{x}) - \frac{d}{ds}F_{\dot{x}}(x,\dot{x}) = 0$$

Now, according to (1), $F_{\dot{x}}(x,\dot{x}) = p$, whence

$$\begin{cases} \dfrac{dp}{ds} = -\dfrac{\partial H}{\partial x} = -\dfrac{\partial L}{\partial x}(x,\xi) = -\dfrac{\partial}{\partial x}(F(x,\xi) + p\cdot\xi) \\ \quad = \dfrac{\partial F}{\partial x}(x,\dot{x}) - \dfrac{\partial \xi}{\partial x}[F_{\dot{x}} - p] = \dfrac{\partial F}{\partial x}(x,\dot{x}) \\ \dfrac{dx}{ds} = \dfrac{\partial H}{\partial p} = \dot{x} \qquad \text{(identity).} \end{cases} \tag{3}$$

Observe that $\frac{\partial}{\partial x} L(x,\xi) = \frac{\partial L}{\partial x} + \frac{\partial L}{\partial \xi} \cdot \frac{\partial \xi}{\partial x}$, but $\frac{\partial L}{\partial \xi}(x,\xi) = 0$ from the stationarity condition (1).

We, therefore, have identified the Euler-Lagrange extremals of the initial problem with the projections on X of the trajectories of the field (3) defined in the space $T^*(X) = (x,p)$ (the covector space). Now (3) is nothing but the classical Hamilton-Jacobi equation

$$X = \frac{\partial H}{\partial p}, \qquad P = -\frac{\partial H}{\partial x} .$$

This shows that there is some incompatibility between the Hamiltonian and the Lagrangian formalisms. For a local Hamiltonian to be defined globally, it is necessary (and sufficient) that $F(x,\dot{x})$ be strictly convex in $\dot{x}$, and one, therefore, has to extremalize a very special functional. Conversely, if $F(x,\dot{x})$ is arbitrary, there is in general a set $(\mathcal{O})$ in the space p, formed by the directions (p) which have an exceptional contact with the graph $u = F(\;,\dot{x}) = m$ (for instance the slopes of the double tangents if $\dim X = 1$). (Fig.2). It, therefore, is generally a matter of determining the projection directions (p) for which the apparent contour of the projection of the graph parallel to (p) (on Ox) is subject to topological variations. This problem is rather easy if $\dim X = 1$, and it has been solved by I. Ekeland. But if $\dim X = 2$, the work of Y. Kergosien gives the generic types of singularities for this set to which one has to add the effect of the parameter (x). In fact, one has in general a section of the global set $(\mathcal{O})$

in the space (x,p) by the plane of directions p,x = const.

In the case of a minimal problem (but no longer extremal), one has a particular type of set : the Maxwell set with 2 types of singularity for the dimension 1 (Fig.3 and Fig.4).

When an extremal crosses a simple arc of the Maxwell set, one recovers the rule of the cosine (the refraction law in optics). Indeed, one has two solutions : ξ_1, ξ_2 ; $H(x,\xi_1)$, $H(x,\xi_2)$ are the corresponding Hamiltonian the values of which coincide at the point of transition $p = \frac{\partial L}{\partial \dot{x}}(\xi_1) = \frac{\partial L}{\partial \dot{x}}(\xi_2)$ (Figs. 4 and 5).

But the equation of the transition locus (Δ) is

$$H(x,\xi_1) = H(x,\xi_2)$$

Hence, a vector ν normal to the straight line (Δ) is defined by

$$\mathrm{Grad}(H(x,\xi_1) - H(x,\xi_2)) = \vec{\nu}$$

Now, the Hamiltonian flows X_1 and X_2 are defined respectively by

$$i \text{ grad } H(x,\xi_1), \quad i \text{ grad } H(x,\xi_2)$$

where i accounts for the rotation of $\pi/2$ which turns the ordinary gradient into the symplectic gradient $(\frac{\partial H}{\partial q_i}, \frac{\partial H}{\partial p_i} \xrightarrow{i} -\frac{\partial H}{\partial p_i}, \frac{\partial H}{\partial q_i})$,

$$\nu \cdot X_1 = \vec{\nu}(i \text{ grad } H(x,\xi_1)),$$
$$\nu \cdot X_2 = \vec{\nu}(i \text{ grad } H(x,\xi_2)),$$

hence $\nu \cdot (X_1 - X_2) = \nu \cdot i\nu = 0$

(Remark : in optics, it is possible to have an extra *reflection*, corresponding to the transformation $\dot{x} \to -\dot{x}$, $p \to -p$).

EXTREMALITY - MINIMALITY.

Starting with a minimizing (resp. maximizing) problem, one is then led to maximizing (resp. minimizing) the Lagrangian $L(x,\dot{x}) = p \cdot \dot{x} - F(x,\dot{x})$ on the space (p,xp considered as the control space, whereas the space $\dot{x}$ is considered as the state space (internal variable). Thus, one knows that there is in general a set K of discontinuities for the Hamiltonian flow in the space (p,x); this set is usually a stratified set which I proposed to call the "Maxwell set" associated with the corresponding minimum problem.

These sets admit local semi-algebraic models defined by some chains of symbols $[\sigma_1^{\mu_1}\sigma_2^{\mu_2}\ldots\sigma_k^{\mu_k}]$, $\mu_i \in N$, where σ_i denotes an algebraic type of minimum (algebraically isolated).

For instance $\sigma_1 = \sum_{i>o} x_i^2$, $\sigma_2 = x_1^4 + \sum_{i>1} x_i^2$,

$\sigma_3 = x_1^4 + x_2^4 = \sum_{i>2} x_i^2$, etc.

Such a chain of symbols defines a stratum of the Maxwell set. One may then give formal rules for the relationships among strata; for instance, if, through a "piccola variazione", a singularity (σ) gives rise to a chain (β) of simpler singularities (all distinct), then every subchain $\beta' \subset \beta$ defines a stratum belonging to the star of the stratum (σ). Likewise, a chain of type $[\beta'\gamma]$ will be in the star of $[\sigma\gamma]$. I refer to the course of the Ecole des Houches "Mathematical Concepts in the Theory of Ordered Media" , Les Houches 1980 (North Holland), for a more detailed study of this Maxwell set. In a situation of this type, one applies the "Maxwell convention" of the catastrophe theory: at each point of the space (x,p), one chooses the point ξ which minimizes the Lagrangian $L(x,p,\dot{x})$.

On the contrary, if one only requires an extremality condition: $L(x,p,\dot{x})$ extremal in $\dot{x}$, then the set $(\mathcal{O})$ of points $(\mathcal{O})$ where the Hamiltonian may undergo some discontinuities is a set called "*obturation* set" according to the terminology of Y. Kergosien (CRAS 1981). Actually, no algorithms are presently available which allow one to define the local models of a generic set $(\mathcal{O})$. For $n = 2$, the question is (for fixed x) of classifying all projections of a surface embedded in $\mathbf{R}^3$, parallelly to a given direction. It is easy to see that one has then a fiber bundle with base PR(2), fiber R, and that the singularities for a *fixed* x are folds, cups and by varying x (of dimension 2) it is possible to obtain swallow tails, butterflies, and an umbilical line (with isolated parabolic umbilics) with, eventually, transversal intersections (multilocal) of these local singularities. Finally, in the extremality problems, there is no reason for choosing one particular Hamiltonian over another one, either at a regular point of the space (p,x), or, if one adopts the convention of perfect delay, at a bifurcation point. This explains why extremality problems with non convex Lagrangians lead to widely undetermined situations. Perhaps one could look along this direction for an interpretation of the quantum indeterminism...

2nd INTERPRETATION OF THE HAMILTONIAN FORMALISM : CONSTRAINTS IN A MEDIUM.

It is known according to Coulomb that if a cylindrical test-bar is subjected to very strong pressure between its flat faces, it will break along a plane section which forms a certain angle (φ) with the stress direction (φ friction angle) (Fig.6). Likewise, in a material medium subject to constraints on the boundary, there exists at each point x a cone Γ_x tangent to the plane directions along which a fracture of the medium can appear, one part of the solid sliding over the other. Hamiltonian formalism provides a tool for dealing with such situations : if at each $x \in X$ there is a cone Γ_x, which is the envelope of the possible fracture plans, then one may construct in the covector space (x,p) a sub-manifold W which is fibered by the field defined by Ker α , where α is the fundamental form $\Sigma p_i dx_i$. Hence, one obtains by projecting on X the possible slidings of one fraction of the medium over the other one (shearing). It is clear that this kind of sliding (which can be found in geology in the theory of faults) offers a particularly well-behaved type of "catastrophe" of a medium, since the topological continuity of the medium is not affected by the sliding of one portion of the medium over the other.

Example : Given a field of quadratic cone Γ_x in a two-dimensional medium, there exist *two* possible directions of sliding. This is a well-known situation in geology, where there are in general two possible directions of faults. For instance, in the massif of Jura, there are two directions of faults slanting with respect to the general direction of sliding, one is approximately North-South, with left-hand sliding, like the Pontarlier accident; the other one, with right-hand sliding (visible for instance near Salins). (Fig.7).

3rd INTERPRETATION : THE PHENOMENOLOGICAL EFFECT OF A CHANGE OF FRAME OF REFERENCE.

Roughly speaking, there are two types of sciences: "macroscopic" sciences, naively realistic, in which one thinks of being in the presence of an "objective" reality, the same for every one, and "non realistic" sciences, in which one thinks that every observation can be perturbed by a change of position, or of kinematic state, of the observer. Since Galileo, physics (mechanics) have turned to be sciences of the second type. Einstein has carried this point of view to its extreme limit : has he not pretended that every phenomenon can be cancelled by choosing an appropriate frame of reference ? To cancel out the effects of gravity, place yourself in a freely-falling elevator...

There is no doubt that the second point of view has become absolutely essential in mechanics, since Galileo has made it clear that inertial motion is not a phenomenon, but the effect of a change of frame of reference. In other words, facing a phenomenon (P), one may try to put in evidence a (moving) frame R(P) in which the phenomenon (P) no longer appears. Thus, in the presence of a mass-point with mass m, with an apparent kinetic momentum p, the observer can cancel the effects of apparent motion, if he puts himself at a speed $\dot{q}$ defined by the relation

$$p = m\dot{q} \quad .$$

Now, this is nothing but the Hamiltonian field associated with the Hamiltonian

$$H = \frac{p^2}{2m} \qquad \dot{q} = \frac{\partial H}{\partial p} = \frac{p}{m} \quad , \qquad \dot{p} = 0$$

We are dealing with a free Hamiltonian and the effect of an external force field can then be expressed through the addition of a potential V(q) to H .

From these considerations, one can draw the following (philosophical) conclusions. In sound logic, one should consider the p_i's as state variables (internal variables) and the q_i's as control variables (external variables). But the characteristic of Hamiltonian formalism is the inextricable mixing of the q's and the p's. In a classical Hamiltonian flow, the change from m(t) to m(t+T) is a global symplectomorphism $S(T): M \to M$ of the phase space M. This is to say that the effect of the evolution is inexistent, since the initial and the final situations are equivalent. In fact, it is a theory of the abolition of motion. But if one goes back to the phenomenology, one observes that the distinction between the (q)'s and the (p)'s cannot be escaped, because in general the material elements related to the variables of position are the room of dissipative phenomena which make them visible. The motion of the planets is a phenomenology, only because the planets are illuminated by the sun, and the solar radiation is an irreversible dissipative phenomenon. But in the covector space $M = T^*(X)$ of a configuration space X, the fibers q_i = const. are Lagrangian manifolds (on which the fundamental form $dp_i \wedge dq_i$ vanishes). The transformation $S(T): M \to M$ transforms these fibers into other Lagrangian manifolds the phenomenological properties of which can be only estimated by a *projection* on the base X . This is to say that the phenomenology is tied to a conflict between an initial fibration (a foliation) and the foliation transformed by S(T). There is a large amount of theoretical work (caustics, Maslov's theories, method of the saddle-point) devoted to the treatment of this projection. But very likely, it is just a special case of a general fact : if every physical phenomenology can be modelized by a foliation, then every time variation of a phenomenology can be described by the intersection of this foliation and the foliation transformed by the time evolution.

It also happens here, again as a miracle, that this time evolution can be described by a flow in a finite dimensionel space ($T^*X = M$) - whereas an evolution described by a partial differential equation should need a flow in an infinite dimensional space. I would not be far from thinking that actually the only really "well-posed" problems can be described by flows in finite dimensional spaces - with the infinite dimensional part quotiented as initial data on the hypersurface $t = 0$ of the Cauchy data. Anyway this is what happens for the Hamilton-Jacobi equation

$\frac{\partial S}{\partial t} + H(q_i \; \frac{\partial S}{\partial q_i}) = 0$, where S denotes the action. In any case,

it would be of interest to characterize the time evolutions which admit in this way a finite dimensional model. Would the extremality condition of a functional be sufficient in higher dimensions... ?

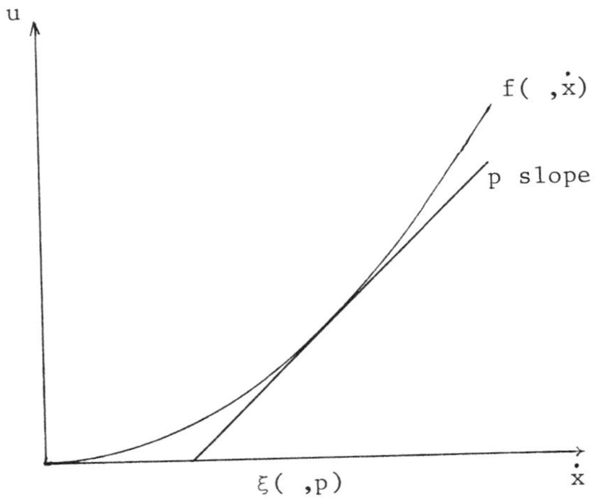

Fig. 1

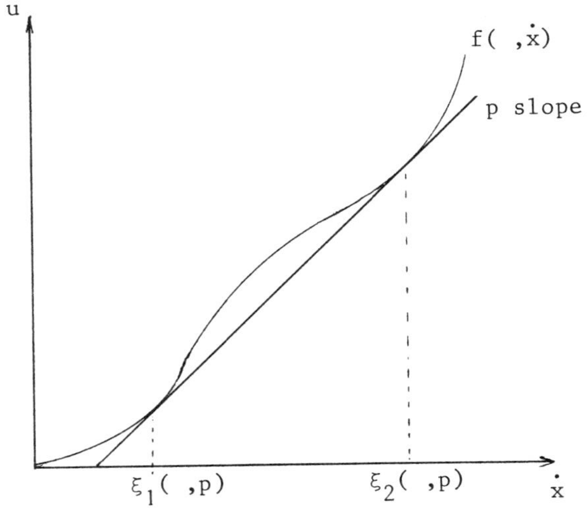

Fig. 2

Stopping-point
Fig. 3

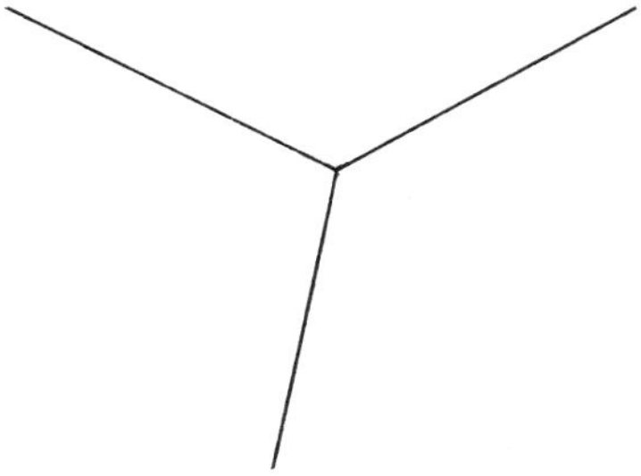

Treble point
Fig. 4

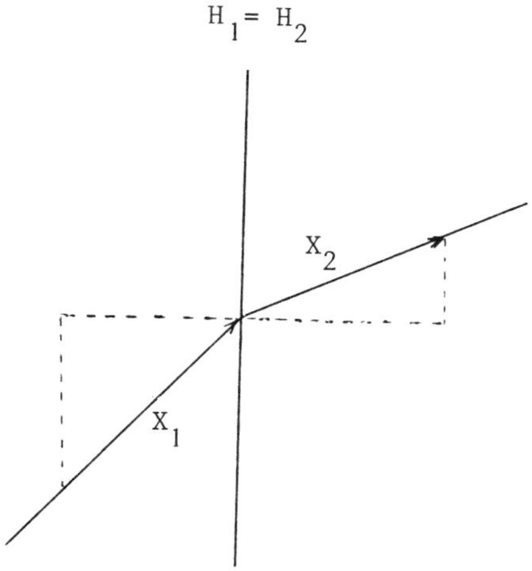

Fig. 5 : Refraction through a line of discontinuity.

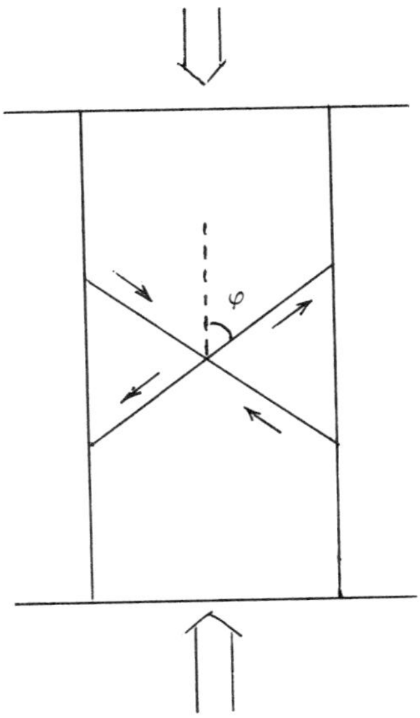

Fig. 6 :
Stress on a test-bar and
the fracture planes

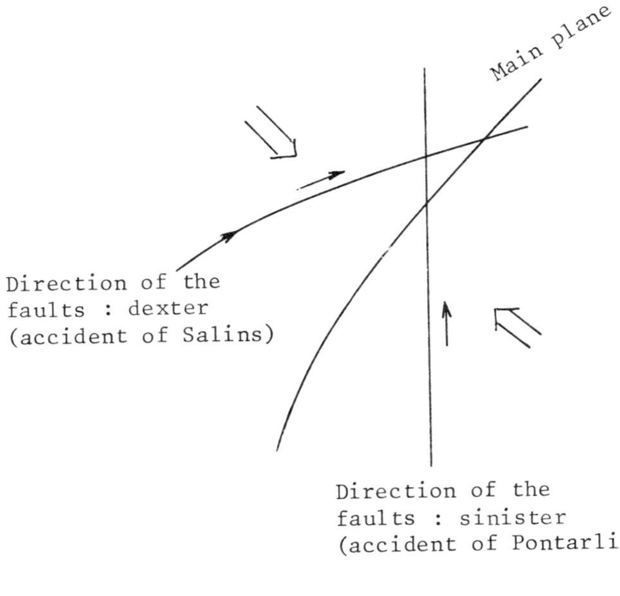

Fig. 7 : Faults in the Jura Franco-Suisse.

REFERENCES

(1) Ekeland, I., Discontinuités des champs hamiltoniens et existence des solutions optimales en Calcul des Variations. Publ. Math. IHES, 47, pp. 5-32, 1977.

(2) Kergosien, Y.L., La famille des projections orthogonales d'une surface et ses singularités. C.R. Acad. Sc. Paris, t. 292, pp. 930-32 (1er Juin 1981).

(3) Thom, R., *Mathematical Concepts in the Theory of Ordered Media*, Cours des Houches 1980 (à paraître chez North - Holland).

ON THE LAGRANGE REPRESENTATION OF A SYSTEM OF NEWTON EQUATIONS

Giacomo Della Riccia

Istituto di Matematica, Informatica
e Sistemistica dell'Università
33100-Udine, Italy

and

Department of Mathematics
The Ben-Gurion University of the Negev
Beer-Sheva, Israel

We give an analytic characterization of integrable systems of Newton equations of motion and a method for the explicit determination of the integrating factors necessary for the solution of the Inverse Problem of classical mechanics.

1. INTRODUCTION

A system of Newton equations of motion

$$F_k = \alpha_{ki}(t, q, \dot{q})\ddot{q}^i + \beta_k(t, q, \dot{q}) = 0 \tag{1}$$

$$\dot{q} = dq/dt, \quad \ddot{q} = d\dot{q}/dt, \quad q = \{q^k(t)\}, \quad k = 1,2, \ldots, n$$

is said to be *Lagrangian* if there exists a function $L(t, q, \dot{q})$ such that

DYNAMICAL SYSTEMS
AND MICROPHYSICS

ISBN 0-12-068720-8

$$F_k \equiv L_k = \frac{d}{dt}\left(\frac{\partial L}{\partial \dot{q}^k}\right) - \frac{\partial L}{\partial q^k}, \quad k = 1,2, \ldots, n \tag{2}$$

It is well-known that $F = \{F_k\}$ admits a Lagrange representation (2) provided the functions $\alpha_{ki}(t, q, \dot{q})$ and $\beta_k(t, q, \dot{q})$, $k, i = 1,2, \ldots, n$, satisfy certain conditions given for the first time by Helmholtz, [3]. When the given system $F = \{F_k\}$ is not Lagrangian the question is whether it is possible to find a Lagrangian system $G = \{G_k\}$ which is equivalent to $F = \{F_k\}$ in the sense that the two systems of Newton equations $F_k = 0$ and $G_k = 0$, $k = 1,2, \ldots n$, admit the same set of solutions. If such $G = \{G_k\}$ exists, $F = \{F_k\}$ is said to be *integrable.*

This so-called Inverse Problem of classical mechanics has been studied with a variety of mathematical approaches, including differential geometry and functional analysis, in configuration space as well as in phase space. A most comprehensive account of the literature on this subject can be found in Santilli, [6], and among more recent papers we can cite the works of Shafir, [7], Oziewicz, [5], and Henneaux, [4].

In this paper we present a straightforward analytic treatment of the Inverse Problem along the lines originated by Helmholtz. Although less elegant than the abstract methods used by the above mentioned authors, our approach has the advantage of being constructive. In fact we shall give the conditions to be satisfied by a system in order to be integrable and the explicit computation of the matrix of integrating factors. We also show that the well-known formal relation between the one-dimensional Inverse Problem and Jacobi's theory of the *last multiplier* persists to some extent in the case of arbitrary number n of degress of freedom.

II. JACOBI'S THEORY OF THE LAST MULTIPLIER AND THE INVERSE PROBLEM WITH ONE DEGREE OF FREEDOM ($n = 1$)

A last multiplier M of a system of m first-order differential equations

$$\frac{dx_1}{X_1} = \frac{dx_2}{X_2} = \ldots \frac{dx_m}{X_m} = \frac{dx}{X} \tag{3}$$

where $(X_1, \ldots, X_m, X)$ are given functions of $(x_1, \ldots, x_m, x)$, is a solution of the partial differential equation (p.d.e.)

$$\frac{\partial}{\partial x_1}(MX_1) + \ldots + \frac{\partial}{\partial x_m}(MX_m) + \frac{\partial(MX)}{\partial x} = 0$$

A last multiplier M plays the role of an integrating factor in the sense that if we already know (m − 1) first-integrals $\Theta_r(x_1, \ldots, x_m, x) = c_r$, $r = 1,2, \ldots, m-1$, for the system (3), then as shown by Jacobi, it is possible to obtain from $M(x_1, \ldots, x_m, x)$ one more first integral, say $\Theta_m(x_1, \ldots, x_m, x) = c_m$. Clearly the general solution $x_r(x, c_1, \ldots, c_m)$, $r = 1,2, \ldots, m$, for (3) is completely determined by these m first integrals.

Let us now consider a Newton equation

$$F \equiv \ddot{q} + g(t, q, \dot{q}) = 0 \tag{4}$$

and suppose that

$$\frac{d}{dt}\left(\frac{\partial L}{\partial \dot{q}}\right) - \frac{\partial L}{\partial q} = \frac{\partial^2 L}{\partial \dot{q}^2}\ddot{q} + \frac{\partial^2 L}{\partial q \partial \dot{q}}\dot{q} + \frac{\partial^2 L}{\partial t \partial \dot{q}} - \frac{\partial L}{\partial q} = 0 \tag{5}$$

is a Lagrange representation of a system equivalent to (4). This implies that (5) must reduce to an identity when $\ddot{q}$ is replaced by $-g(t, q\ \dot{q})$; in other words the Lagrange function L must be a solution of the p.d.e.

$$-\frac{\partial^2 L}{\partial \dot{q}^2} g(t, q, \dot{q}) + \frac{\partial^2 L}{\partial q \partial \dot{q}}\dot{q} + \frac{\partial^2 L}{\partial t \partial \dot{q}} - \frac{\partial L}{\partial q} = 0$$

From this we obtain

$$\frac{\partial}{\partial \dot{q}}\left(-\frac{\partial^2 L}{\partial \dot{q}^2} g\right) + \frac{\partial}{\partial \dot{q}}\left(\frac{\partial^2 L}{\partial q \partial \dot{q}}\dot{q} + \frac{\partial^2 L}{\partial t \partial \dot{q}} - \frac{\partial L}{\partial q}\right) = 0$$

which reduces to

$$\frac{\partial}{\partial \dot{q}}(-Mg) + \frac{\partial}{\partial q}(M\dot{q}) + \frac{\partial M}{\partial t} = 0 \tag{6}$$

where

(7) $$M = \frac{\partial^2 L}{\partial \dot{q}^2}$$

It is clear that (6) is Jacobi's last multiplier p.d.e. for the system of first-order differential equations

$$\frac{d\dot{q}}{-g(t, q, \dot{q})} = \frac{dq}{\dot{q}} = dt$$

corresponding to a Newton equation (4). From a last multiplier $M(t, q, \dot{q})$ we obtain, according to (7), a Lagrangian function

$$L(t, q, \dot{q}) = \int\int M\, d\dot{q}\, d\dot{q} + \xi(t, q)$$

where

$$\xi(q, t) = \int [\{ \dot{q} \frac{\partial}{\partial q} + \frac{\partial}{\partial t} \} \int M d\dot{q} - Mg - \frac{\partial}{\partial q} \int\int M d\dot{q} d\dot{q}] \, dq$$

With this definition of $L(t, q, \dot{q})$ it is easy to check that

$$L_1 = \frac{d}{dt}\left(\frac{\partial L}{\partial \dot{q}}\right) - \frac{\partial L}{\partial q} \equiv M(t, q, \dot{q}) \, [\ddot{q} + g(t, q, \dot{q})]$$

and it is clear that equation $L_1 = 0$ is equivalent to the given equation $\ddot{q} + g(t, q, \dot{q}) = 0$ provided $M(t, q, \dot{q}) \neq 0$. Let us notice that a last multiplier defines a conjugate momentum by $p = \partial L/\partial \dot{q} = \int M d\dot{q}$ and that Jacobi's p.d.e. (6) can be written in the form

(8) $$\frac{dM}{dt} = M \frac{\partial g}{\partial \dot{q}}$$

where the operator

$$\frac{d}{dt} = \{ - g \frac{\partial}{\partial \dot{q}} + \dot{q} \frac{\partial}{\partial q} + \frac{\partial}{\partial t} \}$$

represents *the total derivative of a function along the motion.* We shall see in the following discussion of the n-dimensional Inverse Problem an appropriate generalization of equation (8).

EXAMPLES

a) The harmonic oscillator: $\ddot{q} + q = 0$

We have $g(t, q, \dot{q}) = q$ and $\partial g/\partial \dot{q} = 0$, thus the last multiplier equation becomes $dM/dt = 0$. Obviously $M_1 = 1$ is a solution which generates the Lagrangian $L_1 = 1/2(\dot{q}^2 - q^2)$, a momentum $p_1 = \dot{q}$ and a Hamiltonian

$$H_1 = \frac{1}{2}(q^2 + p^2) = \frac{1}{2}q^2(1 + \frac{\dot{q}^2}{q^2}).$$

$M_2 = \dfrac{1}{q^2 + \dot{q}^2}$ is another solution to which there corresponds the Lagrangian

$$L_2 = \frac{\dot{q}}{q} \text{ arctg } (\frac{\dot{q}}{q}) - \frac{1}{2} \ell n[q^2(1 + \frac{\dot{q}^2}{q^2})],$$

a momentum $p_2 = 1/q \text{ arctg}(\dot{q}/q)$ and a Hamiltonian

$$H_2 = \frac{1}{2} \ell n [\frac{q^2}{\cos^2(pq)}] = \frac{1}{2} \ell n[q^2(1 + \frac{\dot{q}^2}{q^2})].$$

Incidentally $H_2 = 1/2\, \ell n H_1$. Another last multiplier is $M_3 = \text{arctg}(\dot{q}/q) + t$ and, in fact, the general solution of $dM/dt = 0$ is $M = f[q^2 + \dot{q}^2, \text{arctg}(\dot{q}/q) + t]$, where f is an arbitrary function, from which we can generate as many Lagrangian as we want with their corresponding momenta p and Hamiltonian H.

b) The linearly damped oscillator: $\ddot{q} + \varepsilon\dot{q} + q = 0.$

We have $g(t, q, \dot{q}) = \varepsilon\dot{q} + q$ and $\partial g/\partial \dot{q} = \varepsilon$, thus the last multiplier equation becomes $dM/dt = \varepsilon M$. An obvious solution is $M_1 = \exp(\varepsilon t)$ which generates the Lagrangian $L_{1\varepsilon} = 1/2\, e^{\varepsilon t}(\dot{q}^2 - q)$, the momentum $p_{1\varepsilon} = \dot{q}\exp(\varepsilon t)$ and Hamiltonian $H_{1\varepsilon} = 1/2[q^2 \exp(\varepsilon t) + p^2 \exp(-\varepsilon t)]$. Another solution is

$$M_2 = \frac{1}{q^2(1 + U^2)} ,$$

where

$$U = \frac{1}{\omega}\left(\frac{\dot{q}}{q} + \frac{\varepsilon}{2}\right)$$

and $\omega^2 = 1 - \varepsilon^2/4$, with corresponding Lagrangian

$$L_{2\varepsilon} = \omega^2 \, U \, \text{arctg}\, U - \frac{\omega^2}{2} \, \ell n \, [q^2(1 + U^2)] ,$$

momentum $p_{23} = \omega/q$ arctg U and Hamiltonian

$$H_{2\varepsilon} = \frac{\omega^2}{2} \, \ell n[q^2(1 + U^2)] - \frac{\omega\varepsilon}{2} \, \text{arctg}\, U .$$

When $\varepsilon \to 0$, $H_{1\varepsilon}$ and $H_{2\varepsilon}$ become, respectively, the Hamiltonian H_1 and H_2 of the harmonic oscillator.

In Della Riccia, [1], there is a Hamiltonian treatment of the Smith's oscillator. This is a system, described by a Newton equation, whose motion presents a limit cycle in addition to an equilibrium point at $q = \dot{q} = 0$.

The main point of this section is that in general a Newton equation always admits equivalent Lagrangian systems, actually as many as there are non-vanishing solutions M for the last multipliers p.d.e.

III. THE N-DIMENSIONAL INVERSE PROBLEM

We consider a system of Newton equations of motion given in the fundamental form (1). This system is Lagrangian iff the following Helmholtz conditions are satisfied

(A,1) $$|\alpha| = \text{determinant } |\alpha_{ij}(t, q, q)| \neq 0$$

(A,2) $$\alpha_{ij} = \alpha_{ji}$$

$$\left\{ \frac{\partial}{\partial t} + \dot{q}^k \frac{\partial}{\partial q^k} \right\} \alpha_{ij} = \frac{1}{2} \left(\frac{\partial \beta_i}{\partial \dot{q}^j} + \frac{\partial \beta_j}{\partial \dot{q}^i} \right) \tag{A,3}$$

$$\frac{\partial \alpha_{ik}}{\partial \dot{q}^j} = \frac{\partial \alpha_{jk}}{\partial \dot{q}^i} \tag{A,4}$$

$$\frac{1}{2} \left\{ \frac{\partial}{\partial t} + \dot{q}^k \frac{\partial}{\partial q^k} \right\} \left(\frac{\partial \beta_i}{\partial \dot{q}^j} - \frac{\partial \beta_j}{\partial \dot{q}^i} \right) = \frac{\partial \beta_i}{\partial q^j} - \frac{\partial \beta_j}{\partial q^i} \tag{A,5}$$

$$i,j,k = 1,2, \ldots, n.$$

where we use the summation convention over repeated indices. From (A,2), (A,3) and (A,4) one can derive the following relations

$$\frac{\partial \alpha_{jk}}{\partial q^i} - \frac{\partial \alpha_{ik}}{\partial q^j} = \frac{1}{2} \frac{\partial}{\partial \dot{q}^k} \left(\frac{\partial \beta_j}{\partial \dot{q}^i} - \frac{\partial \beta_i}{\partial \dot{q}^j} \right) \tag{A,6}$$

Since we shall consider only so-called regular systems, i.e. those satisfying condition (A,1), we can start without loss of generality with Newtonian systems $G = \{G_k\}$ given in kinematical form

$$G_k \equiv \ddot{q}^k + g^k(t, q, \dot{q}) = 0 \qquad k = 1,2, \ldots, n.$$

Using Helmholtz conditions, we find that G is Lagrangian iff the generalized forces $g = (g^1, \ldots, g^n)^*$ are of the form

$$g^k(t, q, \dot{q}) = \rho^{ki}(t, q)\dot{q}_i + \sigma^k(t, q)$$

$$\rho^{ij} + \rho^{ji} = 0$$

$$\frac{\partial \rho^{ij}}{\partial q^k} + \frac{\partial \rho^{jk}}{\partial q^i} + \frac{\partial \rho^{ki}}{\partial q^j} = 0$$

$$\frac{\partial \rho^{ij}}{\partial t} = \frac{\partial \sigma^i}{\partial q^j} - \frac{\partial \sigma^j}{\partial q^i}$$

$$i,j,k = 1,2, \ldots, n.$$

Let us assume that the given system is not Lagrangian. Since we are considering only equations which are linear with respect to variables $\ddot{q}$, all regular systems $\tilde{G} = \{\tilde{G}_k\}$ equivalent to $G = \{G_k\}$ must be of the form

$$\tilde{G} \equiv M(\ddot{q} + g) = 0$$

where $M = [f_{ij}(t, q, \dot{q})]$ is a nonsingular matrix. $\tilde{G}$ will be Lagrangian, and hence G integrable, iff Helmholtz conditions are satisfied with $\alpha_{ij}(t, q, \dot{q}) = f_{ij}(t, q, \dot{q})$ and $\beta_i(t, q, \dot{q}) = f_{ij}\, g^j$. Therefore the integrating factors f_{ij} must satisfy the following conditions

(B,1) $$|M| = |f_{ij}(t, q, \dot{q})| \neq 0$$

(B,2) $$f_{ij} = f_{ji}$$

(B,3) $$\{\frac{\partial}{\partial t} + q^k \frac{\partial}{\partial q^k}\} f_{ij} = \frac{1}{2}[\frac{\partial}{\partial \dot{q}^j}(f_{ik}\, g^k) + \frac{\partial}{\partial \dot{q}^i}(f_{jk}\, g^k)]$$

(B,4) $$\frac{\partial f_{ik}}{\partial \dot{q}^j} = \frac{\partial f_{jk}}{\partial \dot{q}^i}$$

(B,5) $$\frac{1}{2}\{\frac{\partial}{\partial t} + \dot{q}^k \frac{\partial}{\partial q^k}\}[\frac{\partial}{\partial \dot{q}^j}(f_{ik}\, g^k) - \frac{\partial}{\partial \dot{q}^i}(f_{jk}\, g^k)] = \frac{\partial}{\partial q^j}(f_{ik}\, g^k) - \frac{\partial}{\partial q^i}(f_{jk}\, g^k)$$

$$i,j,k = 1,2, \ldots, n$$

(B,2), (B,3) and (B,4) imply relations

(B,6) $$\frac{\partial f_{jk}}{\partial q^i} - \frac{\partial f_{ik}}{\partial q^j} = \frac{1}{2}\frac{\partial}{\partial \dot{q}^k}[\frac{\partial}{\partial \dot{q}^i}(f_{jk}\, g^k) - \frac{\partial}{\partial \dot{q}^j}(f_{ik}\, g^k)]$$

It is interesting to express these conditions in the following matrix form. Let us denote by $J_x(u) = (\partial u_j/\partial x^i)$ the Jacobian matrix of a column vector $u = (u_1, \ldots, u_n)^*$ with respect to variables $x = (x_1, \ldots, x_n)$. Then $M = (f_{ij})$ is an integrating matrix iff it satisfies

(C,1) $$|M| \neq 0$$

$$M = M^* \tag{C,2}$$

$$\frac{dM}{dt} = \frac{1}{2}(MJ^* + JM) \tag{C,3}$$

$$\frac{\partial M}{\partial \dot{q}^i} = J_{\dot{q}}(M_i), \quad i = 1,2,\ldots,n \tag{C,4}$$

$$\frac{1}{2}\frac{d}{dt}(MJ^* - JM) = MK^* - KM \tag{C,5}$$

where $J = J_{\dot{q}}(g)$, $K = J_q(g)$, M_i is the i-th column of M and the operator

$$\frac{d}{dt} = \{-g^k\frac{\partial}{\partial \dot{q}^k} + q^k\frac{\partial}{\partial q^k} + \frac{\partial}{\partial t}\}$$

represents the total derivative of a function along the motion. The matrix form of (B,6) is

$$\frac{\partial M}{\partial q^i} = J_q(M_i) + \frac{1}{2}J^*_{\dot{q}}[M\gamma_i - JM_i], \quad i = 1,2,\ldots,n \tag{C,6}$$

where γ_i is the i-th column of J^*.

It is clear that for $n = 1$ the matrix p.d.e. (C,3) reduces to Jacobi's last multiplier p.d.e. (8), whereas all other conditions (C,4) to (C,6) are identically satisifed. On the other hand, for $n \geqslant 2$, there will be in general no simultaneous solution M for (C,1) to (C,5) unless certain constraints are imposed on the elements of matrices J and K, that is on the generalized forces $g = (g_1, \ldots, g_n)$. We shall now discuss these constraints.

Let us first substitute in (C,5) dM/dt by its expression (C,3); we find the matrix equation

$$MA^* = AM \tag{C,5$'$}$$

where $A = K - 1/2\ dJ/dt - 1/4J^2$. This equation represents a set of $\frac{n(n-1)}{2}$ linear homogeneous equations for the $\frac{n(n+1)}{2}$ unknown elements of the symmetric matrix M. We then apply the operator $\frac{d}{dt}$ to (C,5)′ and eliminate again dM/dt by using (C,3); we find another set of $\frac{n(n-1)}{2}$ linear homogeneous equations

$$(C,5)'' \qquad MB^* = BM$$

(C,5)′ and (C,5)″ will have a nontrivial solution provided the rank of the coefficient matrix of these $n(n-1)$ homogeneous equations is at most equal to $\frac{n(n+1)}{2} - 1$. This condition on the rank implies a certain number of relations between the elements of A and B. If the rank is equal to $\frac{n(n+1)}{2} - 1$, M can be written in the form $M = f_{11}\ \Lambda$ where $f_{11}\ (t, q, \dot{q})$ is still arbitrary but the symmetric matrix $\Lambda = (\lambda_{ij})$ with $\lambda_{11} = 1$ is completely determined by

$$(C,5)''' \qquad \Lambda A^* = A\Lambda$$
$$\Lambda B^* = B\Lambda$$

Let us call (R,1) the set of algebraic relations implied on the elements of A and B by the value imposed to the rank of (C,5)′ and (C,5)″.

We now substitute in equations (C,4) $M = f_{11}\ \Lambda$ and obtain

$$(C,4)' \qquad \frac{\partial \ln f_{11}}{\partial \dot{q}^i}\Lambda + \frac{\partial \Lambda}{\partial \dot{q}^i} = J_{\dot{q}}(\Lambda_i) + \nabla_{\dot{q}} f_{11} \cdot \Lambda_i^* , \quad i = 1,2,\ldots,n$$

where Λ_i is the i-th column of Λ. Equations (C,4)′ determine the quantities $\frac{\partial \ln f_{11}}{\partial \dot{q}^i}$, $i = 1,2,\ldots,n$, in terms of the elements of Λ and $\frac{\partial \Lambda}{\partial \dot{q}^i}$, $i = 1,2,\ldots,n$. However, in order that equations (C,4)′ be compatible we shall have to write a certain number of algebraic relations between the elements of Λ and $\partial\Lambda/\partial q^i$, $i = 1,2,\ldots,n$. Let us indicate these relations by (R,2).

In a similar way, by substituting $M = f_{11}\ \Lambda$ in equations (C,6) we obtain the quantities $\partial \ln f_{11}/\partial q^i$, $i = 1,2,\ldots,n$, as functions of the elements of Λ, $\partial\Lambda/\partial \dot{q}^i$ and $\partial\Lambda/\partial q^i$, $i = 1,2,\ldots,n$, and their compatibility will imply a certain number of algebraic relations between these elements that we indicate by (R,3).

Since the expressions obtained for $\partial \ln f_{11}/\partial \dot{q}^i$ and $\partial \ln f_{11}/\partial q^i$, $i = 1,2,\ldots,n$, actually represent a set of partial differential equations for f_{11}, they must be compatible in the sense that the second-order partial derivative of $\ell n f_{11}$, with respect to any pair of variables $\dot{q}$ and q is independent of the order in which the differentiation is performed. This leads to certain relations (R,4) between the elements of Λ and their first and second-order partial derivatives.

Finally, by substituting $M = f_{11}\Lambda$ in (C,3) we obtain

$$\frac{d\ell n f_{11}}{dt}\Lambda + \frac{d\Lambda}{dt} = \frac{1}{2}(\Lambda J^* + J\Lambda) \qquad \text{(C,3)}'$$

from which we get

$$\frac{d\ell n f_{11}}{dt} = \sum_{k=1}^{n} \lambda_{1k}\,\gamma_{1k} \qquad (9)$$

and further compatibility conditions (R,5) involving the elements of Λ and $d\Lambda/dt$.

Let us now write (9) as follows

$$\frac{\partial \ell n f_{11}}{\partial t} = \sum_{k=1}^{n} \left(\lambda_{1k}\,\gamma_{1k} + g^k \frac{\partial \ell n f_{11}}{\partial \dot{q}^k} - \dot{q}^k \frac{\partial \ell n f_{11}}{\partial q^k}\right) \qquad (10)$$

where the right-hand side is completely known in terms of g and its partial derivatives with respect to t, $\dot{q}$ and q variables on the basis of the above results obtained from (C,4), (C,5) and (C,6). We can treat (10) as an ordinary differential equation with t as independent variable and determine f_{11} up to an insignificant multiplicative constant and consequently all the matrix $M = f_{11}\Lambda$ has been found.

The conclusion is that in general a system of Newton equations with $n \geqslant 2$ is not integrable unless conditions (R,1), (R,2), . . . , up to (R,5), involving g and its partial derivatives up to the fourth order with respect to the variables t, q and $\dot{q}$, are satisfied. If the system is integrable, the matrix M of integrating factors, hence the equivalent Lagrangian system, is in general unique. There could be more than one solution M only if the rank of equations (C,5)‴ is less than $\frac{n(n+1)}{2} - 1$.

In a future work we shall discuss explicitly conditions (R,1) to (R,5) which represent an analytic characterization of integrable systems.

IV. CONCLUSION

Havas, [2], was able to find a characterization of integrable systems of Newton equations for which the matrix M assumes a diagonal form. We have extended this characterization to all integrable systems. It should be interesting, on the basis of an examination of the explicit expression of conditions (R,1) to (R,5) which can be obtained by long but simple algebraic manipulations, to understand the physical properties of the general forces $g = (g_1, \ldots, g_n)$ which correspond to integrable Newtonian systems.

REFERENCES

[1] Della Riccia, G., "A Hamilton-Jacobi treatment for dissipative systems with one degree of freedom", in Dynamical Systems and Microphysics (Ed. Blaquiére, Fer and Marzollo, CISM Courses and Lectures N. 261, Springer-Verlag, 1980) p. 291-300.

[2] Havas, P., "The range and application of the Lagrange formalism I", Suppl. Nuovo Cimento, Vol. V, 10, (1957), p. 363-388.

[3] Helmholtz, H., Journ. f.d. reine u. angew. Math., 100, (1887), p. 137-

[4] Henneaux, M., "Equations of motion, commutative relations and ambiguities", Université Libre de Bruxelles, Preprint ULB/PHY. No. 722/June 1981.

[5] Oziewicz, Z., "Classical Mechanics: Inverse Problem and Symmetries", Preprint Univ. Genève (1981) UGVA-DPT (# 03-281, # 08-307).

[6] Santilli, R.M., "Foundations of Theoretical Mechanics I: the Inverse Problem in Newtonian mechanics" (Springer-Verlag, 1978).

[7] Shafir, R.L., "The variational principle and natural transformations, I-Autonomous dynamical systems, Preprint, Dept. of Math., King's College, London (1981).

CONSERVATION LAWS AND A HAMILTON-JACOBI-LIKE METHOD IN NONCONSERVATIVE MECHANICS.

B. Vujanović

Faculty of Technical Sciences, University of Novi Sad,
21000 Novi Sad, Yugoslavia

INTRODUCTION

The main objectives of this lecture are:

1. To study the possibility of finding the conservation laws of classical nonconservative dynamical systems.
2. To demonstrate a method for solving the canonical (Hamilton's) differential equations of motion of nonconservative systems, which is, similar to the method of Hamilton and Jacobi.

If the behaviour of a dynamical system is completely describable by means of the Lagrangian function (we conditionally refer to this type of systems as conservative), the determination of its conservation laws is based on the famous theorem of E. Noether |1|, which states that for every infinitesimal transformation of the generalized coordinates and time, which leaves Hamilton's action integral invariant (or gauge invariant), there corresponds a conservation law of the system in question.

For the same type of conservative systems, a well known method for solving canonical equations is based on the theorem of Hamilton-Jacobi (see for example ref. |2|): If a complete solution of the Hamilton-Jacobi partial differential equation is known, then the solution of the canonical system of ordinary equations can be obtained (without any integration) by means of this complete solution.

For nonconservative dynamical systems, neither theorem is applicable due to the existence of nonconservative forces. In these theorems the vital point is the action integral, which does not exist if the system is not conservative.

ISBN 0-12-068720-8

Our approach to the objectives 1 and 2 will be based on D'Alembert's principle of virtual work and its transformation properties with respect to infinitesimal transformations of the generalized coordinates and time.

2. CONSERVATION LAWS OF NONCONSERVATIVE DYNAMICAL SYSTEMS

Let us consider a nonconservative holonomic dynamical system whose position is specified by the n independent coordinates $q = q^1,..,q^n$ which are functions of time t. In accordance with the classical variational calculus, we introduce the following two types of variations:

$$\bar{q}^i(t)-q^i(t) = \delta q^i \ , \quad \delta t = 0, \quad \text{simultaneous variations} \tag{1}$$

$$\bar{q}^i(\bar{t})-q^i(t)= \Delta q^i \ , \quad \bar{t}=t+\Delta t, \quad \text{nonsimultaneous variations} \tag{2}$$

where a bar denotes an infinitesimal variation of the corresponding quantity. For what follows, it is convenient to look at the nonsimultaneous variations as specific infinitesimal transformations of the generalized coordinates and time. We assume that the generators of these transformations have the following structure

$$\Delta q^i= \varepsilon F^i(t,q,\dot{q}), \quad \Delta t = \varepsilon f(t,q,\dot{q}) \tag{3}$$

where ε is a small constant parameter and $\dot{q}=dq^1/dt, dq^2/dt,.,dq^n/dt$ denotes the velocity vector.

The connection between Δq^i and δq^i is given by (see for example |3|)

$$\Delta q^i= \delta q^i+ \dot{q}^i \Delta t \tag{4}$$

By this relation we will define the nonsimultaneous variation of any function, for example

$$\Delta L(t,q,\dot{q}) = \delta L+\dot{L} \ \Delta t \tag{5}$$

where L is the Lagrangian function of the system. Since the symbols Δ and $d/dt= \dot{}$, are noncommutative, the following relation is of interest

$$\Delta\dot{q}^i = (\Delta q^i)^{\cdot}-\dot{q}^i(\Delta t)^{\cdot} \tag{6}$$

The search for the conserved quantities of nonconservative systems will be based on D'Alembert's principle of virtual work. Denoting by $Q_i = Q_i(t,q,\dot{q})$ the nonconservative forces, this principle reads as follow

$$\frac{d}{dt}\left(\frac{\partial L}{\partial \dot{q}^i}\,\delta q^i\right) - \delta L - Q_i\delta q^i = 0 \tag{7}$$

In the eq. (7) and throughout, repeated indices are summed.

Next we transform eq. (7) by the help of a nonsimultaneous infinitesimal transformation and retain the terms linear in the small parameter ε. Using (4) and (5), we write the eq. (7) in the form

$$\frac{d}{dt}\left[\frac{\partial L}{\partial \dot{q}^i}(\Delta q^i - \dot{q}^i\Delta t) + L\Delta t\right] - \left[\Delta L + L(\Delta t)^{\cdot} + Q_i(\Delta q^i - \dot{q}^i\Delta t)\right] = 0 \tag{8}$$

By adding and substracting the time derivative of an arbitrary function εP which depend on the generalized coordinates q, generalized velocities $\dot{q}$ and time t, we write last equation in the form

$$\frac{d}{dt}\left[\frac{\partial L}{\partial \dot{q}^i}(\Delta q^i - \dot{q}^i\Delta t) + L\Delta t - \varepsilon P\right] - \left[\Delta L + L(\Delta t)^{\cdot} + Q_i(\Delta q^i - \dot{q}^i\Delta t) - \varepsilon\dot{P}\right] = 0 \tag{9}$$

We call the function P the gauge-variant function.

From the transformed form of D'Alembert's principle (9) we have immediately that if the relation

$$\Delta L + L(\Delta t)^{\cdot} + Q_i(\Delta q^i - \dot{q}^i\Delta t) - \varepsilon\dot{P} = 0 \tag{10}$$

is satisfied, the dynamical system admits a conservation law, of the form

$$\frac{\partial L}{\partial \dot{q}^i}(\Delta q^i - \dot{q}^i\Delta t) + L\Delta t - \varepsilon P = C = \text{const.} \tag{11}$$

Thus for every infinitesimal transformation of the form (3) and gaugevariant function $P = P(t,q,\dot{q})$ which satisfies the condition (10), there exists a conserved quantity (11).

Using the notation introduced by (3), taking into account eq. (6), we can write the last two equations in terms of space-time generators F^i and f:

$$\frac{\partial L}{\partial q^i}F^i + \frac{\partial L}{\partial \dot{q}^i}(\dot{F}^i - \dot{q}^i\dot{f}) + \frac{\partial L}{\partial t} f + L\dot{f} + Q_i(F^i - \dot{q}^i f) - \dot{P} = 0 \tag{12}$$

$$\frac{\partial L}{\partial \dot{q}^i}(F^i - \dot{q}^i f) + Lf - P = \text{const.} \tag{13}$$

Note that for the case of a conservative dynamical systems, i.e. $Q_i = 0$ eqs. (12) and (13) constitute the classical form of Noethere's theorem (see for example |4|).

Taking into account that the generators of the infinitesimal transformations F^i and f together with the gauge-variant function P are functions of t,q and $\dot{q}$, we can write (12) in the form

$$\frac{\partial L}{\partial t}f + \frac{\partial L}{\partial q^i}F^i + \frac{\partial L}{\partial \dot{q}^i}\left(\frac{\partial F^i}{\partial q^s}\dot{q}^s + \frac{\partial F^i}{\partial t} - \frac{\partial f}{\partial t}\dot{q}^i - \frac{\partial f}{\partial q^s}\dot{q}^i\dot{q}^s\right) + L\left(\frac{\partial f}{\partial t} + \frac{\partial f}{\partial q^s}\dot{q}^s\right) +$$

$$+\ddot{q}^s\left[\frac{\partial L}{\partial \dot{q}^i}\left(\frac{\partial F^i}{\partial \dot{q}^s} - \frac{\partial f}{\partial \dot{q}^s}\dot{q}^i\right) + L\frac{\partial f}{\partial \dot{q}^s} - \frac{\partial P}{\partial \dot{q}^s}\right] + Q_i(F^i - f\dot{q}^i) - \frac{\partial P}{\partial t} - \frac{\partial P}{\partial q^s}\dot{q}^s = 0 \tag{14}$$

This equation represents the condition which must be satisfied for every set of functions F^i, f and P for a given nonconservative dynamical system which is specified by the Lagrangian function $L = L(t,q,\dot{q})$ and the generalized nonconservative forces $Q_i = Q_i(t,q,\dot{q})$.

In practical applications eq. (14) can be used in several different ways. First, due to the fact that the functions F^i, f and P do not depend on the second time derivatives $\ddot{q}$, eq. (14) can be split into a system of linear partial differential equations with respect to F^i, f and P, which are obtained by equating terms of corresponding degrees of $\ddot{q}^i$:

$$f\frac{\partial L}{\partial t} + F^i\frac{\partial L}{\partial q^i} + \left(\frac{\partial F^i}{\partial t} + \frac{\partial F^i}{\partial q^s}\dot{q}^s - \frac{\partial f}{\partial t}\dot{q}^i - \frac{\partial f}{\partial q^s}\dot{q}^i\dot{q}^s\right)\frac{\partial L}{\partial \dot{q}^i} + L\left(\frac{\partial f}{\partial t} + \frac{\partial f}{\partial q^s}\dot{q}^s\right) +$$

$$+Q_i(F^i - f\dot{q}^i) - \frac{\partial P}{\partial t} - \frac{\partial P}{\partial q^s}\dot{q}^s = 0 \tag{15}$$

$$\frac{\partial L}{\partial \dot{q}^i}\left(\frac{\partial F^i}{\partial \dot{q}^s} - \frac{\partial f}{\partial \dot{q}^s}\dot{q}^i\right) + L\frac{\partial f}{\partial \dot{q}^s} - \frac{\partial P}{\partial \dot{q}^s} = 0 \tag{16}$$

We call this system of partial differential equations the generalized Killing equations (see details in |5| and |6|). Thus, if this system of partial differential equations admits a solution, say $f(t,q,\dot{q})$, $F^i(z,q,\dot{q})$ and $P(t,q,\dot{q})$, then a conserved quantity of the form (13) automatically exists. We call every solution of the system (15) and (16) the solution in the "space of variables $q^i, \dot{q}^i, t$".

Second, by using the differential eqs. of motion, we can express all accelerations $\ddot{q}^i$ in terms of q^i, $\dot{q}^i$ and t and substitute these relations into (14). We will call the solution f, F^i and P of the resulting equation "the solution along the trajectory".

Third, if a conserved quantity is given, we can employ (12) and (13) for finding the infinitesimal transformation generators F^i, f and the gauge-variant function P. This procedure constitutes the so called inverse Noether's theorem (see refs. |6|,|7|).

To conclude our considerations, we can formulate the following:

T h e o r e m. For every infinitesimal transformation of generalized coordinates and time (3) and the gauge-variant function $P(t,q,\dot{q})$ which satisfies (12) (in the space of variables $q,\dot{q},t$, or along the trajectory of motion), there exists a conserved quantity, of the form (13), of the given nonconservative dynamical system.

Note that in a slightly different approach based on the invariance of the elementary action (Ldt) and the use of a variational principle with noncommutative rules, the equations (12) and (13) were derived in ref. |6|. The same approach was used in finding the integral invariants of nonconservative systems in |8|. The study of conservation laws in nonconservative field theory, based on D'Alambert's principle, was done in ref. |9|.

As an illustration of the theory, let us consider the motion of a Foucault pendulum in a rectangular coordinate system Oxyz with O a point of the earth's surface, Oz directed vertically upward. Neglecting the small disturbances in the z-direction, the differential equations of motion are

$$\ddot{x} + \Omega^2 x = 2\mu\dot{y}\ , \quad \ddot{y} + \Omega^2 y = -2\mu\dot{x} \qquad (17)$$

where Ω^2 and μ are given constants. Introducing $X=x+iy$, $i=(-1)^{1/2}$

we write the pair of equations as

$$\ddot{X} + \Omega^2 X = -2k\dot{X} \tag{18}$$

where $k = i\mu$. The Lagrangian function and the nonconservative force are $L = (\dot{X}^2 - \Omega^2 X^2)/2$, $Q = -2k\dot{X}$.

Let us suppose that the functions F, f and P are of the form

$$F = u(X)f(t), \quad f=f(t), \quad P=P(t) \tag{19}$$

where u(X), f(t) and P(t) have to be determined in such a way that the relation (12) is satisfied. This relation becomes

$$-\Omega^2 Xuf + \dot{X}(u'f\dot{X} + \dot{f}u - \dot{X}\dot{f}) + (1/2)(\dot{X}^2 - \Omega^2 X^2)\dot{f} - 2k\dot{X}uf + 2k\dot{X}^2 - \dot{P} = 0$$

where the prime denotes differentiation with respect to X. Equating to zero the terms in corresponding degrees of $\dot{X}$ one finds

$$-\Omega^2 Xuf - (1/2)\Omega^2 X^2 \dot{f} - \dot{P} = 0, \quad \dot{f} - 2kf = 0, \quad u'f - (1/2)\dot{f} + 2kf = 0$$

It is easy to show that a solution of these equations are $f = c_1 e^{2kt}$, $u = -kX$ and $P = 0$, $c_1 = \text{const}$. Therefore, the generators of the infinitesimal transformation are $F = -kc_1 X e^{2kt}$, $f = c_1 e^{2kt}$ and according to (13), we have a complex quadratic first integral $(\dot{X}^2 + \Omega^2 X^2 + 2kX\dot{X}) \cdot e^{2kt} = \text{const}$.

Transforming this expression in the real domain, we find the following two first integrals of the equation (17):

$$\begin{aligned} &[(\dot{x}^2 - \dot{y}^2)/2 + \Omega^2(x^2 - y^2) - \mu(x\dot{y} + \dot{x}y)]\cos 2\mu t - \\ &- [\dot{x}\dot{y} + \Omega^2 xy + \mu(x\dot{x} - y\dot{y})]\sin 2\mu t = C_1 = \text{const.} \\ &[(\dot{x}^2 - \dot{y}^2)/2 + \Omega^2(x^2 - y^2) - \mu(x\dot{x} + y\dot{y})]\sin 2\mu t + \\ &+ [\dot{x}\dot{y} + \Omega^2 xy + \mu(x\dot{x} + y\dot{y})]\cos 2\Omega t = C_2 = \text{const.} \end{aligned} \tag{20}$$

It is interesting to note that the system (17) admits an energy integral $(\dot{x}^2 + \dot{y}^2) + \Omega^2(x^2 + y^2) = \text{const.}$ due to the fact that the forces depending of velocities are of a gyroscopis nature. Hence, both quadratic first integrals (20) can serve as an example when quadratic first integrals different from the energy integral exist.

3. A METHOD OF INTEGRATION OF THE NONCONSERVATIVE CANONICAL EQUATIONS

In this section we will briefly demonstratre a method for solving Hamilton's canonical equations in the nonconservative form:

$$\dot{q}^i = \frac{\partial H}{\partial p_i}, \qquad \dot{p}_i = -\frac{\partial H}{\partial q^i} + Q_i(t,q,p) \tag{21}$$

Where H is Hamilton's function and $p=p_1,..,p_n$ denotes the momentum vector.

Our basic supposition is that one component of the momentum vector, say p_1 can be expressed as a continuous function of generalized coordinates q, time t and the rest of the components of the momentum vector $p_2=z_2,..,p_n=z_n$, i.e.

$$p_1 = U(t,q^1,..,q^n,z_2,..,z_n) \tag{22}$$

From what follows, it is necessary to do the following convention. The Latin indices run from 1 to n, small Greek indices go from 2 to n and the capital Latin letters used as indices run from 2 to 2n. Combining (22) with the equation $\dot{p}_1=-\partial H/\partial q^1+Q_1$ and using the rest of eq. (21) we obtain the following quasi-linear partial differential equation:

$$\partial U/\partial t+(\partial U/\partial q^1)(\partial H/\partial U)+(\partial U/\partial q^\alpha)(\partial H/\partial z_\alpha)+\partial U/\partial z_\alpha\ (-\partial H/\partial q^\alpha+Q_\alpha)+$$
$$+\partial H/\partial q^1-Q_1=0, \qquad \alpha=2,..,n \tag{23}$$

We call this equation the basic differential equation. The unknown function U depends of the 2n arguments t, q^i, z_α.

We shall be concerned with the complete solutions of the basic equation (23). We define as a complete solution a relation of the form

$$p_1 = U(t,q^1,..,q^n,z_2,..,C_1,..,C_{2n}) \tag{24}$$

which in addition to the arguments t,q,z contains 2n aribrary constants $C_1,..,C_{2n}$, which when substituted in eq. (23) reduces it to an identity.

Let the initial conditions of the system be given by

$$q^i(0)=\alpha^i, \quad p_i(0)=\beta_i \tag{25}$$

Applying (25) to (24) and expressing one constant, say C_1 in terms of α^i, β_i, and $C_2,..,C_{2n}$, we write (24) in the following conditioned form

$$p_1 = \overline{U}(t,q^1,..,q^n,z_2,..,z_n,\alpha^1,..,\alpha^n,\beta_1,..,\beta_n,C_2,..,C_{2n}) \tag{26}$$

Note that a characteristic feature of our theory is the fact that the constants $C_2,..,C_{2n}$ remain undetermined throughout entire process of finding the motion.

We will show that the motion of the nonconservative dynamical system described by (21) and (25) can be obtained from (26) and the 2n-1 algebraic equations

$$\partial\bar{U}/\partial C_A = 0, \quad A = 2,..,2n \tag{27}$$

for arbitrary values of C_A. We suppose that the following condition is satisfied

$$\det \begin{vmatrix} \partial^2\bar{U}/\partial C_2\partial q^1, .., \partial^2\bar{U}/\partial C_2\partial q^n, & \partial^2\bar{U}/\partial C_2\partial z_2, .., \partial^2\bar{U}/\partial C_2\partial z_n \\ \vdots & \vdots \\ \partial^2\bar{U}/\partial C_{2n}\partial q^1, .., \partial^2\bar{U}/\partial C_{2n}\partial q^n, & \partial^2\bar{U}/\partial C_{2n}\partial z_2, .., \partial^2\bar{U}/\partial C_{2n}\partial z_n \end{vmatrix} \neq 0 \tag{28}$$

in the relevant domain of q, z and C.

To prove this, let us differentiate (27) with respect to time and form the partial derivative with respect to C_A of (23). Forming the difference between these two equations and using (27), we obtain the following equation:

$$(\partial^2\bar{U}/\partial C_A\partial q^1)(\dot{q}^1 - \partial H/\partial\bar{U}) + (\partial^2\bar{U}/\partial C_A\partial q^\alpha)(\dot{q}^\alpha - \partial H/\partial q^\alpha) +$$

$$+ (\partial^2\bar{U}/\partial C_A\partial z_\alpha)(\dot{z}_\alpha + \partial H/\partial q^\alpha - Q_\alpha) = 0$$

Since eq. (28) is satisfied by hypothesis, we find that all equations (21) are satisfied except: $\dot{p}_1 = -\partial H/\partial q^1 + Q_1$. However, this equation is satisfied by (26) and (23), and our statement is proved.

It is clear that the method presented is similar to the method of Hamilton-Jacobi i.e. the solution of the canonical system (21) is available from a complete solution of the basic eq. (23)by simple algebraic operations. At the same time, our method differs conceptually from the method of Hamilton-Jacobi since: a/ It can be applied to nonconservative systems, b/ The partial eq. (23) is quasi-linear and c/ The complete solution of (23) contains 2n-1undetermined constants which remain arbitrary through the whole proces of finding the motion. For more details the reader is refered to the ref. |10|. A slightly different approach to the same question is discussed in |11|.

To give a brief illustration, consider the motion of a heavy particle in a vertical plane with linear friction. If x and y denote the horizontal and vertical axes, the differential equations and initial conditions are

$$\dot{x}=p_1,\ \dot{y}=p_2=z,\ x(0)=0,\quad p_1(0)=v_0\cos\alpha$$
$$\dot{p}_1=-kp_1,\quad \dot{p}_2=\dot{z}=-kz-g,\ y(0)=0,\ z(0)=v_0\sin\alpha \qquad (29)$$

where k,g, and α are given constants.

Taking $p_1= U(t,x,y,z)$, we arrive to the basic equation

$$\partial U/\partial t+U\partial U/\partial x+z\partial U/\partial y-(kz+g)\partial U/\partial z+kU = 0 \qquad (30)$$

Taking $U=f_1(t)x+f_2(t)y+f_3(t)z+f_4(t)$ we find

$$U =\{kC_1x+\left[(C_2/k)+C_3e^{kt}\right]z+gtC_2/k+ge^{kt}C_3/k+C_4\}/(e^{kt}-C_1)$$

Applying the initial conditions and expressing C_4 in terms of C_1, C_2 and C_3 one has the conditioned momentum

$$\overline{U}=\left[kC_1x+C_2y+z(C_3e^{kt}+C_2/k)+gtC_2/k+ge^{kt}C_3/k+\right.$$
$$\left.+v_0\cos\alpha(1-C_1)+v_0\sin\alpha(C_3+C_2/k)-C_3g/k\right]/(e^{kt}-C_1) \qquad (31)$$

Let us form the equations

$$\partial\overline{U}/\partial C_1=0,\ \partial\overline{U}/\partial C_2=0,\ \partial\overline{U}/\partial C_3=0 \qquad (32)$$

However, after finishing the partial differentiation, we may select any particular value for C_1, C_2 and C_3, since (31) and (32) is generating the solution of (29) for the arbitrary values of C_A. Taking for example $C_1= C_2=C_3=0$ in (32),we find respectively

$$x=(v_0\cos\alpha/k)(1-e^{-kt}),\quad y+z/k=-(g/k)t+(v_0\sin\alpha)/k$$
$$z=-(g/k)+e^{-kt}(v_0\sin\alpha+g/k) \qquad (33)$$

For the same particular values of C_A we obtain from (31): $p_1=v_0\cos\alpha e^{-kt}$ which completes the motion of the system.

REFERENCES

|1| *E. Noether, Ges. Wiss, Göttingen, 2, 235 (1918)*
|2| *L.A. Pars, A. Treatise on Analytical Mechanics, Heinemann, London, (1968)*
|3| *A.I. Lure, Analytical Mechanics, "Mir", Moscow, (1961)*
|4| *E.L. Hill, Rev. Mod. Phys., 23, 253, (1951)*
|5| *B. Vujanović, Int. J. Non-Linear Mech. 13, 185 (1978)*
|6| *Dj.Djukić, B. Vujanović, Acta Mech. 23, 17, (1975)*
|7| *________, Archs. Ration.Mech. Analysis, 56,79, (1974)*
|8| *________, Acta Mech. 23, 291, (1975)*
|9| *T.M. Atanacković, Acta Mech., 38, 175, (1981)*
|10| *B. Vujanović, On the integration of the nonconservative Hamilton's dynamical equations. To be publ.Int.J.Eng.Sci. in (1981)*
|11| *__________, Tensor (N.S.), 33(1), 117, (1979)*

ON THE VALIDITY LIMITS OF HAMILTONIAN MECHANICS AND A WAY OF GOING BEYOND THEM.

Francis Fer

Ecole des Mines de Paris

One of the major problems - if not *the* problem - of present-day physics is the microscopic explanation of the irreversibility of macroscopic processes. Until now the mechanical basis of this explanation has been sought in Hamiltonian dynamics. Although all the past and recent attempts along these lines have failed, and that this failure is confirmed by many and prominent scientists[1], yet the general tendency is to continue this line of research.

I personally feel it advisable to look for other ways, and this article is intended to propose such a tentative solution. More precisely, I intend to show

- firstly, that Hamiltonian mechanics, particularly for large assemblies of particles, is a mechanics of stationary phenomena, i.e., a mechanics suited to macroscopic equilibria and, accordingly, incompatible with irreversibility;
- secondly that the explanation of irreversibility can be sought in the interactions between matter and radiation, and more precisely in the delay of propagation of electromagnetic potentials. For this purpose I shall offer a mathematical model based on this point of view and treated within a purely mechanical frame without any recourse to probabilities.

DYNAMICAL SYSTEMS
AND MICROPHYSICS

ISBN 0-12-068720-8

1. SOME PRELIMINARY REMARKS ON IRREVERSIBILITY AND ENTROPY.

First of all, it is absolutely necessary to clarify some questions of language and concepts which have always been a source of constant misunderstanding between the advocates of corpuscular mechanics and those of macroscopic thermodynamics; these concern the notions of irreversibility and entropy.

Let us start with the following indisputable statement: since corpuscular mechanics is intended to join thermodynamics as a firmly established science, it, therefore, follows that we must draw out basic definitions from the latter.

Now, in thermodynamics irreversibility is characterized by a precise and indefeasible criterion, which I shall recall here. To each *individual* system (whether isolated or not) is attached a number, its entropy S, axiomatically endowed with the two following properties: a) *it is a well-defined function of the macroscopic state*† of the system; b) a process is irreversible *if and only if* the increment of S contains *a strictly positive* term, the so-called entropy production d_iS, according to the well-known equation $dS = d_eS + d_iS$, $d_iS > 0$ [2].

For an isolated system, the only one which we shall deal with here, the external contribution d_eS is zero, and the general statement becomes $dS > 0$. The irreversibility of a process is then characterized by the fact that the entropy is a *constantly and strictly increasing function of the time* t.

Now, when thermodynamics is linked to an underlying corpuscular mechanics, what sort of mathematical object becomes entropy ? The answer is: S is a function of the representative point of the mechanical model, namely, when the dynamics employed is Hamiltonian, *S is a well-defined function $S(q,p)$ of the phase-point (or*

† This requires, of course, a previous definition of the macrostate, but thermodynamics has carefully given it (see e.g. [3]).

"micro-state")(q,p). Since this statement is often denied (see below), I shall stress the following argument: given any phase-point *(q,p)*, it determines a perfectly defined macroscopic state and, as *S* is itself a well-defined function of this latter, the conclusion follows. In short, *S is a well-defined function of the micro-state as well as of the macro-state, and irreversibility - i.e., the increasing of S - must be ascertained as well on the trajectory of the micro-state.*

I know that this requirement is not usually accepted. The classical position (1,4) rejects the representation of a system by a single phase-point, and clings to its representation by a statistical ensemble of micro-states. The discussion of this last point of view would need a too long development so that I shall leave it aside. I only want to point out that, *if we pretend to explain thermodynamics on a pure mechanical basis, the representation of a system by a phase-point is a logical obligation*; any other description must call upon non-mechanical extra-hypotheses†.

Just a final remark on these considerations. The word "irreversibility" is often used in a more extensive sense than the one defined above, namely as the "tendency", for a process, to evolve in a direction which cannot be spontaneously reversed. Such a definition is generally far too vague, but it becomes more precise when it refers to ergodicity (ergodicity per se , or mixing, or ergodicity of Markov processes). But even here the word irreversibility is misleading, because ergodicity does not at all imply that there exists a function of *the* micro-state which has a monotonous behaviour (5). The "entropy" (another deceiving term) that one now is accustomed to attach to ergodic processes concerns

† One can even say: conjectures, such as the assumption that "the behaviour of a single system is equivalent to the average behaviour of a suitably chosen ensemble of systems" (4). This is a statement which would need a proof.

an *ensemble* of micro-states and is in fact an entropy of information. Thus, whatever may be, in other respects, the interest of the property of ergodicity, it must not be confused with thermodynamic irreversibility.

2. HAMILTONIAN DYNAMICS IS NOT COMPATIBLE WITH IRREVERSIBILITY.

A.- Thus, if we bear in mind the rules prescribed by thermodynamics, we can see that *it is impossible to assign an entropy to a Hamiltonian system.* This truth is beginning to be accepted today, and is said to date back to Poincaré (1). But the proof offered by Poincaré, though mathematically valid, has no physical significance at all since it deals (6) with an assembly of molecules at total mechanical rest !

In fact, the basic reason for the impossibility stated above lies in the invariance of the measure during the motion. It was the basis for the proof I presented earlier (7). (See also (5) or (8)).

B.- The elementary but fundamental fact is the following one. Let us consider an isolated system of Hamiltonian $H(q,p)$, and let Σ be a surface of constant energy, $H(q,p) = E$, endowed with the usual measure μ. Let $f(q,p)$ be any integrable point-function defined over Σ, and $\Delta f(q,p)$ denote its increment from the point (q,p) to the point reached by the motion during some fixed time Δt. One then demonstrates that $\int_\Sigma \Delta f \, d\mu = 0$, which implies, either that $\Delta f = 0$ almost everywhere on Σ, or that if $\Delta f > 0$ on a subset of Σ of positive measure, then there exists another subset of positive measure on which $\Delta f < 0$. This statement is valid for any measure-preserving motion but it does not give us any information on the relative importance of these two subsets.

On the other hand, more can be said for a Hamiltonian of physical interest. Such a Hamiltonian is an *even* function of the

p's whether it depends on them only through the kinetic energy, or if it takes into account the internal electromagnetic *non-retarded* potentials as in the case of § 4 below. It then can be shown that if $f(q,p)$ is also an *even* function of the p's, the subset of Σ where $\Delta f < 0$ has the same measure as the subset $\Delta f > 0$ (whatever Δt may be).

Making (finite) time-averages on f does not modify these conclusions.

C.- Let us now return to entropy. As we have seen in § 1, the entropy S is a point-function on the surface Σ, for an isolated Hamiltonian system. In addition, thermodynamics asserts that S is a function of the macroscopic state variables. These are the external parameters (such as the volume for a gas), that are kept constant in the present case, and the temperatures, generally non-uniform on the system. If, as is universally done, one accepts that the temperature is proportional to the kinetic energy of agitation, it is readily seen that S is an even function of the p's. Thus the general result applies: either $\Delta S = 0$ all over Σ, and Σ is a surface of macroscopic equilibrium, or the subset of Σ for which $\Delta S < 0$ has the same measure as the subset for which $\Delta S > 0$. *This means, physically, that it is as often possible to build a system whose entropy decreases as the contrary.* It, therefore, is clear that Hamiltonian dynamics is incompatible with irreversibility as characterized by the entropy criterion.

3. A HAMILTONIAN SURFACE OF CONSTANT ENERGY REPRESENTS PRACTICALLY A MACROSCOPIC EQUILIBRIUM.

These facts indicate that, if Hamiltonian dynamics has a link with thermodynamics, it is only in the case where the subsets of Σ for which $\Delta S \neq 0$ have a null or negligible measure. We can verify this view on a particular but frequently used model.

Let us consider a gas composed of N identical particles, enclosed in a thermally insulated vessel D, and submitted to no external forces except those of the walls. The symbols (q,p), H, Σ, μ (this latter here normalized to 1) have the same significance as above. Remember that all the macroscopic states of the gas, whether they are of equilibrium or not, are represented by points of Σ and conversely.

We now ask the question: what is the measure of the part of Σ which represents a macroscopic equilibrium ? This latter term means that the density, as well as the temperature, is homogeneous and, consequently, the kinetic energy of agitation, and the macroscopic velocities are null.

It is immediately evident that the homogeneity just mentioned can be only defined within a certain approximation, determined by the accuracy of our possible measurements.

In order to handle the problem mathematically we adopt the model of "hard spheres", i.e., each particle is represented by a point of mass m and of coordinates xyz, $p_x\ p_y\ p_z$. The interactions reduce in instantaneous collisions and the potential $\Phi(q)$ of the walls is null inside D and gets an infinite or almost infinite value on the boundary.

Let us divide D into r cells D_j $(j=1$ to $r)$ of equal volume; r is supposed to be very large (say 10^{10} to give an order of magnitude). Given a point (q,p) of Σ, there corresponds a determined distribution of the particles in the cells, say N_j in D_j (with $\sum_{j=1}^{r} N_j = N$), and likewise of the energies, say E_j in D_j , with $\sum_{j=1}^{r} E_j = E$ fixed.

Based on the previous assumptions it is easy to calculate the measure $d\mu$ of the subset of Σ for which the gas has a given discrete distribution in densities N_j and a given continuous distribution in energies $(E_j, E_j + dE_j)$. This measure is

$$d\mu = \frac{N!}{r^N} \prod_j \left(\frac{g_j(E_j)}{N_j!}\right) \frac{dE_1 \cdots dE_{r-1}}{g_N(E)} \tag{1}$$

where $g_\nu(u)$ denotes the measure of the 3ν-dimensional sphere $\sum_{\alpha=1}^{\nu} \frac{1}{2m} (p^2_{\alpha x} + p^2_{\alpha y} + p^2_{\alpha z}) = u$.

This formula allows us to calculate the mean values and centered momenta of the N_j's and E_j's. For the mean values one finds as might be expected

$$\overline{N_j} = \frac{N}{r} = n \; ; \quad \overline{E_j} = \frac{E}{r} = \varepsilon \quad \text{for all } j \tag{2}$$

n is also very large ($\sim 10^{15}$) for usual macroscopic gases. As for the momenta, they can be calculated without too much effort up to† the order 6 for the N_j's and the order 4 for the E_j's, which yields

$$\overline{(N_j - n)^6} \simeq 15\, n^3 \quad ; \quad \overline{(E_j - \varepsilon)^4} \simeq \frac{25}{3} \frac{\varepsilon^4}{n^2} \tag{3}$$

the symbol $\simeq$ meaning that the neglected terms are of order $1/r$ or $1/n$ in relative value.

By utilizing the inequality of Bienaymé-Tchebychev, these values of the momenta yields a lower bound of the measure of the part of Σ for which $|N_j - n|$ or $|E_j - \varepsilon|$ remains smaller than a given value for *all* the cells D_j. Consider first the density and then define an *"homogeneity tolerance"* τ as the maximum ratio admitted for $|N_j - n|/n$ *for all* j, and a *"non-equilibrium area"* δ as the measure of the part of Σ for which $|N_j - n|/n > \tau$ for *even a single cell* D_j. One finds, from the first equation (3), $\tau^6 \delta = 15\, N/n^4$.

Let us now calculate the orders of magnitude. For $N = 10^{23}$

† By increasing the order of the momenta, one obtains closer relationships between the quantities τ and δ defined below.

$n = 10^{16}$ (a cube of 0.7 mm in side, or 0.5 μg for air - thus at the boundary of macroscopic measurements), one obtains $\delta = 10^{-4}$ for $\tau = 10^{-6}$.

Let us now define for the energies (total energy : kinetic macroscopic energy plus agitation) a similar ratio τ as $sup|E_j - \varepsilon| / \varepsilon$ for all j, and δ as above. One finds, from the last equation (3) $\tau^4\delta = 25\, N/3n^3$, which yields, for $N = 10^{23}$ and $n = 10^{16}$, $\delta = 10^{-4}$ for $\tau = 10^{-5}$.

Lastly, let us examine the question of macroscopic velocities. They are, of course, the averages of the velocities in each cell D_j . It is obviously found that the mean i.e., the macroscopic velocity is zero. Thus, if k_j denotes the macroscopic kinetic energy contained in D_j, through some calculation and the aid of (3) it is found that the average of $(k_j)^c$ is for small c (say 1 to 10), $(k_j)^c = (E/N)^c = (\varepsilon/n)^c$. If we now define a tolerance ratio τ as $sup(k_j/\varepsilon)$ for all j and δ as above, the inequality of Bienaymé-Tchebychev yields $\tau^c\delta = N/n^{c+1}$. With $N = 10^{23}$, $n = 10^{16}$ $c = 6$, one has $\delta = 10^{-4}$ for $\tau = 6.8\ 10^{-15}$. The macroscopic kinetic energy is then very small compared with the total mean energy, and it is easily seen that, for a gas under ordinary conditions, the last figure corresponds to a macroscopic velocity of 42 μ/s.

Summing up, it follows from these numbers that, for the model of hard spheres, *the largest part of Σ, 99.99 per cent or more, is occupied by macroscopic states which we are not able to distinguish from the perfect equilibrium state* ($N_j = n$, $E_j = \varepsilon$) *with our experimental apparatus.*

One might be tempted to infer that these results are the very proof that Hamiltonian dynamics describes irreversibility, by arguing that a point (q,p) placed on the non-equilibrium part of Σ has a probability nearly equal to 1 to jump in the equilibrium region and to stay in it. *But this is a purely probabilistic view, not a dynamical one*, and, if we cling to this latter, *as logically compelled*, we must admit that the jump towards equilibrium is neither more nor less easy than the opposite process, as we have seen in § 2.

The only reasonable conclusion which can be drawn is that a Hamiltonian surface of constant energy is, save for a small error, a surface of macroscopic equilibrium and that Hamiltonian dynamics represents only stationary processes. If in this domain it has proved very successful, beyond any doubt, we have, for describing irreversible processes, to look for other sort of mechanics.

4. IRREVERSIBILITY AS A CONSEQUENCE OF THE DELAY OF PROPAGATION OF THE ELECTROMAGNETIC FIELD.

For the sake of clarity let us begin with a discrete model: a system (of bounded extension) composed of atoms or molecules which are themselves considered as assemblies of *charged point-masses*. Let us denote by m_α the mass, q_α the charge, $\vec{x}_\alpha$ the position and $\vec{v}_\alpha$ the velocity of the point of index α $(\alpha = 1$ to $N)$. We suppose that the interactions between these points are: firstly the electromagnetic forces they exert one over another, each point exerting no action on itself; secondly (just in case) forces which derive from a potential $G(\vec{x})$ (depending only on the $\vec{x}_\alpha$). We can include in $G(\vec{x})$ the external forces, provided they derive from a potential and that their field is constant in time, which we suppose. Thus the system behaves as if isolated.

Let us consider the - *fictitious* - case when the e.m. potentials are propagated with an *infinite* velocity. It is then easily seen that *the motion of the system is Hamiltonian.* Indeed, we must take for the fictitious e.m. potentials $\tilde{\phi}_\alpha$ and $\tilde{A}_\alpha$ exerted on the point α by all the others the quantities

$\widetilde{\phi}_\alpha = \frac{1}{4\pi\varepsilon_0} \times \sum_{\alpha \neq \beta} \frac{q_\beta}{r_{\alpha\beta}}$, $\widetilde{A}_\alpha = \frac{\mu_0}{4\pi} \times \sum_{\beta \neq \alpha} \frac{q_\beta \vec{v}_\beta}{r_{\alpha\beta}}$ $(r_{\alpha\beta} = |\vec{x}_\beta - \vec{x}_\alpha|)$; the fictitious fields are as usual defined by $\widetilde{E}_\alpha = -\nabla\widetilde{\phi}_\alpha - (\partial\widetilde{A}_\alpha/\partial t)$, $\widetilde{B}_\alpha = \nabla \times \widetilde{A}_\alpha$. Introduce the energies $\Phi = \frac{1}{2}\Sigma_\alpha q_\alpha \widetilde{\phi}_\alpha$, $\mathrm{A} = \frac{1}{2}\Sigma_\alpha q_\alpha \vec{v}_\alpha \cdot \widetilde{A}_\alpha$, and the Lagrangian $L(x_\alpha, \vec{v}_\alpha) = K - \Phi + \mathrm{A} - G$, K being the kinetic energy $\frac{1}{2}\Sigma_\alpha m_\alpha \vec{v}_\alpha^2$. It is then readily seen that the equation of the motion $m_\alpha(d\vec{v}_\alpha/dt) = q_\alpha(\widetilde{E}_\alpha + \vec{v}_\alpha \times \widetilde{B}_\alpha)$ derive from this Lagrangian. In the usual manner we deduce from L the Hamiltonian, which is found to be

$$H = K + \Phi + \mathrm{A} + G \tag{4}$$

H is a constant of the motion.

But in reality the potentials emitted by the charged points travel with the velocity c, and the exact potentials are Liénard-Wiechert ones instead of $\widetilde{\phi}_\alpha$, $\widetilde{A}_\alpha$ written above. The real fields, which we shall denote by $\vec{E}_\alpha$, $\vec{B}_\alpha$, differ from the fictitious ones $\widetilde{E}_\alpha$, $\widetilde{B}_\alpha$. Consequently, in the real motion, H is no longer a constant, and it can be seen from its form (4) and from the true equations of the motion $m_\alpha(d\vec{v}_\alpha/dt) = q_\alpha(\vec{E}_\alpha + \vec{v}_\alpha \times \vec{B}_\alpha)$ that

$$\frac{dH}{dt} = \Sigma_\alpha q_\alpha \vec{v}_\alpha \cdot (\vec{E}_\alpha - \widetilde{E}_\alpha) \tag{5}$$

The right member is of order $1/c$ and our problem is to calculate it.

Here we must complicate the model, because point-masses are a doubtful representation of physical bodies and lead to infinite energies as well as to intricate calculations. It is preferable to substitute a continuous model for the discrete one by replacing each point α by a "fluid" of mass-density $\mu_\alpha(x,t)$ and charge-density $\rho_\alpha(x,t)$, these functions fulfilling the conditions

$\frac{\rho_\alpha}{\mu_\alpha}(x,t) = \frac{q_\alpha}{m_\alpha}$ (a constant which depends upon α but not on x,t); there are as many distinct fluids as we had points previously.

It is obvious to transpose the above equations in this new formalism, but it can be asked whether this transposition is mathematically justified. In fact, it is and I could have begun directly with the fluid model, were it not for the length of this paper. I shall only stress briefly here the main necessary hypotheses.

Several sorts of fluids are allowed, particularly the "quantic fluid" I proposed formerly (9), provided they take into account internal stresses and cohesive forces, and that these latter derive from potentials. The corresponding energy will then replace the term G of the discrete model. The main point is that G have *a lower bound*, which means physically that the fluid representing a particle cannot be reduced to a point, and which is verified for the "quantic fluid" (10). Let us add that, in the case of a gas, one can include in the system the next molecules of the container, provided there exists a boundary to which the velocities of the fluids are tangent.

These assumptions being made, let us set, at each space-time point, $\rho = \Sigma_\alpha \rho_\alpha$, $\vec{j} = \Sigma_\alpha \rho_\alpha \vec{v}_\alpha$. The fictitious potentials of the discrete model thus become

$$\tilde{\phi}(x,t) = \frac{1}{4\pi\varepsilon_0}\int_D \frac{\rho(a,t)}{r}\,da, \quad \tilde{A}(x,t) = \frac{\mu_0}{4\pi}\int_D \frac{\vec{j}(a,t)}{r}\,da \qquad (6)$$

D being the domain of the charges, with $r = |\vec{ax}|$; the Hamiltonian (4) is transposed into

$$H = K + \frac{1}{2}\int_D (\rho\tilde{\phi} + \vec{j}\cdot\tilde{A})\,dx + G$$

$$= K + \frac{1}{2}\int \left(\varepsilon_0 \nabla\tilde{\phi}^2 + \frac{\varepsilon_0}{c^2}\left(\frac{\partial\tilde{\phi}}{\partial t}\right)^2 + \frac{\tilde{B}^2}{\mu_0}\right) dx + G$$

the last integral being extended to the whole space. Clearly

$H - G \geqq 0$ and, since *G has a lower bound, H has also a lower bound.*

Instead of (6), the real potentials ϕ, $\vec{A}$ are now given by

$$\phi(x,t) = \frac{1}{4\pi\varepsilon_0} \int_D \frac{\rho(a, t - r/c)}{r} \, da \qquad (7)$$

$$\vec{A}(x,t) = \frac{\mu_0}{4\pi} \int_D \frac{\vec{j}(a, t - r/c)}{r} \, da$$

On the other hand it is easily shown that the real motion, *which is no longer Hamiltonian* (it is ruled by an hereditary mechanics) satisfies the conservation of energy, namely

$$K + \frac{1}{2} \int (\varepsilon_0 \vec{E}^2 + \frac{\vec{B}^2}{\mu_0}) \, dx + G = E = Ct \qquad (8)$$

from which it results that *K has an upper bound* (since $-G$ has one).

We can now expand the potentials (7) with respect to the parameter $1/c$, which yields a feasible calculation of the right member of (5) transposed in the continuous form $\int_D \vec{j} \cdot (\vec{E} - \tilde{E}) dx$.

Let us begin by the only first term in $1/c$. The result is the following. Set

$$\vec{J} = \int_D \vec{j} \, dx \qquad (9)$$

and

$$H' = H - \frac{1}{4\pi\varepsilon_0 c^3} \frac{d}{dt} \left(\frac{J^2}{2} \right) \qquad (10)$$

We obtain easily

$$\frac{dH'}{dt} = - \frac{1}{4\pi\varepsilon_0 c^3} \left(\frac{d\vec{J}}{dt} \right)^2$$

H', therefore, is constantly decreasing. *I assert it tends towards a finite limit.*

Consider in effect one of the components of the system, an electron for example. Its total current is $\int_D \rho_e \vec{v}_e \, dx$; from Cauchy - Schwartz's inequality we see that

$$\left| \int_D \rho_e \vec{v}_e \, dx \right|^2 \leqq \int_D |\rho_e| dx \int_D |\rho_e| \, \vec{v}_e^{\,2} \, dx = (e^2/m_e) \int_D \mu_e \vec{v}_e^{\,2} \, dx \, .$$

From similar inequalities for all the components, electrons or nuclei, we deduce $\vec{J}^2 \leqq \lambda K$, λ being a suitable constant, and as K is bounded, so is J^2. Hence H', although constantly decreasing, cannot tend towards $-\infty$ since, from (10) and the fact that H has a lower bound, $\frac{d}{dt}\left(\frac{J^2}{2}\right)$ would tend towards $+\infty$ and likewise J^2. Then H' tends towards a finite limit as $t \to +\infty$, as stated above.

As a result, $\frac{dH'}{dt} \to 0$ save on intervals of time whose total duration tends itself towards zero and, therefore, from (11), $\frac{d\vec{J}}{dt}$ does the same; hence, because of (10), *H tends "in measure" towards the limit of H'* (if $d\vec{J}/dt$) were uniformly continuous in $(0, +\infty)$ the result would be more precise: $\frac{d\vec{J}}{dt} \to 0$ and $H \to \lim H'$

The physical interpretation is quite clear. It is easily seen that $\frac{d\vec{J}}{dt} \simeq \sum_\alpha q_\alpha \vec{\gamma}_\alpha$ ($\vec{\gamma}$: acceleration). Thus, $\left(\frac{d\vec{J}}{dt}\right)^2 = \sum_\alpha q_\alpha^2 \vec{\gamma}_\alpha^{\,2} + \sum_{\alpha, \beta \neq \alpha} q_\alpha q_\beta \vec{\gamma}_\alpha \cdot \vec{\gamma}_\beta$. The first term is the well-known Lorentz term, the loss of energy by radiation during a collision; the second term, less known, is the recovery of energy by the particles from this field of radiation. As $t \to \infty$ the two terms balance, and the system reaches a stationary regime characterized by the fact that the fictitious Hamiltonian *H is constant* both in the fictitious motion and in the real one.

We thus have the example of a process which, *in a purely mechanistic and deterministic manner, tends asymptotically towards a stationary regime and whose convergence towards the limit is monotonous*. Of course, we cannot speak of an entropy since the temperature is lacking and H' has the dimension of an energy. Nevertheless, we have found some characteristic features of irreversibility.

Now, this result has been obtained by an expansion limited to the first order. If it is continued up to the second order, it can be seen that, while the equation (10) is modified by the addition to the right member of extra-terms, the new expression of H' still obeys the equation (11). An *estimate* of the extra-terms shows that they are very small in comparison with the main terms of H, and on the other hand that they are fluctuating. It is only an estimate but, if confirmed by more refined treatment, the convergence of H' towards a limit would remain valid. In this case, the fictitious Hamiltonian H would fluctuate around a mean value. The physical interpretation of such a behaviour would be clear: it would represent a fluctuating exchange of energy between particles and their E.M. field at the stationary regime.

While results obtained by means of a limited expansion are not mathematically fully convincing, they hopefully provide a guide towards full proof.

REFERENCES

(1) Prigogine, I., *Physique, temps et devenir*. Masson, Paris, 1980.

(2) Groot, S.R. de, and Mazur, P., *Non-equilibrium Thermodynamics*. North-Holland, Amsterdam, 1963.

(3) Duhem, P., *Traité d'énergétique*. Gauthier-Villars, Paris, 1911.

[4] Tolman, R.C., *The principles of statistical Mechanics.* Oxford Univ. Press, 1962.

[5] Fer, F., *L'irréversibilité, fondement de la stabilité du monde physique.* Gauthier-Villars, Paris, 1977.

[6] Poincaré, H., C.R. Acad. Sc. Paris, 108, 550, 1889.

[7] Fer, F., C.R. Acad. Sc. Paris, 260, 3973, 1965; 260, 4159, 1965; 265, 205, 1967; 265, 289, 1967.

[8] Fer, F., in *Dynamical systems and Microphysics.* C.I.S.M. Courses and Lectures, No 261, Springer-Verlag, Wien, 1979.

[9] Fer, F., C.R. Acad. Sc. Paris, 258, 2983, 3215 and 3435, 1964; 262, 1417, 1966; 263, 103, 1966.

[10] Fargue, D., private communication; see also D. Fargue and F. Fer, C.R. Acad. Sc. Paris, série A, 288, 323, 1979.

PART IV

PERTURBATIONS

ADIABATICAL INVARIANTS IN LINEAR DYNAMICAL SYSTEMS PERIODICALLY DEPENDING ON TIME, WITH AN APPLICATION TO THE STATISTICAL FLUCTUATIONS OF MATHIEU OSCILLATORS

Georges LOCHAK

C.N.R.S. and Fondation Louis de Broglie, Paris, France

José VASSALO PEREIRA

Facultate de Ciêncas de Lisboa, Portugal

I. INTRODUCTION

The fundamental role of adiabatical invariance in quantum mechanics has been mentioned in the two authors' communications at the Ist International Seminar on "Mathematical Theory of Dynamical Systems and Microphysics" (Udine, 1979). This role consists mainly in the fact that a statistical assembly of quantum systems will consist of a mixture of adiabatically invariant states (Lochak G., 1973, 1976, 1981 ; Lochak G. & Alaoui A., 1975). In autonomous systems (i.e. not depending on time), these states are the stationary states, in non autonomous systems periodically depending on time, stationary states non longer exist but adiabatically invariant states can still be defined by using the Floquet theorem : these are the so called *permanent* (or *steady*) states (Young & Deal W.J., Lochak G. 1972, Lochak P.). It is then possible to elaborate a statistical mechanics for assemblies of such non autonomous systems which was successfully used in NMR (Lochak G. & Alaoui A.).

The present communication as well as the subsequent one will be devoted to the problem of adiabatical invariants in *classical dynamical systems* with a perturbation periodic in time. The authors (Lochak G. & Vassalo Pereira J. 1976) have already constructed the

DYNAMICAL SYSTEMS
AND MICROPHYSICS

ISBN 0-12-068720-8

adiabatical invariant of a Mathieu oscillator, but we shall give here an explicit formulation of it and shall make use of it in a theory of *fluctuations* of a Mathieu oscillator immersed in a heat bath.

2. THE INTEGRATION OF THE MATHIEU OSCILLATOR

Let us remind you first that such an oscillator is an absolutely concrete physical being. For instance, in an RLC circuit, the oscillation of an electrical charge q is expressed by :

$$L \frac{d^2q}{dt^2} + R \frac{dq}{dt} + \frac{1}{C} q = 0 \tag{2.1}$$

Let us suppose now that C contains a conveniently oriented cristal of quartz oscillating with a frequency ω. Hence :

$$C = C_0(1 + a \cos\omega t). \tag{2.2}$$

Let us suppose that $R \ll L$ and $a \ll 1$ and let be :

$$\omega_0 = 1/\sqrt{LC_0}\ , \qquad \varepsilon = a\,\omega_0. \tag{2.3}$$

The equation (2.1) is then approximated by the *Mathieu oscillator* :

$$\ddot{q} + \omega_0^2(1 - \frac{2}{\omega_0} \cos\omega t)q = 0. \tag{2.4}$$

With the transformation :

$$q = \xi\, e^{i\omega_0 t} + \eta\, e^{-i\omega_0 t}\ ;\ \dot{q} = i\omega_0(\xi e^{i\omega_0 t} - \eta\, e^{-i\omega_0 t}) \tag{2.5}$$

in the vicinity of the parametrical resonance $\omega = 2\omega_0$, if we neglect the "rapidly oscillating terms", we find from (2.5) :

$$\dot{\xi} = -\, i\, \frac{\varepsilon}{4}\, e^{i(\omega - 2\omega_0)t}\eta\ ;\ \dot{\eta} = i\, \frac{\varepsilon}{4}\, e^{-i(\omega - 2\omega_0)t}\xi \tag{2.6}$$

and we have the solution of (2.6) in the form :

$$\xi = \left[\frac{\Delta - \delta}{2\Delta}\, \xi_0 - \frac{\varepsilon}{8\Delta}\, \eta_0\right] e^{i(\delta + \Delta)t} + \left[\frac{\Delta + \delta}{2\Delta} \xi_0 + \frac{\varepsilon}{8\Delta}\, \eta_0\right] e^{i(\delta - \Delta)t}$$

$$\eta = \left[\frac{\varepsilon}{8\Delta}\, \xi_0 + \frac{\Delta + \delta}{2\Delta}\, \eta_0\right] e^{-i(\delta - \Delta)t} + \left[\frac{-\varepsilon}{8\Delta}\, \xi_0 + \frac{\Delta - \delta}{2\Delta}\, \eta_0\right] e^{-i(\delta + \Delta)t} \tag{2.7}$$

$$\delta = \frac{\omega}{2} - \omega_0\ ;\ \delta^2 > \frac{\varepsilon^2}{16}\ ;\ \Delta = \sqrt{\delta^2 - \frac{\varepsilon^2}{16}}\ . \tag{2.8}$$

We shall suppose that Δ is REAL, which means that ω is, as already said, in the vicinity of the resonance, but OUT of the zone itself. From (2.5) and (2.7), we get the solution of (2.4) in the

Floquet form :

$$\begin{pmatrix} q \\ \dot{q} \end{pmatrix} = T(t)\, e^{iRt} \begin{pmatrix} q_0 \\ \dot{q}_0 \end{pmatrix} \tag{2.9}$$

$$T(t) = \begin{pmatrix} \cos\frac{\omega t}{2} & \frac{1}{\omega_0}\sin\frac{\omega}{2}t \\ -\omega_0 \sin\frac{\omega}{2}t & \cos\frac{\omega}{2}t \end{pmatrix} \begin{pmatrix} \cos n\frac{\omega}{2}t & \frac{\delta - \frac{\varepsilon}{4}}{\omega_0 \Delta}\sin n\frac{\omega}{2}t \\ -\omega_0 \frac{\delta + \frac{\varepsilon}{4}}{\Delta}\sin n\frac{\omega}{2}t & \cos n\frac{\omega}{2}t \end{pmatrix} \tag{2.10}$$

$$R = \begin{pmatrix} 0 & -\frac{\delta - \frac{\varepsilon}{4}}{i\omega_0 \Delta}(\Delta + n\frac{\omega}{2}) \\ -i\omega_0 \frac{\delta + \frac{\varepsilon}{4}}{\Delta}(\Delta + n\frac{\omega}{2}) & 0 \end{pmatrix} \tag{2.11}$$

REMARKS : 1. In these formulae $n \in Z$ because the decomposition of Floquet is not unique, but n will be later fixed by physical conditions. 2. We see that R is a constant matrix. 3. T(t) is periodic but with period $4\pi/\omega$ instead of $2\pi/\omega$: the reason is simply that the equation (2.4) is real.

3. THE TRANSFORMATION OF FLOQUET-LIAPOUNOV AND THE REDUCED EQUATION OF THE MATHIEU OSCILLATOR

Let us write (2.4) in the canonical form :

$$\begin{cases} \dot{q} = \frac{\partial \mathcal{H}}{\partial p} = p \\ \dot{p} = \frac{\partial \mathcal{H}}{\partial q} = -\omega_0^2(1 - \frac{\varepsilon}{\omega_0}\cos\omega t)q \end{cases} ; \quad \mathcal{H} = \frac{1}{2}\left(p^2 + \omega_0^2(1 - \frac{\varepsilon}{\omega_0}\cos\omega t)q^2\right) ; \tag{3.1}$$

and let us introduce the *reduced representation* Q, P by :

$$\begin{pmatrix} q \\ p \end{pmatrix} = T(t) \begin{pmatrix} Q \\ P \end{pmatrix} , \tag{3.2}$$

which is the transformation of Floquet-Liapounov ; from (2.8) and (2.10), it is *canonical* because det T(t) = 1 and we have the *reduced equation* which is autonomous and canonical :

$$\frac{d}{dt}\begin{pmatrix} Q \\ P \end{pmatrix} = iP \begin{pmatrix} Q \\ P \end{pmatrix} = IL \begin{pmatrix} Q \\ P \end{pmatrix} , \tag{3.3}$$

where I is a symplectic matrix ; L = -iIR and, from (2.11) :

$$I = \begin{pmatrix} 0 & -1 \\ 1 & 0 \end{pmatrix} \quad ; \quad L = -\begin{pmatrix} \Omega_1 & 0 \\ 0 & \Omega_2 \end{pmatrix} \quad ; \tag{3.4}$$

$$\Omega_1 = -\frac{\Delta + n\frac{\omega}{2}}{\Delta}\,\omega_0(\delta + \frac{\varepsilon}{4}) \ ; \ \Omega_2 = -\frac{\Delta + n\frac{\omega}{2}}{\Delta}\,\frac{1}{\omega_0}(\delta - \frac{\varepsilon}{4}). \qquad (3.5)$$

The Hamilton function of (3.3) is the *reduced energy*. We have

$$\mathcal{R}(Q,\ P) = -\frac{1}{2}(QP)L\begin{pmatrix}Q\\P\end{pmatrix} = \frac{1}{2}(\Omega_1\ Q^2 + \Omega_2\ P^2). \qquad (3.6)$$

It is obviously a first integral of the system (3.3). But it is also a first integral of (3.1) in the representation q,p :

$$\mathcal{R}(q,p,t) = -\frac{1}{2}(qp)\tilde{T}^{-1}LT^{-1}\begin{pmatrix}q\\p\end{pmatrix} \qquad \text{or :} \qquad (3.7)$$

$$\mathcal{R}(q,p,t) = -\frac{\Delta + n\frac{\omega}{2}}{2\Delta}\Big[\omega_0(\delta + \frac{\varepsilon}{4}\cos\omega t)q^2 - \frac{\varepsilon}{2}\sin\omega t\ qp$$
$$+ \frac{1}{\omega_0}(\delta - \frac{\varepsilon}{4}\cos\omega t)p^2\Big]. \qquad (3.8)$$

Now, let us turn to physics !

4. ADIABATICAL INVARIANCE AND STATISTICAL MECHANICS

If we had a *mixture* of conservative systems (3.3), its equilibrium entropy would be S = k logV, with V = phase volume for a given energy of the system (3.3), so that :

$$V = \oint P\ d\ Q = 2\pi\ J \qquad (4.1)$$

where J is (according a fundamental property of action variable) the *adiabatical invariant* (*) of (3.3) : it is precisely this invariance which makes the preceding expression of S conform with the second principle of thermodynamics (Lochak G., 1973, 1976, 1981). The energy of (3.3), i.e. the reduced energy $\mathcal{R}(Q,P)$, is a linear function of J which is easy to write by putting the solution of (3.3) in the following form, owing to (3.4) and (3.5) :

$$Q = \sqrt{2J\sqrt{\frac{\Omega_2}{\Omega_1}}}\cos\sqrt{\Omega_1\Omega_2}(t - t_0) \ ; \ P = \sqrt{2J\sqrt{\frac{\Omega_1}{\Omega_2}}}\sin\sqrt{\Omega_1\Omega_2}(t - t_0), \qquad (4.2)$$

where appears the angular frequency Ω corresponding to the action

(*) *Let us briefly remind you that this means that* J, *which is a constant of motion of (3.3), remains a constant when a parameter of the system (as* ω_0 *or* ε*) varies infinitely slowly, in other words when we exchange work reversibly with the system.*

variable J, with :

$$\Omega = \sqrt{\Omega_1 \Omega_2} = \left| \Delta + n \frac{\omega}{2} \right| , \tag{4.3}$$

and we have, according to (3.4), (3.5) and (3.6) :

$$\mathcal{R}(Q,P) = J\,\Omega \ ; \ J = \frac{1}{2}\left(\sqrt{\frac{\delta + \frac{\varepsilon}{4}}{\delta - \frac{\varepsilon}{4}}}\, Q^2 + \sqrt{\frac{\delta - \frac{\varepsilon}{4}}{\delta + \frac{\varepsilon}{4}}}\, P^2 \right) . \tag{4.4}$$

Now, let us consider, just for a moment, (3.3) as a given physical conservative hamiltonian system and let us suppose it is immerged in a heat bath. It will be a *canonical system* (in the sense of Gibbs) with a probability of realization of a state given by :

$$d\mathcal{P} = Z^{-1}\, e^{-\frac{\mathcal{R}(Q,P)}{kT}}\, dQdP \ ; \ Z = \int e^{-\frac{\mathcal{R}(Q,P)}{kT}}\, dQ\, dP \ . \tag{4.5}$$

If we had imagined the same statistical problem for a Mathieu oscillator in the initial coordinates q,p, the difficulty would reside in the non conservation of energy, but we have another constant of motion : the *reduced energy* and, perhaps, can we make use of it ? More precisely :

QUESTION : is it possible to consider a Mathieu oscillator in a heat bath as a canonical system in a *modified sense* of Gibbs, with a probability of state $\{q, q + dq \ ; \ p, p + dp\}$ given by (4.5) transformed by (3.2) ? Do we have ? :

$$d\mathcal{P}(q,p,t) = Z^{-1}\, e^{-\frac{\mathcal{R}(q,p,t)}{kT}}\, dq\, dp \ ; \ Z = \int e^{-\frac{\mathcal{R}(q,p,t)}{kT}}\, dq\, dp. \tag{4.6}$$

ANSWER : We shall postulate that we do, with three arguments :

1) The distribution (4.6) is *stationary* because $\mathcal{R}(q,p,t)$ is a constant of motion.

2) The microcanonical distribution corresponding to (4.6) will have the entropy S = k log V with a phase volume V which will be now limited by *a surface of constant reduced energy*, i.e. $\mathcal{R}$ = Const. ; it will be equal to (4.1) with J given by (4.4) in variables Q,P, but we must transform this expression by (3.2), in

a new expression in variables q,p,t ; one finds :

$$J(q,p,t) = \frac{1}{\Delta}\left[\omega_0(\delta + \frac{\varepsilon}{4}\cos\omega t)q^2 - \frac{\varepsilon}{2}\sin\omega t\; qp + \frac{1}{\omega_0}(\delta - \frac{\varepsilon}{4}\cos\omega t)p^2\right]. \quad (4.7)$$

Now, in a preceding paper, we have proved that T(t) in the transformation (3.2), is not only canonical, but in addition, it conserves the adiabatical invariance and so J(q,p,t) is an *adiabatical invariant* for the system (3.1) (Lochak G. & Vassalo Pereira J., 1976 ; a more general proof will be given in this same book in the next communication). Therefore, our statistical distribution will be in accordance with the *2nd Principle* of thermodynamics.

3) There remains to show the *unicity* of the distribution (4.6) which is not evident because there is, a priori, an infinity of transformations T(t) that reduce the system (3.1) to a conservative one. But only the Floquet-Liapounov transformation is adiabatically invariant and so, there remains only an arbitrary number $n \in Z$ which does not appear in J, but only in $\mathcal{R}$ and Ω, in (3.8) and (4.3). We shall now univocally define n by a *linking postulate* : when $\varepsilon \to 0$, the oscillator of Mathieu (3.1) tends to the harmonic oscillator and we shall require that, outside the zone of parametric resonance, *the reduced energy shall tend to the energy of the harmonic oscillator*, i.e. :

$$\lim_{\varepsilon\to 0} \mathcal{R}(q,p,t) = \frac{1}{2}(p^2 + \omega_0^2 q^2) \qquad (\omega \neq 2\omega_0). \quad (4.8)$$

It is easy to find that this implies :

$$\left.\begin{array}{lll} \text{When } \delta > \frac{\varepsilon}{4} & (\omega \text{ on the right of the resonance}) & : n = -1. \\ \text{When } \delta < \frac{\varepsilon}{4} & (\omega \text{ on the left of the resonance}) & : n = +1. \end{array}\right\} \quad (4.9)$$

So, instead of (4.3) and (4.4), we may write the two *positive* expressions :

$$\Omega = n\Delta + \frac{\omega}{2} \;;\; \mathcal{R} = (n\Delta + \frac{\omega}{2})J(q,p,t) \;;\quad n = \pm 1 \;; \quad (4.10)$$

and hence, we will easily find in (4.5) the sum of states :

$$Z = \frac{2\pi}{\Omega} kT = \frac{2\pi}{n\Delta + \frac{\omega}{2}} kT \ ; \quad n = \pm 1. \tag{4.11}$$

These ideas have already been developed in quantum mechanics (Lochak G. et al.) and have been applied in NMR : it has been possible to give a general explanation of the well known hypothesis of Redfield on the spin temperature in the rotating frame.

5. THE STATISTICAL FLUCTUATIONS OF THE MATHIEU OSCILLATOR

Owing to the formulae (4.6), it is possible to calculate the statistical mean value of any dynamical quantity G related to the Mathieu oscillator :

$$\langle G(q,p,t)\rangle = Z^{-1} \int G(q,p,t)\, e^{-\frac{\mathcal{R}(q,p,t)}{kT}}\, dq\, dp. \tag{5.1}$$

But if this statistical mean-value depends on time, we shall take its *time average* $\langle\overline{G}\rangle$. We shall calculate the quadratic mean values of q and p, which is easy because we may write immediately from (4.5) :

$$\langle Q^2\rangle = \frac{kT}{\omega_0(n\Delta + \frac{\omega}{2})}\sqrt{\frac{\delta - \frac{\varepsilon}{4}}{\delta + \frac{\varepsilon}{4}}} \ ; \quad \langle P^2\rangle = \frac{\omega_0 kT}{n\Delta + \frac{\omega}{2}}\sqrt{\frac{\delta + \frac{\varepsilon}{4}}{\delta - \frac{\varepsilon}{4}}} \ ; \tag{5.2}$$

and then, we will get from (3.2) the expressions in terms of q,p which give, by time averaging, the following general formulae which are *always* positive for $n = \pm 1$:

$$\langle\overline{q^2}\rangle = \frac{-kT\ \delta}{\omega_0\Delta(n\Delta + \frac{\omega}{2})} \ ; \quad \langle\overline{p^2}\rangle = \frac{-kT\ \omega_0\ \delta}{\Delta(n\Delta + \frac{\omega}{2})} . \tag{5.3}$$

First we can verify, from (2.7), that *far from the resonance*, on the right as on the left, we have $\Delta \to -n\delta$ and hence, we get the well known result given for the harmonic oscillator :

$$\langle\overline{q^2}\rangle \to \frac{kT}{\omega_0^2} \ ; \quad \langle\overline{p^2}\rangle \to kT \tag{5.4}$$

But the interesting result is obtained *near the parametrical resonance*. If we note η the resonance difference, we have :

$$\delta = \frac{\omega}{2} - \omega_0 = -n(\frac{\varepsilon}{4} + \eta) \quad ; \quad n = \pm 1. \tag{5.5}$$

When η is very small ($\sim\varepsilon^2$), we shall have :

$$\Delta = \sqrt{\delta^2 - \frac{\varepsilon^2}{16}} = \sqrt{\left(\frac{\varepsilon}{4} + \eta\right)^2 - \frac{\varepsilon^2}{16}} \sim \sqrt{\frac{\varepsilon\eta}{2}} \tag{5.6}$$

and introducing this value in (5.3), we find :

$$\overline{\langle q^2 \rangle} = \frac{kT}{\omega_0^2}\sqrt{\frac{\varepsilon}{2\eta}} \quad ; \quad \overline{\langle p^2 \rangle} = kT\sqrt{\frac{\varepsilon}{2\eta}} \quad . \tag{5.7}$$

This is the law we are suggesting for the quadratic fluctuations of p and q (for instance of the tension or of the intensity in the RLC model of section 2), as a function of the temperature, the amplitude of the parametrical perturbation and the resonance difference, supposed here very small. It would be interesting to submit this result to experimentation.

REFERENCES

Alaoui A. and Lochak G., C.R.A.S., 280, 589, 1977.

Coddington E.A. and Levinson N., Theory of ordinary differential equations, Mc Graw-Hill, N.Y., 1955 (useful here for enlightments on the theorem of Floquet).

Lochak G., C.R.A.S. 275 B, 903, 1972 ; 276 B, 809, 1973. Ann. Fond. L. de Broglie, 1, 56, 1976. "The Adiabatical Invariance as One of the Fundamental Principles of Statistical Mechanics", to appear in Russian in the Jubilee Book of Jakob P. Terletsky, 1981.

Lochak G. and Vassalo Pereira J., C.R.A.S., 282 B, 1121, 1976.

Lochak P., Thèse 3e Cycle, "Sur deux généralisations du théorème adiabatique en mécanique quantique", Orsay N° 2964, 1981.

Young R.H. and Deal W.J., J. Math. Phys. II, 3298, 1970.

ADIABATICAL INVARIANTS IN NONLINEAR DYNAMICAL SYSTEMS PERIODICALLY DEPENDING ON TIME, WITH AN APPLICATION TO THE PARAMETRICAL RESONANCE IN A PHYSICAL (NONLINEAR) PENDULUM

Georges LOCHAK

C.N.R.S. and Fondation Louis de Broglie, Paris, France

1. INTRODUCTION

This contribution should be considered as the continuation of the preceding one in this same book, by Lochak G. and Vassalo Pereira J. It will be devoted to a *non linear generalization* of the theory of adiabatical invariants of classical dynamical systems periodically depending on time (See Lochak G., 1979, 1981).

The difficulty lies in the fact that the linear theory is based on the theorem of Floquet and we must begin with a generalization of this result. In a sense, such a generalization may be considered as already performed by Birkhoff in his theory of normal forms. But the latter theory is founded on the formal solutions of Hamilton equations. The generalization which will be presented here does not make use of formal solutions and it seems to me that it will be interesting for a better understanding of the structure and of the geometrical sense of the theorem.

With the help of that, we shall generalize the adiabatical theorem to the case of non linear periodically perturbed systems and we shall give an analytical expression of the adiabatical invariant of a non linear parametrically perturbed pendulum.

2. THE THEOREM OF FLOQUET FROM A MORE GENERAL POINT OF VIEW

Let us consider a non linear periodic differential equation :

$$\dot{\vec{x}} = \vec{v}(\vec{x},t) \ ; \ \vec{v}(\vec{x},t+\tau) = \vec{v}(\vec{x},t) \ ; \ \vec{x} \in \mathbb{R}^n. \qquad (2.1)$$

ISBN 0-12-068720-8

It defines a transformation g_0^t so that :

$$g_{t'}^{t}\, g_0^{t'} = g_0^t \quad ; \quad g_t^{t+\tau} = g_0^t \quad ; \tag{2.2}$$

$$g_0^{t+n\tau} = g_{nt}^{t+n\tau}\, g_0^{nt} = g_0^t\, g_0^{n\tau} \Rightarrow g_0^{n\tau} = (g_0^{\tau})^n \quad ; \tag{2.3}$$

(see Arnold V.). $g_0^{n\tau}$ form a group though g_0^t is not a one parameter group. Now suppose that we are able to *interpolate* the transformation g_0^t by a one-parameter group $g_0^{\star t}$ so that :

$$g_0^{\star t+s} = g_0^{\star t}\, g_0^{\star s} = g_0^{\star s}\, g_0^{\star t} \quad ; \quad g_0^{\star n\tau} = g_0^{n\tau} \; . \tag{2.4}$$

We can now define a new diffeomorphism :

$$T(t) = g^t g^{\star -t} \; . \tag{2.5}$$

From (2.3) and (2.4), we have :

$$T(t+\tau) = g_0^{t+\tau}\, g_0^{\star -t-\tau} = g_0^t g_0^\tau g_0^{\star -\tau} g_0^{\star -t} = g_0^t g_0^{\star -t} = T(t) . \tag{2.6}$$

So, g_0^t may be written as the product of a one-parameter group by a *periodic* transformation (with period τ) :

$$g_0^t = T(t)\, g_0^{\star \tau} . \tag{2.7}$$

Then $g_0^{\star t}$ may be considered as the flow of an *autonomous* equation :

$$\dot{\vec{\xi}} = \vec{v}^{\star}(\vec{\xi}) \quad ; \; \vec{\xi} \in \mathcal{R}^n \; . \tag{2.8}$$

So, we can say that (2.1) is *reducible* to the autonomous system (2.8) by the *periodic* transformation $T(t)$. Therefore, *provided that* $g_0^{\star t}$ *exists*, we have in (2.7) the very essence of the Floquet theorem. The only (but not negligible !) advantage of linearity is to provide us with an expression of $g_0^{\star t}$:

$$g_0^{\star t} = e^{At} \qquad (A : \text{matrix in } \mathcal{R}^n) . \tag{2.9}$$

3. THE CANONICAL REDUCIBILITY OF A PERIODICAL HAMILTONIAN SYSTEM (Lochak G., 1979, 1981)

In the hamiltonian case, we shall express the above statement as follows (we shall drop the indices for the sake of simplicity):

DEFINITION : Given a periodical hamiltonian system :

$$\dot{q} = \frac{\partial H}{\partial p}(q,p,t) \ ; \ \dot{p} = -\frac{\partial H}{\partial q}(q,p,t) \ ; \ H(q,p,t+\tau) = H(q,p,t), \quad (3.1)$$

we shall say it is *canonically reducible* (or a "C.R. system") when it is reducible by a canonical transformation deriving from a generating function $F(q,Q,t)$ such that :

$$F(q,Q,t+\tau) = F(q,Q,t) \ ; \quad (3.2)$$

$$p = \frac{\partial F}{\partial q}(q,Q,t) \ ; \ P = -\frac{\partial F}{\partial Q}(q,Q,t) \ ; \ R(Q,P) = H(q,p,t) + \frac{\partial F}{\partial t} \quad (3.3)$$

$$\dot{Q} = \frac{\partial R}{\partial P}(Q,P) \ ; \ \dot{P} = -\frac{\partial R}{\partial Q}(Q,P) \ . \quad (3.4)$$

In other words, it is stated that in (2.7) T is canonical.

THEOREM : *The system* (3.1) *will be C.R. iff the Hamilton-Jacobi equation* :

$$\frac{\partial S}{\partial t} + H(q,\frac{\partial S}{\partial q},t) = 0 \ ; \ H(t+\tau) = H(t) \ ; \quad (3.5)$$

possesses a complete integral (a principal function) in the form:

$$S(q,\alpha,t) = W(q,\alpha,t) - Et \ ; \ W(q,\alpha,t+\tau) = W(q,\alpha,t). \quad (3.6)$$

Then, the principal function $S^\star$ *of the reduced system (3.4), which is the solution of* :

$$\frac{\partial S^\star}{\partial t} + R(Q,\frac{\partial S^\star}{\partial Q}) = 0 \ , \quad (3.7)$$

will have the form :

$$S^\star(Q,\alpha,t) = W^\star(Q,\alpha) - Et \ ; \ W^\star(Q,\alpha) = W(Q,\alpha,t_0). \quad (3.8)$$

The generating function (3.2) will be :

$$F(q,Q,t) = W(q,\alpha,t) - W(Q,\alpha,t_0). \quad (3.9)$$

PROOF : Using section 2, we will try to interpolate (3.1) by a conservative system. Let then, $S(q,\alpha,t)$ be a complete integral of (3.5) ; the motion will be given by :

$$\begin{cases} p_0 = \dfrac{\partial S}{\partial q}(q_0,\alpha,t_0) \ ; \ \beta = \dfrac{\partial S}{\partial \alpha}(q_0,\alpha,t_0) \\ p \ = \dfrac{\partial S}{\partial q}(q,\alpha,t) \ ; \ \beta = \dfrac{\partial S}{\partial \alpha}(q,\alpha,t). \end{cases} \quad (3.10)$$

Let us consider now a conservative system with a principal function $S^\star(Q,\alpha,t)$, such that :

$$S^\star(Q,\alpha,t) = W^\star(Q,\alpha) - E(\alpha)t \quad (3.11)$$

$$\begin{cases} P_0 = \frac{\partial \overset{\star}{W}}{\partial Q}(Q_0,\alpha) \quad ; \quad \beta = \frac{\partial \overset{\star}{W}}{\partial \alpha}(Q_0,\alpha) - \frac{\partial E}{\partial \alpha}\, t_0 \\ P \;\, = \frac{\partial \overset{\star}{W}}{\partial Q}(Q,\alpha) \quad ; \quad \beta = \frac{\partial \overset{\star}{W}}{\partial \alpha}(Q,\alpha) - \frac{\partial E}{\partial \alpha}\, t \end{cases} \tag{3.12}$$

with the *same* constants α, β.

Let us first suppose that the two systems (3.10) and (3.12) have the same initial conditions :

$$P_0 = p_0 \quad ; \quad Q_0 = q_0 \quad ; \tag{3.13}$$

from which it follows :

$$p_0 = \frac{\partial \overset{\star}{W}}{\partial q}(q_0,\alpha) = \frac{\partial S}{\partial q}(q_0,\alpha,t_0) \tag{3.14}$$

$$\beta = \frac{\partial \overset{\star}{W}}{\partial \alpha}(q_0,\alpha) - \frac{\partial E}{\partial \alpha}\, t_0 = \frac{\partial S}{\partial \alpha}(q_0,\alpha,t_0) \tag{3.15}$$

Hence we have :

$$\overset{\star}{W}(Q,\alpha) = S(Q,\alpha,t_0) + E(\alpha)t_0. \tag{3.16}$$

Because of the periodicity of H, in order to interpolate the system (3.10) by the system (3.12) it's sufficient to suppose that they coincide only at a second time $t_0+\tau$:

$$p(t_0+\tau) = \frac{\partial \overset{\star}{W}}{\partial Q}(q(t_0+\tau),\alpha) = \frac{\partial S}{\partial q}(q(t_0+\tau),\alpha,t_0+\tau) \tag{3.17}$$

$$\beta = \frac{\partial \overset{\star}{W}}{\partial \alpha}(q_0(t_0+\tau),\alpha) - \frac{\partial E}{\partial \alpha}(t_0+\tau) = \frac{\partial S}{\partial \alpha}(q(t_0+\tau),\alpha,t_0+\tau). \tag{3.18}$$

Hence, from (3.16) :

$$\frac{\partial S}{\partial q}(q,\alpha,t_0) = \frac{\partial S}{\partial q}(q,\alpha,t_0+\tau) \tag{3.19}$$

$$\frac{\partial S}{\partial \alpha}(q,\alpha,t_0) - \frac{\partial E}{\partial \alpha}\,\tau = \frac{\partial S}{\partial \alpha}(q,\alpha,t_0+\tau) \tag{3.20}$$

We see on (3.19) that $\frac{\partial S}{\partial q}$ is periodic in t and hence we obtain for S the form :

$$S(q,\alpha,t) = W(q,\alpha,t) + g(\alpha,t) \;\; ; \; W(q,\alpha,t+\tau) = W(q,\alpha,t). \tag{3.21}$$

Introducing this expression in (3.20) it comes :

$$\frac{\partial g}{\partial \alpha}(\alpha,t_0) - \frac{\partial E}{\partial \alpha}\,\tau = \frac{\partial g}{\partial \alpha}(\alpha,t_0+\tau) \;\; ;$$

or : $g(\alpha,t_0) - g(\alpha,t_0+\tau) = E(\alpha)\tau + \text{const.}$

So that g may be written as :

$$g(\alpha,t) = -\,E(\alpha)t + h(t) \;\; ; \quad h(t+\tau) = h(t).$$

The periodic function $h(t)$ may now be included in $W(q,\alpha,t)$ and we see that S *must* have the expression (3.6). From (3.11) and (3.16), we then see that $S^\star$ has the expression (3.8).

In order to find the generating function (3.9), we shall begin with a quite simple but useful remark. Let (q,p) and (Q,P) be two representations of the same system with corresponding hamiltonians H(q,p,t), $H^\star(Q,P,t)$ and let F(q,Q,t) be the generating function of the transformation $(q,p) \to (Q,P)$.

From the principle of least action we have :

$$\sum_i p_i\dot{q}_i - H(q,p,t) = \sum_i P_i\dot{Q}_i - H^\star(Q,P,t) + \frac{dF}{dt}(q,Q,t), \tag{3.22}$$

which means, written in terms of action :

$$\frac{dS}{dt} = \frac{dS^\star}{dt} + \frac{dF}{dt}. \tag{3.23}$$

But S and $S^\star$ are, respectively, complete integrals of the two corresponding Hamilton-Jacobi equations, so :

$$S(q,\alpha,t) = S^\star(Q,\alpha,t) + F(q,Q,t) \tag{3.24}$$

and we obviously have $\frac{\partial F}{\partial \alpha} = 0$ because $\frac{\partial S}{\partial \alpha} = \frac{\partial S^\star}{\partial \alpha} = \beta$.

Owing to the formula (3.24), when we know a complete integral of the Jacobi equation in some representation (q,p), we find a complete integral in the representation (Q,P) provided we know the generating function of the transformation $(qp) \to (QP)$. Conversely if we know S and $S^\star$, we deduce immediately F : in our case, we get (3.9) from (3.6) and (3.8).

Now, suppose we know that (3.5) is C.R., i.e. we have (3.11) and the *periodical* generating function F(q,Q,t). So, the same formula (3.24) gives at once the form of the principal function S :

$$S(q,\alpha,t) = S^\star(Q,\alpha,t) + F(q,Q,t) = W^\star(Q,\alpha) + F(q,Q,t) - Et \tag{3.25}$$

and owing to the periodicity of F, we recognize on this formula the condition (3.6). This concludes the proof.

REMARK : On the formula (3.24) we see at a glance that if two C.R. systems are given with two principal functions :

$$S_1(q,\alpha,t) = W_1(q,\alpha,t) - E_1 t \; ; \; S_2(Q,\alpha,t) = W_2(Q,\alpha,t) - E_2 t$$

they can be transformed into each other through a canonical transformation deriving from :

$$G(q,Q,t) = W_1(q,\alpha,t) - W_2(Q,\alpha,t) - (E_1 - E_2)t$$

This means that the C.R. systems constitute a *class of equivalence* with respect to the canonical transformations deriving from generating functions defined as :

$$G(q,Q,t) = F(q,Q,t) - Ct \; ; \; F(q,Q,t+\tau) = F(q,Q,t).$$

Due to the additivity of generating functions in the composition of canonical transformation, the product of two such transformations is trivially of the same type : they constitute a *group*.

So, in order to establish the C.R. property of a system, we may prove the canonical equivalence -in the sense here defined- of the given system with another one which was already identified as a C.R. system ; and because the group property of the transformations, we may prove this equivalence by constructing a chain of such transformations between the given system and the already known one. We shall use this procedure.

4. THE ADIABATICAL THEOREM

We shall now suppose that the reduced system (3.7) is *integrable* and that we know the action-angle variables $J_1,\ldots,J_n$; $\phi_1,\ldots,\phi_n$. We shall make choice of the J_k as constants α_k in S and $S^\star$.

Then, $W^\star(Q_1,\ldots,Q_n \; ; \; J_1,\ldots,J_n)$ will be a generating function of the transformation from the variables Q,P to ϕ,J, with :

$$P_k = \frac{\partial W^\star}{\partial Q_k}(Q_1,\ldots,Q_n \; ; \; J_1,\ldots,J_n) \; ;$$

$$\phi_k = \frac{\partial W^\star}{\partial J_k}(Q_1,\ldots,Q_n \; ; \; J_1,\ldots,J_n). \tag{4.1}$$

We know, from the general properties of action variables (Goldstein H.) that the reduced energy will take the form :

$$R(Q_1,\ldots,Q_n \; ; \; P_1,\ldots,P_n) = \overline{R}(J_1,\ldots,J_n) = E(J_1,\ldots,J_n). \tag{4.2}$$

The equations of motion are now :

$$\dot{\phi}_k = \frac{\partial E}{\partial J_k}(J_1,\dots,J_n) = \omega_k(J_1,\dots,J_n) \;;$$

$$\dot{J}_k = -\frac{\partial E}{\partial \phi_k}(J_1,\dots,J_n) = 0 \tag{4.3}$$

We shall also need the generating function of the direct transformation $(q,p) \to (\phi,J)$. From the group properties of the canonical transformations, we have :

$$G(q_1,\dots,q_n \;;\; \phi_1,\dots,\phi_n \;;\; t) = F(q_1,\dots,q_n \;;\; Q_1,\dots,Q_n \;;\; t) + W^\star(Q_1,\dots,Q_n \;;\; J_1,\dots,J_n) - \sum_i J_i\phi_i \tag{4.4}$$

$$p_i = \frac{\partial G}{\partial q_i}(q_1,\dots,q_n \;;\; \phi_1,\dots,\phi_n \;;\; t) \;;$$

$$J_i = -\frac{\partial G}{\partial \phi_i}(q_1,\dots,q_n \;;\; \phi_1,\dots,\phi_n \;;\; t). \tag{4.5}$$

In particular, (4.5) gives to us the J_i as constants of motion $J_i(q_k \;;\; p_k \;;\; t)$ of the system (3.1) or (3.5). We see that G is $2\pi/\omega$ periodic in t. But G is also 2π periodic in $\phi_1,\dots,\phi_n$ just as $W^\star - \sum J_i\phi_i$. Really, this function is the "modified generating function" (Goldstein H., Sommerfeld A.) which is periodic in ϕ_k because :

$$J_k = \frac{1}{2\pi}\oint \frac{\partial W^\star}{\partial Q_k}\, dQ_k \;,$$

so that $W^\star$ increases by $2\pi\, J_k$ in one period of the coordinate Q_k.

Let us suppose now that H depends on a parameter a, itself slowly depending on time. Hence, H, $\overline{R}$ and G will be

$$H = H(\dots q_k\dots,\dots p_k,t,a(t)) \;;\; R' = R'(\dots\phi_k\dots,J_k\dots,a(t)) \;;$$
$$G = G(\dots q_k\dots,\dots\phi_k\dots,t,a(t))$$

and, from the classical formulae of canonical transformations :

$$\dot{\phi}_k = \frac{\partial R'}{\partial J_k}(\dots\phi_k\dots J_k\dots a(t)) \;;\; \dot{J}_k = -\frac{\partial R'}{\partial \phi_k}(\dots q_k\dots J_k\dots,a(t)) \tag{4.6}$$

$$R'(\ldots\phi_k\ldots J_k\ldots a(t)) = H(\ldots q_k\ldots p_k\ldots t, a(t)) + \frac{\partial G}{\partial t}(\ldots q_k\ldots\phi_k\ldots, t, a(t)) + \dot{a}\,\frac{\partial G}{\partial a}(\ldots q_k\ldots\phi_k\ldots t, a(t)). \qquad (4.7)$$

Now, due to (4.2) and to the definition of G :

$$H(\ldots q_k\ldots p_k, t, a(t)) + \frac{\partial G}{\partial t}(\ldots q_k\ldots\phi_k\ldots t, a(t)) = \overline{R}(\ldots J_k\ldots, a(t)) = E(\ldots J_k\ldots a) \qquad (4.8)$$

where E does not depend on ϕ_k. So, from (4.3), (4.6), (4.7), (4.8), we shall have :

$$\dot{\phi}_k = \omega_k(J_1\ldots J_n\ ;\ a(t))\ ;$$
$$\dot{J}_k = -\,\dot{a}\,\frac{\partial^2 G}{\partial\phi_k\partial a}(q_1\ldots q_n\ ;\ \phi_1,\ldots,\phi_n\ ;\ t\ ;\ a(t)) \qquad (4.9)$$

Let us suppose, for the sake of simplicity, that $\dot{a}$ = Cnte, so that :

$$J_k(T) - J_k(0) = -\,\dot{a}\int_0^T \frac{\partial^2 G}{\partial\phi_k\partial a}\,dt \qquad (4.10)$$

We shall be content now, with an "intuitive proof for physicists", analogous to the proof given by Sommerfeld in the conservative case (I beg the pardon of mathematicians !).

If a varies slowly ($|\dot{a}| \ll 1$), the ϕ_k will be almost linear on t. Then, we can see on (4.4) and (4.5) that the coefficients of the Fourier expansion of the integrand of (4.10) will be slowly variable, with a constant term equal to zero (due to the derivative on ϕ), provided that :

$$m\omega + \sum_{k=1}^{n} m_k\omega_k \neq 0 \qquad (m, m_k \in Z). \qquad (4.11)$$

So, under these conditions, the integrand in (4.10) will be oscillating, the integral will be bounded and we shall have :

$$|J_k(T) - J_k(0)| = O(\dot{a}) \qquad \forall\ T. \qquad (4.12)$$

This means that $J_k(\ldots q_i\ldots\ ;\ p_i\ldots, a, t)$ *remains a first integral of* (3.1) *when* a *varies infinitely slowly and thus,* J_k *is an adiabatical invariant.*

REMARKS : 1) The theorem makes use explicitely of the periodicity of G.

2) We can see, on (4.11), that the *degeneracy* of the system (3.1) is not the same as the degeneracy of the reduced system (3.4). For the last, it is defined classically by an equality :

$$\sum_{k=1}^{n} m_k \omega_k = 0 \quad ; \quad m_k \in Z,$$

whereas for the initial system (3.1) we see on (4.11) that the degeneracy must be defined by :

$$\sum_{k=1}^{n} m_k \omega_k = 0 \quad (\text{mod.}\,\omega) \quad ; \; m_k \in Z. \tag{4.13}$$

Because of non-linearity, this definition is slightly more complicated than in quantum mechanics (Lochak G., 1974, Lochak G. & Alaoui A., 1977).

3) Due to the dependence of ω_k with regard to the J_k, the whole set of conditions (4.11), even it is true at a given moment, cannot be maintained indefinitely. Then, rigorously speaking, the theorem is surely false just as the corresponding theorem in the conservative case : this problem is related to the well known difficulty of small divisors. But physically, the theorem will be true, in every case where it will be possible to prove that (4.11) remains true up to sufficiently high order in m and m_k and during a time greater that the *relaxation times* of the actual physical system, times which were neglected in making use of Hamilton dynamics, but which cannot be forgotten.

4) Finally, this theorem implies, as particular case, the adiabatical invariance of the expression of J given for the Mathieu oscillator in the formula (4.7) of our previous communication (with J. Vassalo Pereira) in this same book.

5. THE ADIABATICAL INVARIANT OF A NON LINEAR PENDULUM PARAMATRICALLY DISTURBED

We shall consider a "physical Mathieu pendulum", i.e. a pendulum approximated at the first non-quadratic term of the gravitational cosq potential, parametrically disturbed by a periodical force :

$$H(q,p,t) = \frac{1}{2}(p^2 + \omega_0^2 q^2) - \frac{\mu}{4} q^4 - \frac{\varepsilon\omega_0}{2} \cos\omega t \; q^2 \; ; \; \mu = \frac{\omega_0^2}{6}. \quad (5.1)$$

μ is a "small parameter" artificially introduced for the convenience of the perturbation theory. Neglecting the μ^2, $\varepsilon\mu$ and ε^2 terms, we shall now link (5.1) by a chain of transformations of the (3.26) type, to the Mathieu oscillator which was precedingly described (with Vassalo Pereira) and that we shall write again, but with slightly different notations :

$$H'(X,\varpi,t) = \frac{1}{2}(\varpi^2 + \omega_0^2 \kappa^2) - \frac{\varepsilon\omega_0}{2} \cos\omega t \; \kappa^2. \quad (5.2)$$

Let us first apply the canonical transformation giving the set of action-angle variables L,θ :

$$\begin{cases} G_0(\kappa,L) = \dfrac{\omega_0}{2} x\sqrt{\dfrac{2L}{\omega_0} - x^2} + L \operatorname{Arcsin} x\sqrt{\dfrac{\omega_0}{2L}} \\ \varpi = \dfrac{\partial G_0}{\partial \kappa} \; ; \; \theta = \dfrac{\partial G_0}{\partial L} \Rightarrow \kappa = \sqrt{\dfrac{2L}{\omega_0}} \sin\theta \; ; \; \varpi = \sqrt{2L\omega_0} \cos\theta. \end{cases} \quad (5.3)$$

and (5.2) will be written as :

$$H'_0(L,\theta,t) = L\omega_0 - \varepsilon L \cos\omega t \sin^2\theta. \quad (5.4)$$

Now, let us apply, mutatis mutandis, the same transformation to (5.1) with the variables J,ϕ :

$$G_0(q,J) \; ; \; p = \frac{\partial G_0}{\partial q} \; ; \; \phi = \frac{\partial G_0}{\partial J} \Rightarrow q = \sqrt{\frac{2J}{\omega_0}} \sin\phi \; ; \; p = \sqrt{2J\omega_0} \cos\phi. \quad (5.5)$$

and (5.1) will become in turn :

$$H_0(J,\phi,t) = J\,\omega_0 - \frac{\mu}{\omega_0^2} J^2 \sin^4\phi - \varepsilon J \cos\omega t \sin^2\phi. \quad (5.6)$$

We shall then bring back this formula to the same form as (5.4), by making use of the Lindstedt-Poincaré method (see Poincaré H. ; Lochak G. 1981). For that, let us write the shortened Jacobi equation :

$$\omega_0 \frac{\partial G_1}{\partial \phi} - \frac{\mu}{\omega_0^2}\left(\frac{\partial G_1}{\partial \phi}\right)^2 \sin^4\phi = E \quad (5.7)$$

We must now construct a complete integral $G_1(\phi,K)$ expanded in powers of μ, with K representing the constant α of Jacobi. At the first order, we shall find :

$$G_1(\phi,K) = \phi\ K - \frac{\mu}{4\omega_0^3} K(\sin 2\phi - \frac{1}{8} \sin 4\phi). \tag{5.8}$$

It is a new generating function which allows us to go from the variables (J,ϕ) to (K,ψ), such that :

$$\begin{cases} \psi = \dfrac{\partial G_1}{\partial K} = \phi - \dfrac{\mu}{2\omega_0^3} K(\sin 2\phi - \dfrac{1}{8} \sin 4\phi) \\ J = \dfrac{\partial G_1}{\partial \phi} = K - \dfrac{3\mu}{8\omega_0^3} K^2 + \mu \dfrac{K^2}{\omega_0^3} \sin^4\phi \end{cases} \tag{5.9}$$

So, (5.6) becomes (at the first order in μ and ε) :

$$H_1(K,\psi,t) = K\ \omega_0 - \frac{3\mu}{8\omega_0^2} K^2 - \varepsilon\ K \cos\omega t \sin^2\psi. \tag{5.10}$$

Finally, we shall a last transformation $(K,\psi) \to (L,\theta)$ so that :

$$\begin{cases} G_2(\psi,L) = \psi(L + \dfrac{3\mu}{8\omega_0^2} L^2) \\ \theta = \dfrac{\partial G_2}{\partial L} = \psi(1 + \dfrac{3\mu L}{4\omega_0^3}) \ ;\ K = \dfrac{\partial G_2}{\partial \psi} = L + \dfrac{3\mu}{8\omega_0^3} L^2. \end{cases} \tag{5.11}$$

and owing to this transformation, (5.10) becomes identical to (5.4) :

$$H_2(L,\theta,t) = L\ \omega_0 - \varepsilon\ L \cos\omega t \sin^2\theta \tag{5.12}$$

Hence, from the law of group composition of the canonical transformations, (5.1) expressed in terms of variables (q,p) goes to (5.12) written with the variables (L,θ) through a transformation deriving from the following generating function which is independent of time :

$$G(q,L) = G_0(q,J) + G_1(\phi,K) + G_2(\psi,L) - \phi\ J - \psi\ K \tag{5.13}$$

and we will even obtain the oscillator of Mathieu with the help of the generating function :

$$G'(q,\varpi) = G(q,L) - G_0(K.L) + \varpi K \tag{5.14}$$

So, we can assert that (5.1) will be a C.R. system and, for this reason, will possess an adiabatical invariant (owing to the theorem of the section 4) simply if we know that the oscillator of Mathieu is a C.R. system itself. But this is precisely the case, as was directly proved (see Lochak G. & Vassalo Pereira J., 1976 ; Lochak G., 1981). This result may also be established on the

Jacobi equation corresponding to the hamiltonian (5.4) :

$$\frac{\partial \Sigma}{\partial t} + \omega_0 \frac{\partial \Sigma}{\partial \theta}(1 - \frac{\varepsilon}{\omega_0} \cos\omega t \sin^2\theta) = 0. \qquad (5.15)$$

Indeed, this equation has a complete solution of the (3.6) type in virtue of a theorem of Poincaré *simply because it is written on a torus* (Poincaré H. Oeuvres complètes). More concretly, at the first order in μ and ε, in the vicinity of the parametrical resonance, we easily can find the complete solution of (5.15) :

$$\Sigma = \frac{\alpha(\delta - \frac{\varepsilon}{4})}{\Delta} \operatorname{Arctg}\left[\frac{\Delta}{\delta - \frac{\varepsilon}{4}} \operatorname{tg}(\frac{\omega}{2} t - \phi)\right] - \alpha(\delta - \frac{\varepsilon}{4})t \qquad (5.16)$$

with the notations $\delta = \omega/2 - \omega_0$, $\Delta = \sqrt{\delta^2 - \varepsilon^2/16}$, already used in the communication with J. Vassalo Pereira in this book.

In the same communication, we gave (formula (4.7)) the adiabatical invariant of the Mathieu oscillator. So, it is easy to find the invariant of (5.1) by a simple transformation of this formula through (5.5), (5.9), (5.11) and (5.3) and we shall get :

$$J = \frac{p^2 + \omega_0^2 q^2 - \frac{\omega_0^2}{12} q^4}{\Delta(p^2 + \omega_0^2 q^2)}\left[(\delta + \frac{\varepsilon}{4} \cos\omega t) q^2 - \frac{\varepsilon}{2} \sin\omega t \, qp + (\delta - \frac{\varepsilon}{4} \cos\omega t) p^2\right]. \qquad (5.17)$$

REFERENCES

Arnold V., Méthodes mathématiques de la mécanique classique, Mir, Moscow, 1974.

Birkhoff G.D., Dynamical Systems, American Mathematical Society Colloquium Publications, Vol. IX, Providence, 1927.

Goldstein H., Classical Mechanics, Addison-Wesley, Camb. Mass. 1953.

Lochak G., C.R.A.S. 279B, 547, 1974 ; 289 B, 98, 1979 ; Hadronic Journal, 4, 1105, 1981.

Lochak G. & Alaoui A., Ann. Fond. L. de Broglie, 2, 87, 1977.

Lochak G. & Vassalo Pereira J., C.R.A.S. 282 B, 1121, 1981.

Poincaré H., Les Méthodes nouvelles de la Mécanique Céleste, Gauthier-Villars, 1892, 1893, 1899, Paris. Dover, N.Y. 1957. Oeuvres complètes, vol. 1, p. 152 (Sur les courbes définies par les équations différentielles, ch. XV), Gauthier-Villars, Paris, 1951.

Sommerfeld A., Atombau und Spektrallinien, Friedr. Vieweg & Son, Braunschweig, 1951.

A THEOREM ON PHASE-LOCKING IN TWO INTERACTING CLOCKS (THE HUYGHENS EFFECT)

José Vassalo Pereira
Faculty of Sciences, University of Lisbon

The main subject of this paper is a somewhat mysterious phenomenon first noticed by Kristiaan Huyghens in 1665, and which arises when we take two identical clocks and place them near each other. In such circumstances, Huyghens remarked that the phases of oscillation of each clock no longer follow independently from each other but, instead, show an asymptotical tendency towards a certain fixed value of the phase difference.

Huyghens tells in a letter to his father (Huyghens[1]) that in some experiments he had performed there was a tendency of the clocks to come into the coincidence of phases and, in others, to opposite phases. As a matter of fact, Huyghens did not provide an explanation for this phenomenon, and he was not very sure wether the interaction between the two clocks was due to an aerial wave arising from the oscillations of the arms of the two pendulums, or to a material wave propagating along the wall from which the two clocks are hanging. Besides, he repeated many times the same experiment under different conditions - with the clocks far apart, with the clocks in close proximity - and he always arrived at the same conclusion: there must be an interaction, even if he could

DYNAMICAL SYSTEMS
AND MICROPHYSICS

ISBN 0-12-068720-8

not be very sure about the asymptotical phase difference of the two clocks.

Now, there are at least two reasons that make very interesting the Huyghens Effect:

The first is that it is the first example, chronologically speaking, of interactions between two self-oscillationg systems. In fact, a clock is a self-oscillating system, and since 1936 we have at our disposal a very beautiful theory of the clock, which is due to the Russian mathematician. A.A.Andronov (Andronov, Vitt, Khaikin[2]). So we may except that by making use of Andronov's model for the clock, we may perhaps arrive to a satisfactory explanation of the Huyghens Effect.

The second reason is that Huyghens Effect is intrinsically an effect of phase locking, very much alike those which have been developed in Electronics in the last few years. In these problems there have been a few trials to describe the transient states, enough to allow us to assert that it is an extremely important but, unfortunately, extremely difficult task. Now, we think that there is a chance to prevail over these difficulties because, as it will be seen in the sequel, at least for the Huyghens Effect a painless mathematical description can be given.

We shall start with a few words about the dynamical model for the clock which we make use here:

Let us take a pendulum. Of course, a real pendulum is always enduring the action of some kind of friction, so that the oscillations are always damped. In order to maintain them, some energy must be supplied from the exterior by means of some mechanism coupled to the pendulum. Now, what distinguishes a clock is that this mechanism (which is called here "the escapment") supplies energy to the pendulum in a very peculiar way, which is a function of the state of the pendulum itself. Let us denote by x the distance of the pendulum from its (stable) equilibrium position, and by $y = \dot{x}$ its velocity. We shall suppose, for the sake of simplicity, that the supply of energy by the escapment to the pendulum is performed by jumps, and just once by period, when the arm of the pendulum goes from the left to the right ($\dot{x} > 0$), and at the maximum value of the velocity.

Now it has been proved by Andronov that, in order to this system to actually have the properties of a real clock, one must require that

1. The friction which acts upon the pendulum must be a "dry" friction, that is, the pendulum undergoes a friction of the form

(1) $$f = f(\dot{x}) = - f_o \text{ sign } \dot{x} .$$

2. In the collision between the escapment and the arm of the pendulum, when energy is supplied to it, the amount of energy is always the same whatever may be the state of the pendulum before the collision:

(2) $$y_2^2 - y_1^2 = \text{constant} = h^2.$$

The dynamical evolution of the pendulum then comes given by

(3) $$m \frac{d^2x(t)}{dt^2} = - \frac{m\ g}{1} x(t) - f_o \text{ sign } \frac{d\ x(t)}{dt}$$

(m = mass of the pendulum ; g = acceleration of gravity ; 1 = = length of the arm of the pendulum). If we now define the positive constants

(4) $$\alpha = g/1 \qquad \beta = 1\ f_{o/m\ g},$$

and if we adopt a new time scale t'which is related to t by means of

(5) $$t' = + (\alpha)^{1/2}\ t,$$

equation (3) then takes the form

(6) $$\ddot{x} + x = - \beta \text{ sign } \dot{x}$$

(Of course, we still call <u>t</u> the new time scale, but this is unimportant, and it is just to avoid a cumbersome notation). Now we see that this equation splits in two:

(7a) $\ddot{x} + x = -\beta$ (for $\dot{x} > 0$); (7b) $\ddot{x} + x = +\beta$ (for $x < 0$).

Let us see what happens in the $x, y = \dot{x}$ plane (phase plane):

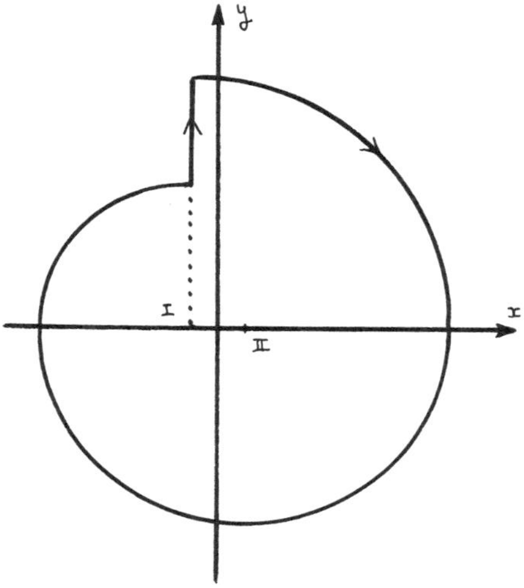

In the upper half plane (equation (7a) the representative point (rp) of the system describes circles centered in I = $(-\beta, 0)$ in the clockwise sense. In the lower half plane (eq. (7b)), the rp describes circles centered in II = $(+\beta, 0)$, in the clockwise sense too. Both in the upper and the lower half plane the angular velocity of the rp is constant and equal to + 1, so that the period is 2π.

If the pendulum starts, for instance, in the position $x<0$, $y = 0$, when it reaches the maximum value of the velocity(in $x=-\beta$) then, as we have said, it undergoes a sudden supply of energy and its velocity changes instantaneously from the value y_1 to y_2, according to (2). It is now almost evident that the rp of this system tends asymptotically to a closed discontinuous curve, which is a limit cycle (a doubly stable limit cycle).

Before we start using this dynamical model of the clock, we must still say a few words about the meaning of the word "phase" when we deal with in the theory of limit cycles:

Let us consider in the x,y plane a doubly stable cycle, that is, an isolated periodic solution of a system of two order autonomous ordinary differential equations

(8) $$\dot{x} = P(x, y) \qquad \dot{y} = Q(x, y).$$

Let us denote by

(9) $$x = \Phi(t) \qquad y = \Psi(t)$$

that periodic solution (with period τ) whose rp is thus describing the limit cycle. Since the system is autonomous, any other solution differing by Δt from the origin of time, also describes the limit cycle, that is, $x = \Phi\ (t - \Delta t)$, $y = \Psi(t - \Delta t)$ is also a periodic solution of (8), where $\Delta t \varepsilon (0,\tau)$. Let us now take a rp starting at some $t = t_o$ in a point P_o in the vicinity of the limit cycle. Since the limit cycle is stable, the rp will wind around it and for $t = \infty$ it will be upon the limit cycle and coincident with of the rp's whose evolution is given by $x = \Phi(t-\Delta t)$, $\Psi(t-\Delta t)$, with a certain value of Δt, and which has always been describing the limit cycle. That is, when a rp starts from an initial condition P_o near the limit cycle and at a certain instant t_o, it is always possible to find, starting at the same instant t_o and upon the limit cycle another rp such that for $t = \infty$ they will both coincide. We then say that they are "in phase". The definition of the phase difference of two rp's evolving in the vicinity of the limit cycle is then evident: it is just the time difference of two other rp's upon the limit cycle, each one of which is in phase with one of the former two.

Now in the case of the limit cycle of a clock such notions are easily introduced: as we have said above, in the upper half plane x,y the rp describes circles centered in $I = (-\beta, 0)$ and always with angular velocity equal to +1 (the same thing happens in the lower half plane, but centered in $II = (+\beta,0)$). So we see that the rp's which are in phase with a rp starting in P_o at $t=t_o$ upon the limit cycle are those which start, at the same instant, from the straight line joining P_o and I (or II, if P_o is on the lower half plane). Besides, the phase difference between two rp's, P_1 and P_2, in the vicinity of the limit cycle, is just the angle centered in I (if P_1, P_2 are both in the upper half plane) and defined by P_1 and P_2. Of course, if the two rp's are not in the same half plane, then the phase difference is given by the sum of

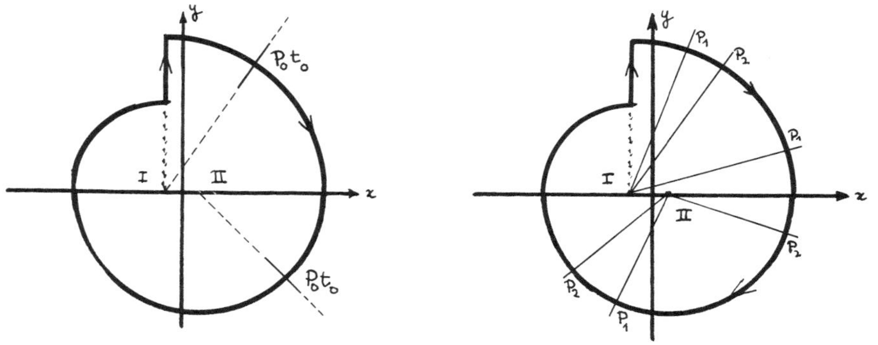

two angles: one centered in I and the other in II (see figure).

Let us go back to the Huyghens Effect.

The model we give to describe the interaction is the following: Let us have two identical clocks, C_1 and C_2, oscillating in the same vertical plane and hanging from an upper horizontal wall at two points P_1 and P_2, with P_1 at the left of P_2. We shall follow in the x, y plane the motion of the rp's of the clocks.

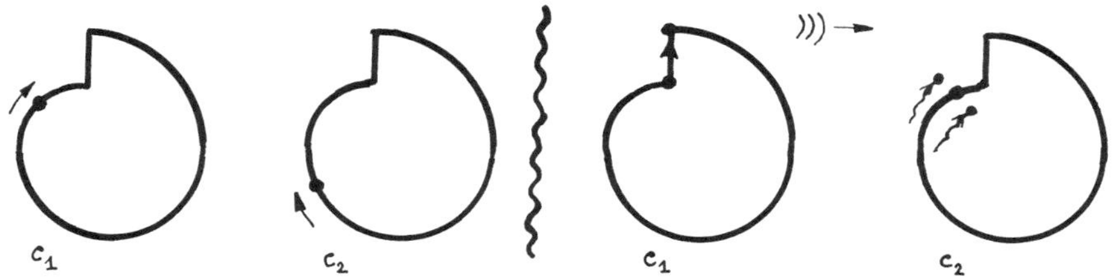

Let us supopose that there is a phase difference τ_0 between the two clocks, or, more precisely, that the rp of C_1 goes τ_0 units of time in advance over that of C_2, with $\tau_0 \varepsilon (0,\pi/2)$. Now when the rp of C_1 reaches and this collision gives rise to a perturbation propagating all around, that is, a progressive wave which travels along the wall with compressions and expansions. When this wave arrives over P_2, the fixed point of the other pendulum, this one undergoes a very quick swelling motion and this perturbation will then travel down along the rigid arm of the pendulum until it reaches its lower extremity. For the sake of simplicity, we may admit that the propagation of the disturbances (along the wall and along the arm of the pendulum) is instantaneous. Besides we may even consider that the perturbation is given, mathematically speaking, by a very small sinusoidal force acting very quickly during only one period. This means that during this very short interval of time, C_2 is no longer ruled by (7a) but, instead, by

$$\ddot{x}(t) + x(t) = -\beta + \Theta \sin \omega(t-t_o) \quad (\Theta, \omega^{-1} \ll 1). \tag{10}$$

When this force ceases to act, the rp of C_2 is <u>out</u> of the limit cycle, at a state which is different from the one it would have if the force had not acted. This is the way the disturbance gives rise to a variation of the phase difference: if the motion of C_2 had not been disturbed, its rp would pursue its motion <u>upon</u> the limit cycle, and the phase difference between C_1 and C_2 would remain the same; but the disturbance will certainly throw C_2 out of the limit cycle, giving rise to a small phase difference with regard to the unperturbed motion, and so the phase difference of the two clocks will change. We find that this variation is equal to

$$\Delta^1 = \frac{\Omega}{2} \cos \tau_o \text{, with} \tag{11}$$
$$\Omega = 4\Pi\Theta\,(\omega^2 - 1)^{-1},$$

where τ_o is the phase difference of the two clocks before the perturbation. This result is easily obtained by integrating with the same initial conditions, both equations (7a) and (10), that is, the unperturbed and perturbed equations; from this we can calculate the position of both rp's after the perturbation, and then obtain the angle defined by both, which is the variation of the phase difference.

After this, the motion of the two clocks goes on, and time will come when it is C_2 that reaches its maximum value for the velocity. It then receives a sudden supply of energy, by collision with its own escapment. We now have another progressive wave generated by this sudden perturbation, travelling from the right to the left, which will arrive upon P_1 and disturb the pendulum of C_1 exactly in the same way, and giving rise another small variation of the phase difference. We can proceed exactly as before and find for this (second) variation

$$(13) \qquad \Delta'^1 = \frac{\Omega}{2} \cos\,(\tau_o - \Delta^1) \cong \frac{\Omega}{2} \cos \tau_o ,$$

where $\tau_o - \Delta^1$ is still the phase difference between the two clocks before the perturbation. We thus see that during a single period of the clocks, the phase difference, which was τ_o when we started, becomes less than $\tau_o - \delta_1$, where

$$(14) \qquad \delta_1 \equiv 2\Delta_1 = \Omega \cos \tau_o$$

Let us recall that we have admitted that $\tau_o \varepsilon (0,\pi/2)$. We could undertake the same reasoning for $\tau_o > \pi/2$, and we would then arrive at the conclusion that the phase difference always <u>decreases</u> δ_1 unities of time, with δ_1 given by (14); but since $\cos \tau_o$ may now be negative, this means that for some values of the phase difference there may be an <u>increasing</u> of the phase difference. More precisely, we see that for τ_o (=advance of C_1 over C_2) ε $(0,\pi/2)$, $(3\pi/2)$, $(3\pi/2, 2\pi)$, δ_1 (that is, the variation of the phase difference during one period) is then <u>decreasing</u>; and for $\tau_o\ \varepsilon\ (\pi/2\ ,3\pi/2)$, then δ_1 is <u>increasing</u>.

To put it briefly, this means that no matter the value we have for the initial phase difference between C_1 and C_2, this will always tend asymptocally to what we might call a "<u>stable equilibrium</u>" value for the phase difference, equal to $3\pi/2$. We see that there is also an opposite situation: the one where the initial phase difference is equal to $\pi/2$; this is an "<u>unstable equilibrium</u>" for the phase difference since the slightest disturbance will place it wether in $\tau_o\ \varepsilon\ (0,\pi/2)$ or in $\tau_o\ \varepsilon\ (\pi/2,\pi)$, and in both cases we still find the same asymptotical phase difference $3\pi/2$.

This is the qualitative theory of the Huyghens Effect. We

are now going to follow quantitatively the transient states and seek for some information about the time it takes to have a certain variation for the phase difference.

We have seen that during a period of the two clocks, that is, during an interval of time $(t_o, t_o + 2\pi)$, the phase difference has changed by δ_1, gives by (14). More generally, during an interval of time $(t_o, t = 2\pi N)$, with $N >> 1$, the total variation of the phase difference must be

(15) $$\sum_{k=1}^{v} \delta_k = \sum_{k=1}^{v} \Omega \cos (\tau_o - \sum_{s=1}^{k-1} \delta_s).$$

Let us call τ (t) the phase difference between the two clocks at instant t (Of course, we have $\tau(t_o) = \tau_o$). It then comes

(16) $$\tau(t) = \tau_o - \sum_{1}^{v} \delta_k.$$

From this we may calculate $\tau(t = 2\pi(k+1))$ and $\tau(t = 2\pi k)$; if we take their difference and divide by 2π, what we get, in the limit, is just

(17) $$\dot{\tau} (t) = - \frac{\Omega}{2 \pi} \cos \tau(t).$$

This is the equation that rules the variation of the phase difference in time. It is an hereditary equation: The phase difference between the two clocks is a functional of the whole past story of the system.

Of course, the integration of (17) is not an easy task, but nevertheless we may get some information from it, by some approximate calculations. Let us suppose that the starting value of the phase difference is $\tau_o \in (0, \pi/2)$; then, as we have said, it will <u>tend</u> to $3\pi/2$, a value that is reached asymptotically, that is, for $t = \infty$. But in order to have an estimate about how strong is this asymptotical tendency, let us determine, for instance, the time needed for the two clocks to get into coincidence of phases (NB: of course, $\tau(\infty) = 3\pi/2 = - \pi/2$.). Now since Ω is very small, we are going to replace equation (17) by the aproximate one

$$\dot{\tau} = \frac{\Omega}{2\pi}\left(\frac{2}{\pi}|\tau| - 1\right). \tag{18}$$

Now the integration is obvious and we get

$$\tau(t) = \begin{cases} (\tau_o - \pi/2)\exp. \frac{\Omega}{\pi^2}(t - t_0) + \pi/2 & \text{for } \tau\varepsilon(o,\tau_o) \\ \pi/2.\ \exp. -\frac{\Omega}{\pi^2}(t - t') - \pi/2 & \text{for } \tau\varepsilon(-\pi/2, 0), \end{cases} \tag{19}$$

where t' is precisely the instant of time where $\tau(t') = 0$, that is, where we have the coincidence of phases, and the calculation gives for it the following expression

$$t' = t_0 - \frac{\pi(\omega^2-1)}{4\,\Theta}\log\left(1 - \frac{2\tau_o}{\pi}\right). \tag{20}$$

This value thus provides an approximate value of the time needed for the two clocks to come into the coincidence of phase. It is worth nothing that this value depends only upon ω, Θ, that is, the amplitue and the period of the small disturbance, and that it is independent of the inner structure of the clocks (the constants α, β , etc.).

BIBLIOGRAPHY

(1) Huyghens, K., Oeuvres Complètes, Vol. V, La Haye, 1893, pp. 243-4

(2) Andronov, A.A., Vitt, A.A., Khaikin, S. E., Theory of Oscillators, Pergamon, 1963.

SIMPLE CLOSED GEODESICS ON SURFACES

André AVEZ

University Pierre et Marie Curie (Paris).

1. INTRODUCTION

Any geodesic $\gamma : R \to S$ of a compact surface S is defined for all $t \in R$. If $\gamma(p) = \gamma(o)$, $\dot{\gamma}(p) = \dot{\gamma}(o)$ for some $p > o$, γ is called a closed geodesic. If, moreover, $\gamma(u) = \gamma(v)$ and $o \leqslant u < v \leqslant p$ imply $u = o$, $v = p$, then γ is called a simple closed geodesic (no double point).

Geodesics of the standard sphere are simple and closed. This is exceptional and one may wonder if there exists at least a simple closed geodesic on a surface S diffeomorphic to a sphere. This problem deserved a lot of works (see [5]). The most recent one revives a heuristic method of Poincaré and its Authors [2] proved the following: suppose the Gaussian curvature $K > o$, and $|\operatorname{grad} K| < 2.K^{3/2}$ on S, then there exists a simple closed geodesic. Their proof is a hard one and makes strong use of $K > o$. My aim is to prove a similar result under weaker assumptions and with more elementary tools.

We begin, in Parts 1 to 7, to collect some useful and well-known facts.

2. THE UNIT TANGENT BUNDLE OF A SURFACE S .

This is the union $M = (TS)_1$ of the unit tangent vectors to the 2-dimensional Riemannian manifold S .

ISBN 0-12-068720-8

The natural projection $\pi : M \to S$ maps a vector u of M tangent at $x \in S$ to its origin x .

M has a natural structure of 3-dimensional manifold, with respect to which π is smooth : given isothermal coordinates (x^1, x^2) on S, the Riemannian metric reads $g = e^{2A} \cdot [dx^1 \otimes dx^1 + dx^2 \otimes dx^2]$, where $A \in C^\infty(S)$. Therefore the components of $u \in M$ satisfy $e^{2A} \cdot [(u^1)^2 + (u^2)^2] = 1$ and we may write $u^1 = e^{-A} \cdot \cos\theta$, $u^2 = e^{-A} \cdot \sin\theta$. Then (x^1, x^2, θ) are coordinates on M, in which π reads : $(x^1, x^2, \theta) \rightsquigarrow (x^1, x^2)$.

2.1. Example

If S is diffeomorphic to the standard sphere $S^2 = \{(x,y,z) : x^2 + y^2 + z^2 = 1\}$, then $(TS)_1$ is diffeomorphic to the rotation group SO(3), i.e. the 3-dimensional real projective space S^3/Z_2 .

Sketch of proof : We may assume $S = S^2$. Fix $y_o \in S^2 \subset R^3$ and a unit vector u_o tangent at y_o . Given $(y,u) \in (TS^2)_1$ there exists a unique rotation $r \in SO(3)$ which maps the orthonormed frame $(y_o, u_o, y_o \times x_o)$ to $(y, u, y \times x)$.

2.2. The Sasaki's metric (see [1])

If ∇ is the covariant derivative of g and $u \in M$, the Sasaki's metric is $G = g + g(\nabla u, \nabla u)$ (see Appendix).

For each $b \in S$ the fiber $\pi^{-1}(b)$ is the set of the unit vectors tangent at b to S ; it is diffeomorphic to the circle. A vector field on M is called horizontal if it is always G-orthogonal to the fibers. It is easy to prove that G is characterized by the following property : the derivative map of π preserves lengths of horizontal vectors. In other words G makes π a Riemannian submersion [8].

It can be proved that the unit tangent bundle M of the standard sphere, endowed with G, has constant curvature 1/4. Therefore its universal covering $\widetilde{M}$ is actually isometric to the usual sphere S(o, 2) of radius 2 of R^4.

3. THE GEODESIC FIELD.

Given a unit vector u tangent to S at x, there is a unique geodesic γ such that $\gamma(o) = x$, $\dot{\gamma}(o) = u$. Since conservation of energy implies $|\dot{\gamma}(s)| = |\dot{\gamma}(o)| = 1$, s is the arc-length of γ counted from x .

The map $M \times \mathbb{R} \to M$, which sends $((x,u),s)$ to $(\gamma(s), \dot{\gamma}(s))$, is a one-parameter group of diffeomorphisms of M, called the geodesic flow. Its infinitesimal generator is a vector field X on M called the geodesic field.

<u>3.1. Example</u>

The equations of the geodesics, parametrized by arc-length, of the standard plane $\{(x^1,x^2)\}$ are : $\dot{x}^1 = u^1$, $\dot{x}^2 = u^2$, $\dot{u}^1 = \dot{u}^2 = 0$.

In the coordinates (x^1, x^2, θ) we saw that $u^1 = \cos\theta$, $u^2 = \sin\theta$. Hence $\dot{x}^1 = \cos\theta$, $\dot{x}^2 = \sin\theta$, $\dot{\theta} = 0$ and the geodesic field is $X = (\cos\theta, \sin\theta, o)$.

4. GEODESIC FIELD AND STRUCTURAL EQUATIONS.

Write the Riemannian metric of S in isothermal coordinates : $g = e^{2A} \cdot [dx^1 \otimes dx^1 + dx^2 \otimes dx^2]$. The equations of the geodesics are the Lagrange equations for $L = g(\dot{x}, \dot{x})$:

$$\frac{d}{ds}(e^{2A} \cdot \dot{x}^i) - \partial_i A = 0, \qquad i = 1,2 . \tag{4.1}$$

If s is the arc-length, conservation of energy gives $L = 1$. Therefore $\dot{x}^1 = e^{-A} \cdot \cos\theta$, $\dot{x}^2 = e^{-A} \cdot \sin\theta$ and (4.1) gives

$$\dot{\theta} = e^{-A} \cdot (\cos\theta \cdot \partial_2 A - \sin\theta \cdot \partial_1 A) .$$

Then the components of the geodesic field are

$$X = (e^{-A}\cdot\cos\theta,\ e^{-A}\cdot\sin\theta,\ e^{-A}(\cos\theta\cdot\partial_2 A - \sin\theta\cdot\partial_1 A)). \tag{4.2}$$

Assume <u>S is orientable</u>. We may define an action of the circle group S^1 on M : to each unit vector u tangent to S at x and to $t \in \mathbb{R}/2\pi\mathbb{Z}$ we make correspond the tangent vector at x obtained from u by counter-clock rotation of t radians. In coordinates : $(x^1,x^2,\theta) \rightsquigarrow (x^1,x^2,\theta + t)$. The generator of this flow is

(4.3) $T = (o, o, 1)$

and its orbits are circles, which are nothing but the fibers $\pi^{-1}(x)$ of the unit tangent bundle.

If we change θ into $\theta + \frac{\pi}{2}$ in (4.2), we get that L.W. Green called the second geodesic field :

(4.4) $Y = (-e^{-A}.\sin\theta, e^{-A}.\cos\theta, -e^{-A}(\sin\theta.\ \partial_2 A + \cos\theta.\ \partial_1 A))$

Since the Gaussian curvatuve K of S is -div (grad A) is isothermal coordinates, direct computations yield the Lie brackets

(4.5) $[Y, T] = X, \quad [T, X] = Y, \quad [X, Y] = K\cdot T$

They are the structure equations of the Riemannian connection written in vector notation.

4.6. Example : the standard sphere S^2

Here K = 1 and (4.5) becomes the multiplication table of the Lie algebra of SO(3). The unitary tangent bundle M may be identified with SO(3) and its universal covering $\widetilde{M}$ with the sphere S(o,2). The circle orbits of T are the fibers of $\pi : M \to S^2$ and the π - projection of the circle orbits of X are the geodesics of S^2, i.e., the great circles.

5. REMARK

From the very definition of X the π - projection of an orbit of X is a geodesic of S, and closed orbits of X correspond bijectively to closed geodesics of S . Let $\gamma : [o,p] \to S$ be a closed geodesic of S, of minimal period $p > o$. It is the π - projection of a closed orbit $\widetilde{\gamma} : [o,p] \to M$ of X : $\widetilde{\gamma}(s) = (\gamma(s), \dot{\gamma}(s))$. If γ self-intersects at $a \in S$, $\widetilde{\gamma}$ cuts the fiber $\pi^{-1}(a)$ in several distinct points and conversely. Therefore : closed simple geodesics γ of S correspond bijectively to closed orbits $\widetilde{\gamma}$ of X such that $\widetilde{\gamma}$ cuts each circle orbit of T at most in one point.

Warning : changing the orientation of γ reverses the unit tangent vector $\dot{\gamma}(s)$: $-\gamma$ is obtained from γ by the action $\theta \rightsquigarrow \theta + \pi$ of S^1 . Therefore γ and $-\gamma$ cut a circle orbit of T in two antipodal points.

6. HOPF FIBRATION.

A point M of mass +1 of the plane $\{q = (q_1, q_2)\}$ is attracted by the origin o proportionately to the distance $|oM|$. The equations of the motion in the phase space $R^4 = \{(q_1, p_1, q_2, p_2)\}$ are $\dot{q}_1 = p_1$, $\dot{p}_1 = -q_1$, $\dot{q}_2 = p_2$, $\dot{p}_2 = -q_2$. Since the Hamiltonian $H = \frac{1}{2}[\,|q|^2 + |p|^2\,]$ is a first integral, the spheres $S(o,r) = H^{-1}(r^2/2)$ are invariant and the generator

$$T' = (p_1, -q_1, p_2, -q_2) \tag{6.1}$$

induces a vector field on $S(o,r)$.

Let us identify $\mathbf{R}^4$ with $\mathbf{C}^2 = \{(u = q_1 + ip_1, v = q_2 + ip_2)\}$. The equations of the motions become $\dot{u} = -iu$, $\dot{v} = -iv$; therefore the orbits are the great circle $u = u(o)\cdot e^{-it}$, $v = v(o)\cdot e^{-it}$ of $S(o,r)$. This is the so-called Hopf fibration of $S(o,r)$.

We define two other vector fields on $R^4 = \mathbf{C}^2$:

$$\begin{aligned} X' &= (q_2, -p_2, -q_1, p_1) = (\bar{v}, -\bar{u}), \\ Y' &= (p_2, q_2, -p_1, -q_1) = (i\bar{v}, -i\bar{u}). \end{aligned} \tag{6.2}$$

Since $T' = -i(u,v)$, X' is obtained from T' by the rotation $(u, v) \rightsquigarrow (i\bar{v}, -i\bar{u})$ of $\mathbf{R}^4$. Then X' defines also a Hopf fibration of $S(o,r)$.

It is readily seen that (om, T', X', Y') is an orthonormed frame at $m \in S^3 = S(o,1)$.

Direct computations yields

$[X', Y'] = 2T'$, $[T', X'] = 2Y'$, $[Y', T'] = 2X'$.

Now we put $2X = X'$, $2Y = Y'$, $2T = T'$. Since X, Y, T are invariant under the antipodal map $om \rightsquigarrow -om$, they define three vector fields on $SO(3) = S(o,2)/Z^2$. Their orbits are great circles on $S(o,2)$ and, since $|X| = |Y| = |T| = 1$, their period is 4π on $S(o,2)$ and 2π on $SO(3)$. Obviously, $[X,Y] = T$, $[T,X] = Y$, $[Y,T] = X$.

This allows us to identify (see [4]) SO(3) with the unit

tangent bundle of the standard sphere S^2 and X, Y, T respectively with the first, the second geodesic fields and the circle field of S^2.

7. B. FULLER'S RESULTS ([3], [4], [7]).

Let T_1 be another vector field on S^3. We may think of T_1 as a perturbation of T'. Does this perturbation destroy the periodicity of the orbits of T' ? Improving a Seifert's result Fuller proved that if $|T_1(m) - T'(m)| < \frac{1}{3}$ for all $m \in S^3$, then T_1 has at least a periodic orbit, the period of which is close to 2π.

Of course the some results holds if we replace S^3 by S(o,2) and T' by T (warning : then, the period is close to 4π).

8. NEW FORMULATION OF THE PROBLEM.

We want to prove that a surface S, diffeomorphic to a sphere and C^2 close enough to S^2, possesses a simple closed geodesic.

Let be *can* the canonical metric of the standard sphere. The uniformization theorem says that an arbitrary Riemannian metric g' on S^2 is conformally equivalent to *can* : $g' = e^{2v} \cdot can$, where $v \in C^\infty(S^2)$.

Therefore isothermal coordinates for *can* are still isothermal for g' and g' is obtained from $can = e^{2A} \cdot [dx^1 \otimes dx^1 + dx^2 \otimes dx^2]$ by substitution $A \rightsquigarrow A + v$. Consequently, changing A into A + v in (4.2) gives the geodesic field X' of g' :

$X' = e^{-v}[X + e^{-A} \cdot (\cos\theta \, . \, \partial_2 v - \sin\theta \cdot \partial_1 v) \cdot T]$.

Using (4.4) we may rewrite

$X' = e^{-v}[X + Y(v) \cdot T]$. Since $e^{-v} > o$, X' and $\overline{X} = X + Y(v) \cdot T$ have the same orbits. Now, using Remark 5, our problem becomes : given $v \in C^\infty(S^2)$ [or, equivalently, a smooth function v on S^3, such that $v(om) = v(-om)$ and $T(v) = \frac{\partial v}{\partial \theta} = o$] does the vector field $\overline{X}$ possesses a closed oriented orbit $\tilde{\gamma}$ which cuts each orbit of T at most in one points ?

9. THEOREM.

Let v be a smooth function on the standard sphere (S^2, can), such that $|\operatorname{grad} v| < \frac{1}{3}$. Then $(S^2, e^{2v}.can)$ possesses a closed geodesic, which is simple if, moreover, v is C^2-close enough to zero.

Proof. (4.2) and (4.4) imply

$|X(v)|^2 + |Y(v)|^2 = |\operatorname{grad} v|^2$. Since $|T| = 1$,

$|\overline{X} - X| = |Y(v)| \leqslant |\operatorname{grad} v| < \frac{1}{3}$ and Fuller's result shows that $\overline{X}$ has a closed orbit γ' on $S(o,2)$.

Now, let us assume more : v is C^2-close enough to zero. Then (e.g. [7]) we may suppose that the length of γ' is close to 4π, and we argue by contradiction to prove that γ' is simple.

Suppose γ' cuts an orbit of T in two distinct points a and b, and let γ be the orbit of X issuying from a. We introduce $f(t) = |\gamma(t) - \gamma'(t)|$. Since

$\gamma = \gamma(t) = (q_1, p_1, q_2, p_2)$,

$\gamma' = \gamma'(t) = (q_1', p_1', q_2', p_2')$,

$\dot{\gamma}(t) = X(\gamma) = \frac{1}{2}(q_2, -p_2, -q_1, p_1)$,

$\dot{\gamma}'(t) = \overline{X}(\gamma') = X(\gamma') + a'\cdot T(\gamma')$, where $a' = Y(v)(\gamma')$, and

$T(\gamma') = \frac{1}{2}(p_1', -q_1', p_2', -q_2')$, we find

$\frac{d}{dt} f^2 = 2\cdot < \gamma - \gamma', \dot{\gamma} - \dot{\gamma}' > = -2\cdot a' \cdot < \gamma - \gamma', T(\gamma') >$. Then, $|T| = 1$, $|Y(v)| < \delta$ and the Schwarz' inequality give $|\dot{f}| \leqslant \delta$. Since $f(o) = o$, we get $f(t) \leqslant \delta \cdot |t|$, and, if $M = \operatorname{Max}(4\pi, p)$, $f(t) \leqslant \frac{1}{2}\delta\cdot M$. Therefore γ' lies in the tubular neighbourhood of γ of radius $\delta\cdot M/2$ on $S(o,2)$.

Let us call P the plane of the great circle γ, and ν the orthogonal projection onto P. Starting at a and following γ' until b we get an arc ab. Elementary Geometry shows that the projection $\nu(ab)$ lies in the annulus of P bounded by γ and a circle of radius $r = |\nu(b) - o| > 2\cdot\left(1 - \frac{\delta^2 \cdot M^2}{32}\right)$.

Therefore length (ab) $\geqslant$ length $\nu(ab) \geqslant 2\pi r$. Let us keep following γ' from b until going back to a : we get an arc ba, with the same estimate. Finally

$p = \text{length } \gamma' = \text{length}(ab) + \text{length}(ba) \geqslant 4\pi r \geqslant 4\pi\cdot\left(2 - \frac{\delta^2 \cdot M^2}{16}\right)$.

Since $M = \operatorname{Max}(4\pi, p)$, this is impossible if $\delta\cdot M < 4$; in particular if δ is small enough and if p is bounded from above by a number close to 4π .

APPENDIX : GEOMETRY OF THE UNIT TANGENT BUNDLE.

1. THE SASAKI'S METRIC (1).

In isothermal coordinates (x^1, x^2) the Riemannian metric of the surface S has components $g_{11} = g_{22} = e^{2A}$, $g_{12} = o$; $g^{11} = g^{22} = e^{-2A}$, $g^{12} = o$ and their Christoffel symbols are $\gamma^1_{11} = -\gamma^1_{22} = \gamma^2_{12} = \partial_1 A$, $\gamma^1_{12} = -\gamma^2_{11} = \gamma^2_{22} = \partial_2 A$.
This allows to compute $\nabla u^i = du^i + \gamma^i_{rs} \cdot u^r \cdot dx^s$, where $u^1 = e^{-A} \cdot \cos\theta$, $u^2 = e^{-A} \cdot \sin\theta$ are the components of $u \in M = (TS)_1$. Therefore the components of the Sasaki's metric G are

$$\text{(A.1)} \quad \begin{array}{lll} G_{11} = e^{2A} + (\partial_2 A)^2, & G_{12} = -\partial_1 A \cdot \partial_2 A, & G_{13} = -\partial_2 A \\ G_{22} = e^{2A} + (\partial_1 A)^2, & G_{23} = \partial_1 A, & G_{23} = 1 \end{array}$$

and the contravariant components are

$$\text{(A.2)} \quad \begin{array}{lll} G^{11} = e^{-2A}, & G^{12} = o, & G^{13} = e^{-2A} \cdot \partial_2 A, \\ G^{22} = e^{-2A}, & G^{23} = -e^{-2A} \cdot \partial_1 A, & G^{33} = 1 + \Delta_1 A, \end{array}$$

where $\Delta_1 A = e^{-2A} \cdot [(\partial_1 A)^2 + (\partial_2 A)^2] = |\text{grad } A|^2$.

We may compute the Christoffel symbols of G. For instance, if K is the Gaussian curvature,

$$\text{(A.3)} \quad \Gamma^1_{13} = \Gamma^1_{33} = o, \qquad \Gamma^1_{23} = K/2 \; .$$

For the sake of simplicity we shall put $G(a,b) = \langle a,b \rangle$ and $|a| = \sqrt{\langle a,a \rangle}$.

2. THE VECTOR FIELDS X, Y, T .

Formulas (4.2), (4.3) and (4.4) give the components of X, Y, T in the coordinates (x^1, x^2, θ); with the help of (A.1) we see that (X, Y, T) is an orthonormed frame of M, which is direct because $X = Y \times T$.

Using the well-known formula $\operatorname{div} Z = G'^{-1/2} \cdot \partial_i(\sqrt{G'} \cdot Z^i)$, where $G' = \text{dét}(G_{ij}) = e^{4A}$, we obtain

(A.4) $\operatorname{div} X = \operatorname{div} Y = \operatorname{div} T = o.$

From the classical formula

$\operatorname{curl}(a \times b) = [b,a] + a\cdot \operatorname{div} b - b\cdot \operatorname{div} a$, and with (4.5) and (A.4) we deduce

(A.5) $\operatorname{curl} X = -X, \quad \operatorname{curl} Y = -Y, \quad \operatorname{curl} T = -K\cdot T$

The first one is the Bernoulli formula for the velocity vector X of a perfect fluid. This leads to a generalization of the closed geodesic problem : given a vector field X on a compact 3-dimensional manifold, such that $|X| = 1$ and $\operatorname{curl} X = -X$, does there exist periodics orbits ?

3. THE ORBITS OF X, Y, T ARE GEODESICS OF M.

Let us denote by $\tilde{X}$ the 1-form $< X,\cdot >$, by $*$ the Hodge operator and by v the volume element of G. Formulas (A.5) imply $* d\tilde{X} = \widetilde{\operatorname{curl} X} = -\tilde{X}$; then $i_X d\tilde{X} = -i_X * \tilde{X} = -i_X i_X v = o$. In coordinates : $X^r(\nabla_r X_s - \nabla_s X_r) = o$. Since $|X| = 1$, we deduce

(A.6) $\nabla_X X = o$, and similarly $\nabla_Y Y = o$.

On the other hand the components of G in (x^1, x^2, θ) do not involve θ. Therefore $\theta \rightsquigarrow \theta + t$ is an isometry group of G and its generator T satisfies $\nabla_r T_s + \nabla_s T_r = o$. Since $|T| = 1$, we deduce

(A.7) $\nabla_T T = o$

Hence the orbits of T are simple closed geodesics of M.

4. THE INVARIANT DERIVATIVES OF X, Y, T.

Computation of $\nabla_X T$: Since $|T| = 1$, $< \nabla_X T, T> = 1$. Since $\nabla_X X = o$, $<T, X> = o$, we get $<\nabla_X T, X> = X <T,X> - < T, \nabla_X X> = o$. Therefore $\nabla_X T = a\cdot Y$. But $T = (o,o,1)$, and (A.3) implies $(\nabla_X T)^1 = X^r \partial_r T^1 + \Gamma^1_{rs} X^r T^s = \frac{K}{2}\cdot X^2 = -\frac{K}{2}\cdot Y^1$. Finally $a = 1$ and $\nabla_X T = -\frac{K}{2}\cdot Y$.

Computation of $\nabla_T X$: From the previous formula and (4.5) : $\nabla_T X = [T,X] + \nabla_X T = (1 - \frac{K}{2})\cdot Y$

Computation of $\nabla_Y T$: Similar to that of $\nabla_X T$: $\nabla_Y T = \frac{K}{2}\cdot X$

Computation of $\nabla_T Y$: Similar to that of $\nabla_T X$: $\nabla_T Y = (\frac{K}{2} - 1)\cdot X$

Computation of $\nabla_X Y$: Since $|Y| = 1$, $< \nabla_X Y, Y > = o$. Since $\nabla_X X = o$, $<Y,X> = o$, we get $< \nabla_X Y, X> = X<Y,X> - <Y, \nabla_X X> = o$. Finally $< \nabla_X Y, T > = X < Y,T > - < Y, \nabla_X T >$; and since we know $\nabla_X T$ and $< Y,T > = o$, $\nabla_X Y = \frac{K}{2} T$.

Computation of $\nabla_Y X$: Similar to that of $\nabla_T X$: $\nabla_Y X = -\frac{K}{2}\cdot T$.

To summarize we have

$$(A.8)\quad \begin{cases} \nabla_X X = o, & \nabla_Y X = -\frac{K}{2}\cdot T, & \nabla_T X = (1-\frac{K}{2})\cdot Y, \\ \nabla_X Y = \frac{K}{2}\cdot T, & \nabla_Y Y = o, & \nabla_T Y = (\frac{K}{2} - 1)\cdot X, \\ \nabla_X T = -\frac{K}{2}\cdot Y, & \nabla_Y T = \frac{K}{2}\cdot X, & \nabla_T T = o . \end{cases}$$

5. THE CURVATURE OF $(TS)_1$.

We compute the curvature in the orthonormed frame (X,Y,T) with (A.8). Recall that the Riemann curvature tensor of M is defined by $R(X_1,X_2,X_3,X_4) = < \nabla_{X_3}\nabla_{X_4}X_2 - \nabla_{X_4}\nabla_{X_3}X_2 - \nabla_{[X_3,X_4]}X_2 , X_1 >$. Taking into account the well-known properties of R, we need only the significant components.

Computation of R(X,Y,X,Y) :

$$R(X,Y,X,Y) = < \nabla_X\nabla_Y Y - \nabla_Y\nabla_X Y - \nabla_{[X,Y]} Y, X > = < -\nabla_Y(\frac{K}{2}T) - \nabla_{KT}Y, X > =$$

$$< -\frac{K}{2}\cdot\nabla_Y T - K\cdot\nabla_T Y, X > =$$

$$< -\frac{K^2}{4}X - K(\frac{K}{2}-1)\cdot X, X > = K - \frac{3}{4}K^2 .$$

Similar computations yields

$$R(X,Y,X,T) = \frac{1}{2}\cdot\nabla_X K ,$$

$$R(X,Y,T,Y) = -\frac{1}{2}\cdot\nabla_Y K,$$

$$R(X,T,X,T) = K^2/4$$

$$R(X,T,Y,T) = o$$

$$R(Y,T,Y,T) = K^2/4 .$$

Consequences :

1. M has constant curvature $\frac{K^2}{4}$ if, and only if, K = o or 1.

The "if" part was proved by Klingenberg and Sasaki [6].

2. M has positive curvature if, and only if, $o < K < 2$ and $|\operatorname{grad} K|^2 < K^3 \cdot (2 - K)$. Observe that Berger-Bombieri's assumptions are fullfilled if M has positive curvature.

REFERENCES.

[1] Avez, A. and Buzzanca, C., Flusso geodetico sul fibrato unitario tangente di una superficie, Rend. Circ. Mat. Palermo, 25, 1976, pp.176-182

[2] Berger, M.S. and Bombieri, E., On Poincaré isoperimetric problem for simple closed geodesics, J. funct. Analysis, 42, 1981, pp.274-298

[3] Fuller, F.B., *The existence of periodic orbit*, Institut Math. Strasbourg, 1968, I.R.M.A., 67 rue R. Descartes, Strasbourg (France).

[4] Fuller, F.B., An index of fixed point type for periodic orbits, Amer. J. Math. 89, 1967, pp.133-148.

[5] Klingenberg, W., *Lectures on closed geodesics*, Springer-Verlag (Berlin), 1978.

[6] Klingenberg, W. and Sasaki, S., On the tangent sphere bundle of the 2-sphere, Tohoku Math. J. 27, 1975, pp.49-56.

[7] Moser, J., Periodic orbits near an equilibrium and a theorem by Alan Weinstein, Comm. pure Appl. Math. 29, 1976, pp. 727-747.

[8] O'Neill, B., The fundamental equations of a submersion. Michigan Math. J. 13, 1966, pp. 459-469.

PART V

SOME PROBLEMS IN QUANTUM MECHANICS

SOME TOPICS IN THE QUANTUM THEORY OF MEASUREMENT

GianCarlo Ghirardi

Istituto di Fisica Teorica dell'Università, Trieste, Italy

and

International Centre for Theoretical Physics, Trieste, Italy

and

Alberto Rimini

Istituto di Fisica dell'Università, Salerno, Italy

1. INTRODUCTION

The quantum theory of measurement, according to the general formulation which has been given by von Neumann[1], can be described in the following way:

a) Suppose the quantity M to be measured is associated with the self-adjoint operator M. For simplicity we assume that M has a purely discrete non degenerate spectrum so that we write the corresponding eigenvalue equation as

$$M \,|\, m\rangle = m \,|\, m\rangle \tag{1}$$

b) A measurement of M is performed on a statistical ensemble of identically prepared quantum systems, i.e. we deal with a pure case, associated to the state vector $|\psi\rangle$

c) The probability of getting the result m in a measurement

ISBN 0-12-068720-8

of M is given by

$$P_{M=m} = |\langle m \mid \psi \rangle|^2 , \tag{2}$$

and those systems for which the same result m is obtained constitute, after the measurement a pure case associated with the state $|m\rangle$. This obviously implies that for the measurement on an ensemble associated to $|\psi\rangle$, the ensemble becomes, after the measurement, a statistical mixture of the pure states $|m\rangle$ with weights $|\langle m|\psi\rangle|^2$. We can then describe formally the measurement process using the language of the density operators,by saying that the system-apparatus interaction induces the following change in the density operator

$$\rho_{before} = |\psi\rangle\langle\psi| \rightarrow \sum_m P_m |\psi\rangle\langle\psi| P_m = \rho_{after} \tag{3}$$

Here P_m is the projection operator on state $|m\rangle$:

$$P_m = |m\rangle\langle m| . \tag{4}$$

Up to this point, we have considered the description of measurement process within the Hilbert space of the system. However actual measurements consist of the interaction of the measured system with a macro object, i.e. the measuring apparatus. It is then necessary to take into account an ensemble of systems and measuring apparatuses and to describe the process in the Hilbert space $H = H^S \times H^A$ which is the direct product of the Hilbert spaces of the system and of the apparatus. Moreover, in order that the apparatus acts for the purpose it is being used one should assume that those apparatuses which have found the systems in state $|m\rangle$ go into a quantum state $|A_m\rangle$ which is uniquely correlated to the obtained result m, so that they record the desired information about the outcome of the measurement. This leads to the idea of an ideal measurement process which takes place according to the following scheme:

i. There exists a macro-system (measuring device) which is such that when properly prepared in a state $| A >$, if a system in state $|m>$ interacts with it, the quantum mechanical evolution describing the interaction leads to a final state in which the system is still in the state $| m >$ and the apparatus is in state $| A_m >$. Denoting by V the unitary time evolution operator for the process we then assume

$$V \, |m \; A > = | m \; A_m > \tag{5}$$

ii. The states $|A_m >$ corresponding to different values of m must be orthogonal so that they can be identified by a subsequent measurement on the apparatus.

Let us make some remarks. First of all, one could realease the (unphysical) requirement that the apparatus be in a pure quantum state before and after the measurement process. However as well known no essential changes occur, both from practical and the conceptual point of view if we allow for mixtures. Secondly, we remark that the ideal measurement scheme summarized in (5) is non distorting in the sense that when the system is in an eigen state of the measured quantity it remains in the same state after the measurement.

When the initial state of the system is not in an eigenstate of the measured quantity,

$$| \psi > \; = \; \sum_m Q_m \, |m > \tag{6}$$

the scheme (5) implies, due to the linearity of the quantum evolution, the following relation:

$$V \mid \psi A > = \sum_m Q_m \mid m A_m > \tag{7}$$

The result of the measurement, as expressed by eq. (6) is different from what one would like to get, i.e. a statistical mixture with weights $|Q_m|^2$ of the final states $|m A_m>$.

From the above discussion it appears that the key problems of the quantum theory of measurement are

1. To justify and investigate the possibility of the occurence of the ideal scheme (5)
2. To justify the transition from the state at the r.h.s. of (6) to a statistical mixture, i.e. the so called"reduction of the wave pocket".

The second problem is probably the most controversial and crucial problem in the foundations of quantum mechanics. We will not discuss it here, we only mention that a clear understanding of the first problem could be useful also for a clarification of the second one. The main subject of this work will be a detailed discussion of the first problem. As we shall see, in fact, there are various situations in which the scheme (5) in conceptually untenable since it leads to contradictions.

2. LIMITATIONS TO THE IDEAL MEASUREMENT SCHEME

We shall first of all quote a general theorem which has been proved by various authors[2-4] following an original suggestion by Wigner[5], proving that there are many instances in which the ideal measurement scheme cannot hold.

Theorem: If there exist an addive conserved quantity $N^S + N^A$ of the system plus apparatus, and N^S is a bounded operator of H^S the ideal measurement scheme (5) is impossible, unless the measured quantity M commutes with N^S.

Formally

$$[N^S + N^A, V] = 0, \text{ and } [M, N^S] \neq 0$$

are incompatible with eq. (5).

The now quoted theorem is very important and actually puts severe limitations to the possibility of an ideal measurement scheme. In fact, in nature there are invariance principles which are thought to be valid for any type of phenomena and in particular they must govern also the interaction between the quantum systems and the measuring apparatuses. Associated with the invariance there are conservation laws for physical quantities, and, when these quantities are additive, they imply limitations on quantum measurements. A typical example is the one arising from the assumed isotropy of space implying rotational invariance and consequently the conservation law of total angular momentum. In this case if we denote by $\underline{S}$ the spin operators of the quantum system and by $\underline{L}$ the angular momentum of the apparatus (including in it also the orbital angular momentum of the quantum system) then an ideal measurement of a spin component, say S_3, is forbidden by the above theorem since $\underline{S} + \underline{L}$ is additive and conserved and S_3 does not commute with the bounded operators S_1 or S_2. There follows that no apparatus can measure S_3 according to the ideal scheme.

When this situation occur, one is compelled to consider, in

place of the ideal measurement scheme an approximate scheme. Of course, one assumes that the ideal scheme can still be valid at least in an approximate way, otherwise we would have to change completely our ideas concerning the measurement. As we will see shortly, the use of an approximate measurement scheme introduces the possibility of distortions of the initial state and of errors in the results of the measurements. All the terms appearing in thefinal state which are not present in the ideal scheme (5) are related to a "malfunctioning" of the apparatus. The problem is then that of minimizing this malfunctioning, compatibly with the existence of the additive conserved quantity, by relating it to some physically meaningful parameter which can be monitored to make very small the undesired terms. This problem has been discussed for the first time by Yanase in ref. (6), but only for the case of a spin ½ system. Moreover the derivation of the lower bound for the malfunctioning obtained in (6) is not mathematically rigorous. In ref. (7) we have given a rederivation of the Yanase bound which is much simpler and rigorous and in ref. (8) we have generalized the result to arbitrary spin values.

3. LIMITATIONS FOR SPIN MEASUREMENTS

Suppose we have a quantum system of spin S, and we want to measure the spin component S_3 of the system. The eigenvalue equation for S_3 is

$$S_3 \,| m > = m \hbar \,| m > , \; - S \leq m \leq S$$

and the ideal measurement scheme would be the same as (5) with the now specified meaning of the states $| m >$ and limitations on the values of m. Due to the conservation of total angular momentum the ideal scheme is forbidden and we have to introduce an approximate measurement scheme:

$$V|A\ m> = |A_m\ m> + \sum_{m \neq m'}^{S}{}_{-S} |\varepsilon_{mm'}, m'> \qquad (8)$$

with $<A_m | A_{m'}> = 0$ for $m \neq m'$. In order that the scheme (8) be a good approximation of the ideal scheme (5) and could then be an acceptable basis for the description of the measurement, we have to assume that the norms of the states $|\varepsilon_{m\,m'}>$ are much smaller than the norms of the states $|A_m>$ and can be made vanishingly small under some suitable limit. We remark that (8) is not the most general approximately ideal scheme due to the assumed orthogonality of the states $|A_m>$. This limitation could be easily removed and would not change the final result we are going to derive. We note that the scheme (8) contains distorting terms. As a parameter representing the non-ideality of the process (8) it is very natural to assume the total amount of distortion ε^2, defined as

$$\varepsilon^2 = \sum_{\substack{m,m' \\ m \neq m'}} || \, |\varepsilon_{m\,m'}> ||^2 \qquad (9)$$

To derive a lower bound for ε^2 we start by defining the set of (2 S + 1) x (2 S + 1) linearly independent (2 S + 1) dimensional matrices I_{ij}, (i,j = - s, ... + S) having only one element different from zero

$$(I_{ij})_{lk} = \delta_{il}\ \delta_{jk} \qquad (10)$$

Obviously, any operator acting in the spin space S can be written as a linear combination of these matrices. Since V is an operator of the Hilbert space $H^S \times H^A$, introducing properly defined operators C_{ij} on H^A we can write:

$$V = \sum_{i,j=-s}^{+s} C_{ij} I_{ij} \tag{11}$$

Unitarity of V implies

$$\sum_{r=-s}^{s} C^{+}_{ri} C_{rl} = \delta_{il} \tag{12}$$

Expressing the spin operators in terms of the matrices I_{ij} and imposing rotational invariance

$$[V, L_\alpha + S_\alpha] = 0 \qquad \alpha = 1,2,3 \tag{13}$$

one easily gets the commutation relations of the operators C_{ij} with the operators L_α

$$[C_{lk}, L_{\frac{1}{2}}] = \frac{\hbar}{2} \quad A(s,l)\, C_{l-1,k} \pm A(s,l+1) C_{l+1,k}$$

$$\mp A(s,k) C_{l,k-1} - A(s,k+1) C_{l,k+1} \tag{14}$$

$$[C_{lk}, L_3] = \hbar (l-k)\, C_{lk}$$

where $A(s,m) = \sqrt{(s+m)(s-m+1)}$
Moreover the apparatus states $|A_m>$ and $|\varepsilon_{m m'}>$ appearing in (8) are easily seen to be related to the operators C_{ij} and the initial apparatus state $|A>$ by

$$|A_m> = C_{m m} |A> , |\varepsilon_{m m'}> = C_{m' m} |A> \tag{15}$$

Let us evaluate the matrix element

$$\langle A,m\,|(L_1 + S_1)|\,A,\ m+1\rangle = \frac{\hbar}{2}\,A(s,\ m+1)$$

This matrix element equals the expression $\langle A,m\,|\,V^+(L_1+S_1)V\,|\,A,\ m+1\rangle$ due to eq. (13) and the unitarity of V. Using eq. (8) to express the states $V\,|\,A,\ j\rangle$ after some formal manipulations one then gets the equality[(8)]

$$\frac{\hbar}{2}\,A(s,m+1) = \sum_{m'} \langle A\,|\,C^+_{m'\,m}\,L_1\,C_{m'\,m+1}\,|A\rangle + R \tag{16}$$

where R is a quantity which vanish in the limit of all norms $\|\,|\,\varepsilon_{m\,m'}\rangle\,\|$ going to zero. Eq. (16) can also be written

$$\begin{aligned}\frac{\hbar}{2}\,A(s,m+1) = &\sum_{m'\leq m} \langle A|\,L_1 C^+_{m'\,m} C_{m'\,m+1}\,|\,A\rangle \\ &+ \sum_{m'\geq m+1} \langle A\,|C^+_{m'\,m} C_{m'\,m+1} L_1\,|A\rangle + R + R^1\end{aligned} \tag{17}$$

where R^1 contains the contributions coming from having commuted the C operators with L_1 in going from (16) to (17). Also these contributions vanish in the limit of vanishingly small $\|\,|\varepsilon_{m\,m'}\rangle\|$. Using the fact that the modulus of a sum is always smaller or equal to the sum of the moduly of its terms, applying the Schwarz inequality and using the unitarity relations (12) one can easily deduce from (17) the inequality (for the details of the derivation the reader is referred to ref. (8))

$$\begin{aligned}\frac{\hbar}{2}\,A(s,m+1) \leq \{&\sum_{m'\leq m} \|\,|\varepsilon_{m+1,m'}\rangle\,\|^2 + \sum_{m'\geq m+1} \\ &\|\,|\varepsilon_{m\,m'}\rangle\,\|^2\}^{1/2} \cdot \sqrt{2\langle A\,|\,L_1^2\,|\,A\rangle} + |\,R + R^1\,|\end{aligned} \tag{18}$$

This inequality shows that the norms of all $|\varepsilon_{m\,m'}\rangle$ states can be made arbitrarily small only if the quantity $< A | L_1^2 | A >$ is increased correspondingly. In this limit, one can disregard the term $| R + R^1 |$ in (18) obtaining

$$\sum_{m' \leq m} \| |\varepsilon_{m+1,m'} > \|^2 + \sum_{m' \geq m+1} \| |\varepsilon_{m\,m'} > \|^2 \geq \frac{\hbar^2 A^2(s,m+1)}{8 <A | L_1^2 | A>} \quad -S \leq m \leq S-1 \tag{19}$$

From eq. (19) one immediately gets a lower bound for the total amount of distorsion ε^2 defined in eq. (9)

$$\varepsilon^2 \geq \frac{2}{12 <A | L_1^2 | A >} s(s+1)(2s+1) \tag{20}$$

The bound (20) expresses the limitations occurring in the measurement of the third component of the spin for a system of spin s, imposed by the conservation law of the first component $L_1 + S_1$ of the total angular momentum. The conservation law for $L_2 + S_2$ implies an analogous bound while conservation of $L_3 + S_3$ does not give any lower bound for ε^2 since S_3 commutes with the measured quantity. For a spin -½ system eq. (20) gives the bound obtained in refs. (6, 7).

4. CONSTRUCTION AND STUDY OF A MODEL FOR A SPIN MEASUREMENT

In this section we will introduce a model for a spin measurement of a spin -½ particle. The reasons for which the study of an explicit model is useful are various. First of all even though we have obtained lower bounds for the parameter ε^2, and we could also have deduced from the previous derivation mathematical equations for the operators C_{ij} and the state $|A>$ guaranteeing that the bound can be actually attained (see ref. 7), we cannot prove that an operator V and a state $|A>$ satisfying these relations exists and correspond to a physically sensible apparatus. Secondly the mathematical study of the previous section does not give informations about the expansion of the states $|\varepsilon_{m\,m'}>$ in terms of the states $|A_m>$ and it can be easily seen that the features of the "malfunctioning" of the apparatus depend precisely on the form of such an expansion. Analogously we did not get any information about the terms (which may be present) which destroy the orthogonality of the states $|A_m>$. Finally, from the previous analysis we know that the deviations from the ideal scheme are imposed by the conservation laws but we do not know very much about the physical mechanism through which such deviations are generated. The construction of an explicit, physically reasonable model will help us in understanding all the questions quoted above.

Let us define now our model. We consider a spin -½ particle and denote by S_i its spin operators. We want to measure the third component S_3 of the spin of the particle. Let us write the eigenvalue equation for S_3:

$$S_3 |u_\pm> = \pm \frac{\hbar}{2} |u_\pm> \qquad (21)$$

We start by describing an apparatus working according to the ideal measurement scheme. Of course it will violate rotational invariance. Be M_i the operators associated to the orbital angular momentum of the particle. If a suitable interaction is introduced we can use the orbital motion of the particle as an apparatus to measure S_3.

This is just what happens for instance in a Stern-Gerlach experiment. Let us assume that the evolution of the particle is governed by the Hamiltonian

$$H = K_p + H_I \tag{22}$$

where K_p is the kinetic energy of the particle and H_I is a spin-orbit interaction term of the form:

$$H_I = \gamma S_3 M_3 \tag{23}$$

Let us assume that the operator $M^2 = \Sigma M_i^2$ has a fixed value equal to $\hbar^2 1(1 + 1)$, and denote by $| M = 1,0,-1>$ the three eigenstates of M_3. Assume that initially the orbit is in the state

$$|A> = \tfrac{1}{2} |M = 1> + \frac{1}{\sqrt{2}} |M = 0> + \tfrac{1}{2} |M = -1> \tag{24}$$

The time evolution of the initial states $| u_\pm A>$ is then given by

$$V(t) |u_\pm A> = |u_\pm> (\tfrac{1}{2} e^{\mp iet/\hbar} |M = 1> + \frac{1}{\sqrt{2}} |M = 0> + \tfrac{1}{2} e^{\pm iet/\hbar} |M = -1>) \tag{25}$$

where $e = \gamma \hbar^2/2$. If the interaction time T is chosen according to the relation

$$\frac{e\,T}{\hbar} = \frac{\Pi}{2} \tag{26}$$

we have

$$V |u_\pm A> = |u_\pm A_\pm> \tag{27}$$

where

$$|A_\pm> = \mp \frac{i}{2} |M = 1> + \frac{1}{\sqrt{2}} |M = 0> \pm \frac{i}{2} |M = -1>, \quad < A_+ |A_-> = 0 \tag{28}$$

The time evolution (27) corresponds to an ideal measurement scheme. Obviously this is possible only because the Hamiltonian H_I is not rotationally invariant. Since we believe that space is isotropic, the lack of rotational invariance can be due only to the interaction with an external object. The full rotational invariance can then be recovered by taking into account the rotational degrees of freedom of the external object. The Hamiltonian we have used so far should then correspond to the limit of the rotationally invariant Hamiltonian when the external object becomes infinitely massive. We then assume that the interaction between the spin and the orbit takes place via the interaction of both of them with a massive object which we shall call <u>the magnet</u>. The total Hamiltonian is then assumed to be

$$H_{tot} = K_p + K_m + H_I \tag{29}$$

where K_p and K_m are the kinetic energy operators for the particle and the magnet, respectively and

$$H_I = \gamma \, \underline{S} \cdot \underline{\underline{H}} \cdot \underline{M} \tag{30}$$

where $\underline{\underline{H}}$ is a rank - 2 tensor built up with the dynamical variables of the magnet. Note that the total Hamiltonian is rotationally invariant so that denoting with $\underline{N}$ the angular momentum of the magnet, the total angular momentum $\underline{S} + \underline{M} + \underline{N}$ is conserved. We take again $2\hbar^2$ as the fixed value of $\underline{M}^2$, and we observe that we can fix the values of K_p, $\underline{M}^2$, forget about K_p and use simply $K_m + H_I$ as the Hamiltonian. We further assume that $\underline{\underline{H}}$ depend only on the orientation ω of the magnet and that there is an orientation, say ω_o for which the value of $\underline{\underline{H}}$ is

$$\underline{\underline{H}}_o \equiv \underline{\underline{H}}(\omega_o) = \begin{pmatrix} 0 & 0 & 0 \\ 0 & 0 & 0 \\ 0 & 0 & 1 \end{pmatrix} \tag{31}$$

If the magnet is massive enough so that we can assume it is at rest in the orientation ω_o then H_I reduces to $\gamma\, S_3\, M_3$ and we go back to the evolution (27) corresponding to the ideal scheme. On the contrary if the moments of inertia of the magnet are finite we must take into account its motion and use the complete Hamiltonian $K_m + \gamma\, \underline{S} \cdot \underline{\underline{H}} \cdot \underline{M}$.

Let $|\psi\rangle$ be the initial state of the magnet. The total Hamiltonian induces during the time T an evolution which can be written

$$\begin{aligned} U\,|u_+ A\, \psi\rangle &= |u_+ A_+ X_+\rangle + |u_+ A_- Y^-_+\rangle + |u_+ A^1 Y^1_+\rangle + \\ &\quad + |u_- A_+ Z^+_+\rangle + |u_- A_- Z^-_+\rangle + |u_- A^1 Z^1_+\rangle \\ U\,|u_- A\, \psi\rangle &= |u_- A_- X_-\rangle + |u_- A_+ Y^-_+\rangle + |u_- A^1 Y^1_-\rangle \\ &\quad + |u_+ A_- Z^-_-\rangle + |u_+ A_+ Z^+_-\rangle + |u_- A^1 Z^1_-\rangle \end{aligned} \tag{32}$$

where $|A_\pm\rangle$ are the states of the orbit defined in eq. (28), $|A^1\rangle$ is a state of the orbit orthogonal to both $|A_+\rangle$ and $|A_-\rangle$,

$$|A^1\rangle = \frac{1}{\sqrt{2}}\,|M = 1\rangle + \frac{1}{\sqrt{2}}\,|M = -1\rangle \tag{33}$$

and $|X\rangle$, $|Y\rangle$, $|Z\rangle$ with various apices are proper apparatus states. Obviously in the conditions in which the apparatus is supposed to work the norms of the $|X\rangle$ states are expected to be of the order of 1, while the norms of the $|Y\rangle$ and $|Z\rangle$ states are expected to be very small. The final state of the magnet is never detected:just as in the ideal limiting case the result of the measurement is obtained by looking at the orbit. Note that the final state of the particle, due to the interaction with the magnet is an improper mixture.

The problem we are discussing consists then in determining the composition of such a mixture,i.e. to evaluate the squared norms of the states $|Y>$ and $|Z>$. For instance we will define

$$P^{(+)}_{u_- A_+} = <Z^+_+ | Z^+_+>$$

and similarly for the other terms. Note that $P^{(+)}_{u_- A_+}$ represents the probability that after the interaction of a state u_+ with the apparatus, the orbital state of the particle be A_+ while its spin state be u_-. This probability refers then to a distorsion of the initial state. Analogously $P^{(+)}_{u_+ A_-} = <Y^-_+ | Y^-_+>$ refers to the probability that even though the spin state is u_+ both before and after the measurement it is recorded by the orbital state $|A_->$ which is associated in the correctly functioning apparatus to a state u_- . Therefore $P^{(+)}_{u_+ A_-}$ represents a probability of an error in the measurement.

It can be useful to compare our model of measurement with the Stern-Gerlach apparatus. We note that in both cases

i) The orbit of the particle is a part of the apparatus

ii) The detection of the spin state is done by looking at the orbital state

iii) The spin-orbit interaction takes place via a massive object.

The differences between the two apparatuses are instead the following

I) The interaction is much simpler in our case

II) $[M^2, H_I] = 0$ in our case, so that H_I does not couple states belonging to different eigenmanifolds of M^2 and the Hilbert space of the orbit is effectively finite dimensional

III) The orthogonality of the final states is obtained in our case only for precise interaction times rather than for all times larger than a certain minimum.

We came now to the construction of the explicit expression of the second rank tensor $\underline{\underline{H}}$ entering eq. (30). The assumptions that $\underline{\underline{H}}$ is a tensor pertaining to the magnet and that for a fixed orientation of the magnet has the expression $\underline{\underline{H}}_o$ of eq. (31) determine completely its expression:

$$\underline{\underline{H}}(\omega) = R(\omega, \omega_o)\, \underline{\underline{H}}(\omega_o)\, R^{-1}(\omega, \omega_o)$$

where $R(\omega, \omega_o)$ is the orthogonal matrix describing the rotation from ω_o to ω. Due to the massive character of the magnet we can assume that its orientation is very near to ω_o throughout the interaction time and we can accordingly specify ω by the three small rotation angles Θ_1, Θ_2, Θ_3 around the three axes of the fixed reference frame corresponding to the orientation ω_o. The initial state of the magnet $|\psi\rangle$ is a quantum state corresponding to the classical state of the magnet at rest in the position ω_o. Therefore $|\psi\rangle$ will satisfy the relations

$$\langle\psi|\Theta_i|\psi\rangle = \langle\psi|N_i|\psi\rangle = 0 \tag{34}$$

On the other hand

$$\begin{aligned} \langle\psi|\Theta_i^2|\psi\rangle &\equiv (\Delta\Theta_i)^2 \neq 0 \\ \langle\psi|N_i^2|\psi\rangle &\equiv (\Delta N_i)^2 \neq 0 \end{aligned} \tag{35}$$

Since ω is always near to ω_o we have

$$R(\omega, \omega_o) = \begin{pmatrix} 1 & -\theta_3 & \theta_2 \\ \theta_3 & 1 & -\theta_1 \\ -\theta_2 & \theta_1 & 1 \end{pmatrix} \tag{36}$$

so that the Hamiltonian can be written

$$H = K_p + K_m + \gamma S_3 M_3 + V \tag{37}$$

$$V = V_n + V_d$$

where

$V_n = \gamma(\theta_2 S_3 M_1 - \theta_1 S_3 M_2)$ is a non distorting term

$V_d = \gamma(\theta_2 S_1 M_3 - \theta_1 S_2 M_3)$ introduces distorsions

The interaction term V which originates the deviations from the ideal functioning of the apparatus would be zero if we could assume $\theta_i = 0$. This is impossible for two reasons

1. The magnet recoils during the interaction process
2. The angles θ_i cannot have perfectly sharp values in the initial state due to quantum uncertainty.

Which is the relative importance of these two effects? To discuss this point we introduce the following parameters (we suppress for simplicity indices relative to the various axes)

$$a = T\frac{\hbar}{I}, \quad b = \frac{\hbar}{\sqrt{\langle\psi|N^2|\psi\rangle}} \tag{38}$$

when I denotes the moment of inertia of the magnet. Note moreover that the angular momentum transferred during the interaction from the particle to the magnet is of the order of $\hbar$. There follows that $\frac{\hbar}{I}$ gives the order of magnitude of the rate of the recoil rotation of the magnet and consequently a gives the order of magnitude of the recoil rotation during the time T. On the contrary since $\sqrt{\langle \psi | N^2 | \psi \rangle} = \Delta N(0)$, b represents roughly the initial spread $\Delta\Theta(0)$. A rough estimate for $\Delta\Theta(T)$ is then

$$[\Delta\Theta(T)]^2 \simeq [\Delta\Theta(0)]^2 + \frac{\langle \psi | N^2 | \psi \rangle}{I^2} T^2 = b^2 + \frac{a^2}{b^2} \qquad (39)$$

In order that $\Delta\Theta$ be small throughout the interaction time one must then have

$$a \ll b, \quad b \ll 1 \qquad (40)$$

which means that the initial spread of the angles must be very small compared with unity, but much larger than the magnitude of the recoil movement, in order that the angular position of the magnet do not spread completely during the interaction time. For any reasonable macroscopic instrument and interaction time both the above inequalities can be largely satisfied.

The above inequalities allow also to prove that, in order to solve the quantum problem we are discussing we can use first order perturbation theory. We do not give here the details of the solution of the dynamical problem. For this the reader is referred to ref. (9). The results of the calculations are

$$P^{(+)}_{u_+A_+} = P^{(-)}_{u_-A_+} = P^{(+)}_{u_-A^1} = P^{(-)}_{u_+A^1} = 0$$

$$P^{(+)}_{u_+A^1} = P^{(-)}_{u_-A^1} = \frac{1}{2}[(\Delta\Theta_1)^2 + (\Delta\Theta_2)^2] \qquad (41)$$

$$P^{(+)}_{u_-A_+} = P^{(-)}_{u_+A_-} = P^{(+)}_{u_-A_-} = P^{(-)}_{u_+A_+} = \frac{1}{4}[(\Delta\Theta_1)^2 + (\Delta\Theta_2)^2]$$

From this we get the probabilities for the various types of malfunctioning.

Probability of wrong answer:

$$P^{(+)}_{u_+A_-} + P^{(+)}_{u_-A_-} = \frac{1}{4}\,[(\Delta\,\Theta_1)^2 + (\Delta\,\Theta_2)^2\,]$$

Probability of no answer:

$$P^{(+)}_{u_+A1} + P^{(+)}_{u_-A1} = \frac{1}{2}\,[(\Delta\,\Theta_1)^2 + (\Delta\,\Theta_2)^2\,]$$

Probability of distorsion:

$$P^{(+)}_{u_-A_+} + P^{(+)}_{u_-A_-} + P^{(+)}_{u_-A1} = \frac{1}{2}\,[(\Delta\,\Theta_1)^2 + (\Delta\,\Theta_2)^2\,]$$

and analogous formulae for the case in which the initial spin state is u_- instead of u_+. We see from the above results that the various probabilities of malfunctioning are directly related to the quantum uncertainty in the initial position of the magnet, while the recoil movement is irrelevant.

To compare our result with the general theorem of sect. 3 we remark that the total amount of distorsion is, for our model

$$\varepsilon^2 = (\Delta\,\Theta_1)^2 + (\Delta\,\Theta_2)^2 \tag{42}$$

To compare this with the previous result we can use the uncertainty relation, assuming that the initial state is of minimal uncertainty

$$\Delta\,\Theta_1 = \frac{\hbar}{2} \,/\, \surd <\psi\,|\,N_1^2\,|\,\psi> \tag{43}$$

Noting further that from $b \ll 1$ it follows $<N_i^2> \ggg \hbar^2 \;\tilde{\sim} <M_i^2>$ so that one has essentially $<N_i^2> \simeq <L_i^2>$ one can finally write

$$\varepsilon^2 = \frac{\hbar^2}{4}\left(\frac{1}{<L_1^2>} + \frac{1}{<L_2^2>}\right) \tag{44}$$

which is twice the sum of the two Yanase bounds corresponding to the conservation laws of $L_1 + S_1$ and $L_2 + S_2$.

Summarizing, our explicit model, which is physically sensible, shows that the theoretical bounds for the malfunctioning can be almost attained by realistic measuring apparatuses so that an almost ideal measurement process is practically possible. It is worth noting that according to our model we cannot make ε^2 vanishingly small by choosing the initial state of the magnet due to the relation $a \ll b$. The parameter ε^2 can be made to approach zero only by letting the moment of inertia of the magnet to become very large. Finally we remark that our detailed analysis of the considered model has revealed the physical mechanism which give rise to the unavoidable deviations from the ideal measurement scheme. First, the physical requirement of rotational invariance calls into play the rotational degrees of freedom of the massive part of the apparatus. Secondly, the quantum uncertainty, making it impossible to fix such a massive part of the apparatus in the precise orientation corresponding to the ideal functioning, introduces the deviations. On the other hand the recoil movement of the massive apparatus, which is obviously present, is practically negligible and plays no role in giving origin to the nonideal terms

REFERENCES

1. J. von Neumann: Matematische Grundlagen der Quantenmechanik (Berlin 1932); English Edition Princeton, N.J., 1955.

2. H. Araki and M.M. Yanase: Phys. Rev. 120, 622 (1960).

3. H. Stein and A. Shimony: Proc. SIF, course IL (New York, N.Y., 1970), p. 56

4. M.M. Yanase: Proc. SIF., Course IL (New York, N.Y., 1970), p.77.

5. E.P. Wigner, Z. Phys., 131, 101 (1952).

6. M.M.Yanase: Phys. Rev. 123, 666 (1961).

7. G.C.Ghirardi, F. Miglietta, A. Rimini and T. Weber, Phys. Rev. 24D, 347 (1981).

8. F.Crisciani, G.C. Ghirardi, A. Rimini and T. Weber, Nuovo Cim. 64B, 338 (1981).

9. G.C. Ghirardi, F. Miglietta, A. Rimini and T. Weber, Phys.Rev. 24D, 353 (1981).

LIGHT-CONE QUANTIZATION OF GAUGE THEORIES WITH PERIODIC BOUNDARY CONDITIONS

V.A. Franke
Department of Theoretical Physics, University of Leningrad

Yu. V. Novozhilov
Department of Theoretical Physics, University of Leningrad and Science Sector, UNESCO, 75700 Paris

E.V. Prokhvatilov
Department of Theoretical Physics, University of Leningrad

INTRODUCTION

The idea of quantization on a plane tangential to the light-cone ("light-cone quantization") was advocated by Dirac as early as 1945 [3]. It was found later [9, 10] that in the resulting non-covariant perturbation theory vacuum diagrams are suppressed so that the bare Fock vacuum is the ground state interacting fields. Such simplicity of the vacuum state is a very attractive feature of the light-cone quantization approach in view of the complex structure of vacuum states in the covariant approach or in the standard canonical quantization scheme.

However, the light-cone quantization presents some difficulties: (a) the Lorentz-invariance of formalism should be checked after quantization; (b) such formalism involves both first and second-class constraints in the Dirac classifications [4], and (c) there are additional infra-red divergences. The two latter difficulties are related to zero-modes of fields A_μ in the longitudinal momentum $P_- = 1/\sqrt{2}(P_0 - P_3)$ (when $x^+ = 1/\sqrt{2}(x^0 + x^3)$ plays the role of time). Zero-modes of fields are very important for an understanding of the vacuum structure, because, in fact, vacuum diagrams contribute only to zero-modes. These modes are also necessary for a construction of gauge invariant states in a non-abelian theory, because they are the only field modes which transform inhomogeneously

DYNAMICAL SYSTEMS AND MICROPHYSICS

ISBN 0-12-068720-8

under a gauge transformation in the light-plane formulation. In this situation it appears important to deal correctly with zero-modes. This can be achieved if we separate them by considering fields in x^--box of length 2L and imposing periodic boundary conditions on fields A_μ, field strengths $F_{\lambda\nu}$ and their derivatives $\partial_\lambda F_{\mu\nu}$: $A_\mu(L) = A_\mu(-L)$ etc. The periodic boundary conditions are interesting also by themselves, because periodic structures might be important in the physics of confinement [8]. As far as periodic boundary conditions are assumed, one cannot choose, in a general case, the usual light-cone gauge $A_- = 1/\sqrt{2}(A_0 - A_3)$. This gauge should be replaced by $\partial_- A_- = 0$, and the function $A^a_- = v^a(x^+, \mathbf{x})$ describes topological charge [5]. At the same time field A_μ may become singular in the points or regions defined by $v(x^+, \mathbf{x})$. In the singularity free case the light-cone quantization is self-consistent [6]. The second-class constraint can be solved in principle, so that the Hamiltonian is finally well defined in terms of canonical variables. One can also show that in the case of non-abelian fields the gauge-invariance of physical states (in the Fock space) is incompatible with nonvanishing eigenvalues of quantized version of $v(x^+, \mathbf{x})$. However, in the case of abelian fields (i.e. quantum electrodynamics) gauge-invariant states may be constructed out of redefined gauge-invariant fermion operators so that $v \neq 0$ [7].

We use for fields both matrix A_μ and isovector A^a_μ forms, $A_\mu = 1/2i\, \tau^a A^a_\mu$, the gauge group is SU(2); $F_{\mu\nu} = \partial_\mu A_\nu - \partial_\nu A_\mu + g[A_\mu, A_\nu]$. The covariant derivative is $D_\mu G = \partial_\mu G + g[A_\mu, G]$. The metric is chosen as $g_{+-} = g_{+-} = 1$, $g_{kk} = -1$, $k = 1, 2$. We use the notation $x = (x^+, x^-, \mathbf{x})$, $\mathbf{x} = (x^1, x^2)$, $r = |\mathbf{x}|$. For any function $f(x)$ we define $\overset{\circ}{f}(x^+, \mathbf{x})$ by

$$\overset{\circ}{f} = \frac{1}{2L} \int_{-L}^{L} f\, dx^- .$$

PERIODICITY, LIGHT-CONE GAUGE CONDITION AND TOPOLOGICAL CHARGE

We start with fields A_μ that are regular in the whole x^μ-space including the spacial infinity $r \to \infty$ and obey periodic boundary conditions in x^-. The fields are assumed: (a) to be periodic in x^- in the direct sense

$$A_\mu(\bar{x} = L) = A_\mu(\bar{x} = -L) \tag{1}$$

(b) to be continuous together with their derivatives $\partial_\nu A_\mu$ in the whole x^μ-space including hypersurfaces $x^- = \pm L$, and (c) to have a finite energy and a finite momentum. It follows that at the spacial infinity $r \to \infty$, field strengths F_{-k}, F_{kj} and F_{+-} should vanish as r^{-2}, while fields should asymptotically become pure gauge fields.

Fields A_μ do not satisfy any particular gauge condition. Now we want to eliminate a gauge freedom by a suitable gauge transformation bringing fields A_μ to a standard form A'_μ and preserving at the same time the periodicity of A'_μ.

Theorem Let A be a field satisfying conditions (a), (b) and (c). By means of a gauge transformation the field A_μ can always be brought to the form A'_μ with the properties

$$\partial_- A'^a_- = 0 , \quad A'^a_-(x) = v^a(x^+, \mathbf{x}) ,$$
$$A'_k \to 0 , \qquad r \to \infty , \tag{2}$$

while the function $v^a(x^+, \mathbf{x})$ admits the representation

$$v^a(x^+, \mathbf{x}) = v(x^+, \mathbf{x})\, n^a(x^+, x) , \qquad n^a n^a = 1 ,$$

where an isoscalar function $v(x^+, \mathbf{x})$ is continuous in the whole $(x^+, \mathbf{x})$-space, $v(x^+, \mathbf{x}) \to v_\infty(x^+)$ at $r \to \infty$, and an isovector field $n^a(x^+, \mathbf{x})$ is continuous everywhere with the exception of a discrete number of points $(x_i^+, \mathbf{x}_i)$. In these singularity points of $n^a(x^+, \mathbf{x})$

$$v(x_i^+, \mathbf{x}_i) = \frac{\pi N_i}{gL} , \qquad N_i = 0, \ \pm 1, \ \pm 2 \ldots \tag{3}$$

If $v(x^+) = \pi N/gL$, then $n^a(x^+, \mathbf{x}) \to n^a(x^+)$ at $r \to \infty$, where $n^a(x^+)$ does not depend on x^k/r.

Remarks (i) If $v(x^+, \mathbf{x}) = \pi N/gL$, $N = 0;\ \pm 1,\ \pm 2 \ldots$ on contours, surfaces or, even, in three-dimensional regions, nevertheless the isovector $n^a(x^+, \mathbf{x})$ can be extended by continuity to these contours, surfaces and regions in such a way that there remains only a discrete number of singularity points of $n^a(x^+, \mathbf{x})$.

(ii) It is not possible in a general case to eliminate sigularities of fields $A'_\mu(x)$ in the gauge $\partial_- A'_- = 0$. Fields $A'_\mu(x)$ are singular on the lines $x^+ = x_i^+$, $\mathbf{x} = \mathbf{x}_i$, where where $n^a(x^+, \mathbf{x})$ has singularity points. The proof of the Theorem is given in [5].

Consider now fields $A'_\mu(x)$ in the light-cone periodic gauge $\partial_- A'_- = 0$, $A'^a_- = v^a(x^+, \mathbf{x})$. (to avoid confusion with a general field $A_\mu(x)$ we continue to denote fields in this gauge by $A'_\mu(x)$). The appearance of an isovector v^a is a consequence of the periodic boundary condition (1). The isovector v^a may not vanish at special infinity $r \to \infty$, so that in this case the unit vector $n^a = v^a/v$ defines a preferred direction in the gauge space; this relation between periodicity and an appearance of a preferred direction has already been noted in another context [1]. As was stated in the Theorem, the unit vector $n^a(x^+, \mathbf{x})$ may have singular points in the $(x^+, \mathbf{x})$-space. These points are related to topological charge. To show this fact we consider a Minkowski space expression for the topological charge, or Pontryagin index

$$\nu = \frac{1}{16\pi^2 g^2} \int d^4x \, \mathrm{Tr}\, F_{\mu\nu} \, {}^*F^{\mu\nu} , \tag{4}$$

where ${}^*F^{\mu\nu} = 1/2\, \varepsilon^{\mu\nu\lambda\sigma} F_{\lambda\sigma}$ is the dual of $F^{\mu\nu}$; $\varepsilon_{+-12} = 1$. As was shown in [2], for time-periodic fields this quantity gives the winding number of the gauge transformation of time periodicity, while for the dyon configurations it gives the monopole strength. One can expect that similar relations would exist in the light-cone frame too.

The integrand in (4) is well known to be a total divergence

$$\begin{aligned} F^a_{\mu\nu} \, {}^*F^{\mu\nu a} &= 4\, \partial_\lambda K^\lambda , \\ K^\lambda &= \frac{1}{2}\, \varepsilon^{\lambda\nu\rho\sigma} A'^a_\nu \left(\partial_\rho A'^a_\sigma + \frac{1}{3}\, g\, \varepsilon^{abc} A'^b_\rho A'^c_\sigma\right) , \end{aligned} \tag{5}$$

so that ν is represented by a surface integral over K^μ. The contribution from the K^--part vanishes because of the periodic boundary condition, the contribution from the K^j-part vanishes because fields A_k decrease like $(1/r)$ at large distances. Therefore,

$$\nu = \frac{g^2}{8\pi^2} \int \{K^+(T) - K^+(0)\}\, dx^- d^2x , \tag{6}$$

where $x^+ = T$ and $x^+ = 0$ are limits of integration over x^+ in Eq. (4).

The quantity K^+ is not conserved, and in order to calculate K^+ one should know the time development of the system. We evaluate ν in the case when fields are periodic in time x^+ up to a gauge transformation:

$$A'_\mu (x^+ + T) = G(x^+) A'_\mu (x^+) G^{-1} (x^+) + \frac{1}{g} G(x^+) \partial_\mu G^{-1} (x^+) , \tag{7}$$

where all quantities depend also on x^- and x^k. Then

$$\nu = \frac{1}{16\pi^2} \int dx^- d^2 x \, \varepsilon^{abc} (G^{-1} \partial_- G)^a (G^{-1} \partial_1 G)^b (G^{-1} \partial_2 G)^c , \tag{8}$$

where we have taken into account the regularity of G at spacial infinity as well as vanishing of A_k at $r \to \infty$. It follows from (8) that $\nu = 0$, if G does not depend on x^-. Now, the field $A'_\mu(x)$ should be periodic in x^- and satisfy the gauge condition $\partial_- A'_- = 0$ at any light-cone moment x^+, including $x^+ = T$. Thus, the transformation $G(x^+, x^-, x)$ in Eq. (7) should be periodic in x^- up to a sign and preserve the gauge $\partial_- A'_- = 0$. A particular form for G is

$$G(x^+, x^-, x) = \exp (i \frac{\pi N}{L} \tau^a n^a x^-) \tag{9}$$

With the expression (9) for G the quantity ν in Eq. (8) is

$$\nu = \frac{N}{4\pi} \int \varepsilon^{abc} n^a \partial_1 n^b \partial_2 n^c d^2 x . \tag{10}$$

The length of the longitudinal box 2L has disappeared from Eq. (10); ν , does not depend on the limit $L \to \infty$, but only on the properties of the unit vector $n^a (x^+, x) = v^a/\nu^v$. It follows that the limit $L \to \infty$ cannot be taken without considering global quantities.

In the light-cone coordinates the unit isovector n^a depends on x^+ and x^k. We introduce functions $f_1(x^+, x)$ and $f_2(x^+, x)$ by

$$n^a \tau^a = \tau^3 \frac{a - i\bar{f}}{a + i\bar{f}} , \qquad \bar{f} = \tau^1 f_1 + \tau^2 f_2 , \tag{11}$$

where a may depend on x^+. For a particular choice $f_k = x^k$

$$\int d^2 x \, \varepsilon^{abc} n^a \partial_1 n^b \partial_2 n^c = 4\pi \tag{12}$$

Thus, the quantity ν takes on integer values $\nu = 0, \pm 1, \pm 2 \ldots$ for fields periodical in time (up to a gauge transformation) in the light-cone periodical gauge. Therefore it defines the mapping in this case and may be called the Pontrjagin index, or topological charge. ν is associated with such light-cone time development which is described for the periodicity time T by the matrix (9). Note that within the usual light-cone gauge $A_- = 0$ one cannot have field configurations with nonvanishing ν. But if we start with $A_- = 0$ at $x^+ = 0$ and (with periodicity T in x^+) reach $A_- = \pi N/gL$ at $x^+ = T$, then $\nu = N$. We came to the conclusion that for the description of fields with a topological charge the time-dependent functions $v^a(x^+, \mathbf{x})$ are indispensable.

CANONICAL FORMALISM AND QUANTIZATION [6]

We consider fields A_μ in the periodic light-cone gauge

$$\partial_- A_- = 0, \qquad A^a_- = \delta_{a3}\, v(x^+, \mathbf{x}) \tag{13}$$

Eq. (13) differs from the condition (1) by an x^--independent gauge transformation. We assume that fields A_k vanish at least as $1/r$ at $r \to \infty$; fields may have singularity strings where $v(x_i^+, \mathbf{x}) = \pi N_i/gL$, $N_i = \pm 1, \pm 2, \ldots$ We shall consider only a singularity free case when $v \neq \pi Ni/gL$.

To develop the canonical formalism, it is helpful to introduce, in addition to field variables A_k, a new one, namely Π_{+-} , in such a way that the variational principle leads to the usual equations. We write the action as

$$W = \int d^2x dx^+ \int_{-L}^{L} dx^- \mathcal{L} = - \int d^2x\, dx^+ \int_{-L}^{L} dx^- \operatorname{Tr} \{2\Pi_{+-}\, \partial_+ v + \\ + 2 (F_{-k}\, \partial_+ A_k) - (\Pi^2_{+-} + F^2_{12}) + 2A_+ (D_- \Pi_{+-} + D_k F_{-k})\} \tag{14}$$

The definition of canonical momenta π_k for fields A_k constitutes, in fact, a second class constraint

$$\pi^a_k = F^a_{-k} \tag{15}$$

This constraint can be easily solved when we take into account the periodic boundary condition and introduce the Fourier-decomposition of fields A_k:

$$A_k^3 = \frac{1}{2\sqrt{2L}} \sum_{n\neq 0} e^{-ip_n x^-} \frac{Q^3_{k|n|} + i\,\varepsilon(p_n)\,\Pi^3_{k|n|}}{\sqrt{|p_n|}} + \mathring{A}_k^3 \tag{16}$$

$$A_k^{1+i2} = \frac{1}{2\sqrt{2L}} \sum_{n} e^{-ip_n x^-} \frac{Q_{kn} + i\,\varepsilon(p_n - gv)\,\Pi_{kn}}{\sqrt{|p_n - gv|}}\,, \tag{17}$$

where all coefficients and $\mathring{A}_k^3$ are functions of (x^+, x); $p_n = \pi n/L$, $n = 0$, ± 1, $\pm 2 \ldots$ The sign function $\varepsilon(p_n - gv)$ is necessary in order to bring the action (14) into the canonical form. We see that fields $A_k^{1+i2} = 1/\sqrt{2}(A_k^1 + iA_k^2)$ contain the function $v(x^+, x)$.

We now express the action W in terms of new variables

$$W = \int dx^+ d^2x [\partial_+ v\, \Pi^{(v)} + \sum_{n>0} \partial_+ Q^3_{kn}\, \Pi^3_{kn} + $$

$$+ \sum_n \partial_+ Q_{kn}\, \Pi_{kn} - \mathcal{H}]\,, \tag{18}$$

$$\mathcal{H} = -\int_{-L}^{L} dx^- \,\mathrm{Tr}\{\Pi_{+-}^2 + F_{12}^2 - 2A_+(D_- \Pi_{+-} + D_k F_{-k})\} \tag{19}$$

where $\Pi^{(\nu)} = \int_{-L}^{L} dx^- (\Pi^3_{+-} - \partial_k A_k^3)$.

The canonical variables are Q^3_{kn} and Π^3_{kn}, Q_{kn} and Π_{kn}, v and $\Pi^{(v)}$. The variables $\mathring{A}_k^3$ and $\Pi^3_{+-} = \Pi^3_{+-} - \Pi^3_{+-}$ should be expressed as functions of the canonical ones as follows. The variation of A_+ in (18) gives us the constraint $D_- \Pi_{+-} + D_k F_{-k} = 0$. From this equation the quantity $\tilde{\Pi}^3_{+-}$ can be found if the condition

$$\rho^3(x^+, x) \equiv \int_{-L}^{L} dx^- (D_k F_{-k})^3 = 0 \tag{20}$$

is satisfied. The function $\mathring{\Pi}^3_{+-}$ drops out of this equation and remains an independent variable, which is replaced later by the variable $\Pi^{(v)}$. The variation of $\mathring{A}_k^3$ in (18) leads to the constraint

$$\Lambda_k^3(x^+, \mathbf{x}) \equiv \int_{-L}^{L} dx^- (D_k F_{+-} + D_j F_{jk})^3 = 0 \tag{21}$$

which is second class in the Dirac classification [8]. Expressed in terms of canonical variables, this constraint is quite a complicated equation for $\mathring{A}_k^3$ which we cannot solve in a closed form. But one can show that this constraint uniquely defines the functions $\mathring{A}_k^3$. Let us rewrite Eq. (21) in terms of canonical variables using (16) and (17). The resulting equation has a form $R\,\mathring{A}_k^3 + T_k = 0$, where R and T_k do not contain $\mathring{A}_k^3$. We look for solutions $\mathring{A}_k^3(x^+, \mathbf{x})$ vanishing at $r \to \infty$. Evidently, the condition (21) defines $\mathring{A}_k^3$ uniquely, if the equation $R\,\mathring{A}_k^3 = 0$ has no solutions vanishing at $r \to \infty$. Therefore, we need to study the operator R which is given explicitly by

$$R = -2L\partial_k \partial_k + g^2 \sum_n \frac{\Pi_{kn}\Pi_{kn} + Q_{kn}Q_{kn}}{2\,|p_n - gv|} \tag{22}$$

The second term in (22) is a positive definite function. This ensures the nonexistence of solutions of $R\,\mathring{A}_k^3 = 0$ regular in $(x^+, \mathbf{x})$-space and vanishing at $r \to \infty$. Therefore, the function $\mathring{A}_k^3(x^+, \mathbf{x})$ can be expressed uniquely in terms of Q_{kn} and Π_{kn}, Q_{kn}^3 and Π_{kn}^3, and v.

Thus, the Hamiltonian (19) is indeed a function of only the canonical variables; the quantities ρ^3 (Eq. 20) are generators of gauge transformations leaving the gauge condition (13) invariant. We rewrite the Hamiltonian (19) as

$$\begin{aligned}\mathcal{H} &= \frac{1}{2}\left(\frac{1}{\sqrt{2L}}\Pi^{(v)} + \sqrt{2L}\,\partial_k \mathring{A}_k^3\right)^2 + \lambda\rho^3 + \\ &+ \frac{1}{2}\int_{-L}^{L} dx^- \left[(\Pi_{+-}^1)^2 + (\Pi_{+-}^2)^2 + (\tilde{\Pi}_{+-}^3)^2 + F_{12}^a F_{12}^a\right],\end{aligned} \tag{23}$$

where the constraint $\rho^3 = 0$ enters with the Langrangian multiplier $\lambda = \mathring{A}_+^3$.

The quantization is obtained through a standard procedure by replacing the Poisson brackets of canonical variables on the surface $x^+ = \text{const.}$ by the commutation relations. We postulate

$$[Q_{kn}(x^+, x), \Pi_{k'n'}(x^+, x')] = [Q^3_{kn}(x^+, x), \Pi_{k'n'}(x^+, x')] =$$

$$= i\, \delta_{kk'}\, \delta_{nn'}\, \delta^2(x - x')\,, \tag{24}$$

$$[v(x^+, x'), \Pi^{(v)}(x^+, x)] = i\delta^2(x - x')$$

We shall denote quantum quantities by the same symbol as classical ones.

The physical states $|\phi>$ are characterized by the condition

$$\rho^3(x^+, x)\,|\phi> = 0\,, \tag{25}$$

expressing their gauge invariance.

Let us turn now to the definition of basic states in the large Hilbert space, i.e. when (25) is not taken into account. In the light-cone formulation basic states are defined as eigenstates of the longitudinal momentum P_- which has non-negative spectrum. We introduce creation and annihilation operators

$$a^3_{kn} = \frac{1}{\sqrt{2}}(Q^3_{kn} + i\Pi^3_{kn})\,, \quad a_{kn} \pm \frac{1}{\sqrt{2}}(Q_{kn} + i\Pi_{kn}) \text{ etc} \tag{26}$$

and express P_- by means of (16), (17) and (26)

$$P_- \equiv \int d^2x dx^- T_{--} = -\int d^2x dx^- \, 2\mathrm{Tr}\, F^2_{-k} =$$

$$= \int d^2x \,\{\, 2L\partial_k v \partial_k v + \sum_{n>0} P_n a^{3+}_{kn} a_{kn} + \sum_n |p_n - gv|\, a^+_{kn} a_{kn} \tag{27}$$

The operator ordering in (27) ensures that P_- is non-negative.

In the same way we get for the generator ρ^3 (Eq. 20)

$$\rho^3 = -2L\partial_k \partial_k v - g \sum_n \varepsilon(p_n - gv)\, a^+_{kn} a_{kn}\,. \tag{28}$$

The operator v commutes with creation and annihilation operators, so that all terms in the r.h.s. of (27) can be diagonalized independently. We define eigenstates $|v'>$ of v:

$$|v'> = \exp\{-i\int d^2x\, v'(x)\, \Pi^{(v)}(x)\}\,|0> \tag{29}$$

and then build the Fock-type states by applying creation operators to states $|v'>$. We introduce also the vacuum state $|0>$ with properties

$$a_{kn}|0> = 0, \quad a^3_{kn}|0> = 0, \quad v|0> = 0 \tag{30}$$

The typical basic state in the large Hilbert space will then be

$$a^{3+}_{kn}(\mathbf{x}) \ldots a^{+}_{k'n'}(\mathbf{x}') \ldots |v'> \tag{31}$$

Consider a gauge transformation of states (31) with a gauge function $\omega^3(\mathbf{x})$. It is induced by an operator $U = \exp i \int d^2 x\omega^3(\mathbf{x})\rho^3(\mathbf{x})$ containing the generator ρ^3: ρ^3:

$$v|v'> = \exp\{i \int d^2x\, \omega^3(\mathbf{x}) 2L\partial_k\partial_k v'(\mathbf{x})\}\, |v'> , \tag{32}$$

$$va^{+}_{kn} v^{-1} = \exp\{-ig\omega^3(\mathbf{x})\,\varepsilon(p_n - gv)\}\, a^{+}_{kn} \tag{33}$$

The gauge invariance of a physical state of the form (31) should be the result of overall compensation of gauge changes due to creation operators and the state $|v'>$. But with quite different transformation laws (32) and (33) such compensation can be achieved only when $v' = 0$. Take a state $a^{+}_{kn}(\mathbf{x}_1)|v'>$; it is gauge invariant if $v'(\mathbf{x})$ satisfies the equation

$$2L\partial_k\partial_k v'(\mathbf{x}) + g\delta^2(\mathbf{x}-\mathbf{x}_1)\,\varepsilon(p_n - gv'(\mathbf{x})) = 0 \tag{34}$$

In the region where $\varepsilon = +1$ the solution is $4\pi Lg\, v'(\mathbf{x}) = -g^2 \ln|\mathbf{x}-\mathbf{x}_1| + \text{const.}$, so that $gv'(x) \to \infty$ at $|\mathbf{x}-\mathbf{x}_1| \to 0$, or $\varepsilon = -1$ in contradiction with the initial assumption $\varepsilon = +1$. The contradiction remains if one starts with the region $\varepsilon = -1$ or if one considers a basic state (26) with any number of creation operators. We conclude that $v' = 0$ for physical states. Thus the assumption of singularity free fields, $v' \neq \pi N/gL$, together with the boundary condition $v' \to 0$ at $r \to \infty$ lead to $v' = 0$ in the physical subspace.

The physical subspace then is nothing but a Fock space where all states are obtained by applying gauge invariant combinations of creation operators with $n \neq 0$ to the unique vacuum state $|0>$; This result was obtained earlier [10] directly in the gauge $A_- = 0$, where one assumes that $v' = 0$ in the large Hilbert space of states

(31). In the gauge $A_- = 0$ the first and second class constraints of the classical theory take the form $\rho'^a = 0$ and $\Lambda_k^a = 0$.

If we assume that $v'(x)$ is finite at $r \to \infty$ (instead of $v'(x) \to 0$) then the gauge invariance condition (25) will give $v' =$ const. The relativistic invariance then requires $v' \sim 1/L$, when $L \to \infty$.

In conclusion, let us discuss the role of zero-modes $\overset{\circ}{A}{}_k^3$ (in case of $v' \neq kN/gL$) and $\overset{\circ}{A}{}_k^a$ (in case of $v' = 0$) for the construction of gauge invariant states. Unlike independent variables a_{kn}, $\overset{+}{a}_{kn}$ and a_{kn}^3, $(a_{kn}^3)^+$, $n \neq 0$ which transform as field strengths under the gauge transformation $G(x)$, zero-modes transform as fields (or potentials), i.e. with an inhomogeneous term. This is just what is needed to build up a phase factor (in case of $v = 0$)

$$\Phi(1,2) = P\exp\left(\frac{1}{2} i \int_1^2 g\,\tau^a \overset{\circ}{A}{}_k^a \, dx^k\right)$$

which transforms according to

$$\Phi(1,2) \to \Phi'(1,2) = G(x_1)\,\Phi(1,2)\,G^{-1}(x_2)$$

Closing the contour integral in $\Phi(1,2)$ we get a gauge invariant and contour dependent operator $\Phi_r = T_r\Phi(1,1)$ or the Wilson loop. The phase factors $\Phi(1,2)$ enter into gauge invariant combinations of creation operators, e.g.

$$\Omega(1,2) = \mathrm{Tr}\,\{\tau^b a_{kn}^{b+}(x_1)\,\theta(\rho_n - gv)\,\Phi(1,2)\,\tau^c \overset{+}{a}{}_{k'n'}^{c}(x_2)\,\theta(-p_n + gv)\,\Phi(2,1)\}$$

(in case of $v' = 0$). These combinations $\Omega(1,2,\ldots N)$ together with loops Φ_r constitute a complete set of gauge invariant operators necessary for construction of physical states. Thus, zero-modes contain important information about physical states.

The most difficult problem for practical non-perturbative calculations is that zero-modes are dependent variables which should be found in terms of other modes from operator constraints (21) $\Lambda_k^3 = 0$ (in case $v \neq \pi N/gL$) or $\Lambda_k^a = 0$ (in case $v = 0$). Note also that phase factors $\Phi(1,2)$ do not commute with creation operators so that an operator ordering is essential in the construction of physical states.

REFERENCES

[1] Ambjorn, J., Felsager, B., and Olesen, P., The Niels Bohr Institute report NBI-HE-80-22, 1980.

[2] Christ, N., and Jackiw, R., Phys. Lett., **91B**, 228, 1980.

[3] Dirac, P.A.M., Rev. Mod. Phys., **81**, 392, 1949.

[4] Dirac, P.A.M., "Lectures on Quantum Mechanics", Yeshiva University, New York, 1964.

[5] Franke, V.A., Novozhilov Yu. and Prokhvatilov, E.V., Lett. Math. Phys., 5, 239-245, 1981.

[6] Franke, V.A., Novozhilov Yu. and Prokhvatilov, E.V., Lett. Math. Phys., **5**, 437-444, 1981.

[7] Franke, V.A., Novozhilov Yu. and Prokhvatilov, E.V., Theoret. Math. Fizika (to be published).

[8] t'Hooft, G., Nucl. Phys., **B138**, 1, 1978; **B153**, 141, 1979. Mandelstam, S., Mandelstam, S., Phys. Rev., **D19**, 2391, 1979.

[9] Klauder, J.R., Leutwyler, H., Streti, L., Nuovo Cim. **66A**, 539, 1970. Kogut, J., and Susskind, L., Phys. Rep., **86**, 75, 1973.

[10] Tomboulis, E., Phys. Rev. **D8**, 2736, 1973. Casher, A., Phys. Rev. **D14**, 452, 1976.

PART VI

CONTRIBUTED PAPERS

SOLUTIONS OF GENERALISED THEORIES OF GRAVITATION DERIVED FROM A MODIFIED DOUBLE DUALITY CONDITION

I.M. Benn, T. Dereli, R.W. Tucker

Department of Physics, University of Lancaster, Lancaster, U.K.

Motivated by the desire to find a unification among the fundamental interactions, it is worthwhile to explore some of the alternatives to Einstein's theory of gravity that are consistent with the classical experimental tests.

We use the formalism of complex quaternionic differential forms; essentials of which may be found in reference (1). If the linear connection on space-time frames is restricted to be metric compatible, taking values in the algebra of SL(2,C), then the connection 1-form may be written $\hat{\omega} = -\frac{1}{2}(i\omega^{o}{}_{k} + \frac{i}{2}\varepsilon_{kij}\omega^{i}{}_{j})\hat{e}_{k}$ where $\omega_{ab} = -\omega_{ba}$ are the usual real components. (Tetrad indices are raised and lowered by the Minkowski metric $g_{ab} = \mathrm{diag}(-+++)$). The algebraic elements $\hat{e}_k$, k = 1,2,3 satisfy $\hat{e}_i\hat{e}_j = -\delta_{ij} + \varepsilon_{ijk}\hat{e}_k$. Quaternionic conjugation is defined by $\hat{e}_k \to \overline{\hat{e}_k} \equiv -\hat{e}_k$, while $\dagger$ denotes Hermitian conjugation, meaning quaternion conjugation followed by complex conjugation. $\mathcal{H}$ denotes the Hermitian, and A the anti-Hermitian part of any quaternionic form. The (anti-Hermitian) basis 1-form $e = ie^{o} + \sum_{k=1}^{3} e^{k}\hat{e}_{k}$ is expressed in terms of the real orthonormal tetrad 1-forms e^{a}, a = 0,1,2,3. The

ISBN 0-12-068720-8

structure equations $\hat{R} = d\hat{\omega} + \hat{\omega}\wedge\hat{\omega}$ and $T = de + \hat{\omega}\wedge e + e\wedge\hat{\omega}^{\dagger}$ define the (complex q-vector valued) curvature 2-form $\hat{R}$ and the (anti-Hermitian) torsion 2-form T of space-time.

The Einstein-Hilbert action 4-form may be written

$$\Lambda = - \text{Re } S\left[i\hat{R}\wedge e\wedge\bar{e}\right] \tag{1}$$

where the gravitational constant has been set to unity. Independent frame and connection variations result in the vacuum Einstein equations

$$\mathcal{H}(\hat{R}\wedge e) = 0 \tag{2}$$

$$D(e\wedge\bar{e}) \equiv T\wedge\bar{e} - e\wedge\bar{T} = 0 \tag{3}$$

(It may be noted that the 24 real equations (3) have the unique solution T = 0.) It is well known that the only static, spherically symmetric solution to (2) and (3) yields the Schwarzschild metric.

An alternative theory replaces the Einstein-Hilbert action by one that is analogous to the theory of free electromagnetism[2]:

$$\Lambda = - \alpha \text{ Re } S\{\hat{R}\wedge *\hat{R}\} \tag{4}$$

where * takes the Hodge dual with respect to a Minkowski signatured metric tensor. The corresponding field equations are

$$D*\hat{R} = 0 \tag{5}$$

$$T^{\alpha} \equiv \text{Re } S\left\{i_{X_\alpha}\hat{R}\wedge *\hat{R} - \hat{R}\wedge i_{X_\alpha} *\hat{R}\right\} = 0 \qquad \alpha = 0,1,2,3 \tag{6}$$

where the interior multiplication is with respect to a set of four vector fields X_α that are dual to a g-orthonormal frame e^{β}, i.e. $i_{X_\alpha}(e^{\beta}) = \delta^{\beta}{}_{\alpha}$. The field equations (5) and (6) admit a large class of classical solutions. In particular there is no Birkhoff theorem that determines a unique spherically symmetric solution. The conformally flat zero torsion solution with metric

$$ds^2 = -\frac{dt^2}{\left[1+\frac{c}{r}\right]^2} + \frac{dr^2}{\left[1+\frac{c}{r}\right]^2} + r^2 d\Omega^2 \tag{7}$$

found by Pavelle and Thompson[3] satisfies *both* sets of equations and must compete with Schwarzschild solution for interpretation in the zero torsion spherically symmetric static sector of solutions. Here we wish to point out that a simple double duality ansatz may be used to solve both sets (5) and (6), and in particular generate the Schwarzchild solution and (7). We first note that the double (anti-) self dual curvature condition:

$$*\hat{R} = \pm i\ \hat{R} \tag{8}$$

automatically satisfies the field equations (5) and (6) by virtue of the Bianchi identity $D\hat{R} = 0$ and the properties of interior multiplication. Furthermore, all torsion free solutions of the vacuum Einstein equations with arbitrary cosmological term:

$$\hat{R}\wedge e = \Lambda\ e\wedge\bar{e}\wedge e \tag{9}$$

imply that the curvature is double *self-dual* ($*\hat{R} = i\hat{R}$) and hence obeys the field equations (5) and (6). Thus the sector of double *anti-self dual* solutions ($*\hat{R} = -\ i\hat{R}$), one of which is generated by (7), is consequently not vacuum Einstein.

In an attempt to constrain the theory of gravity, an action that also includes the Einstein action has been suggested by many authors[4]:

$$\Lambda(\omega,e) = -\ \mathrm{Re}\ S\left[\alpha R\wedge *\hat{R} + i\ \hat{R}\wedge e\wedge\bar{e}\right] \tag{10}$$

In this manner both the field equations for the connection and metric take on a more dynamical status:

$$2\, D{*}\hat{R} = -\, i\,(T\wedge\bar{e} - e\wedge\bar{T}) \tag{11}$$

$$2\mathcal{H}(\hat{R}\wedge e) = -\, i\alpha\, \tau \tag{12}$$

where the stress form $\tau = -\, i\tau^{0} + \sum_{a=1}^{3} \tau^{a}\hat{e}_{a}$ is defined from equation (6). For this theory once again the double self-duality condition ensures that all torsion free vacuum Einstein solutions to equation (9) with $\Lambda = 0$ also satisfy the systems (11) and (12). Debney, Fairchild and Siklos[5] show that these solutions are now the unique torsion free solutions. Ramaswamy and Yasskin[6] furthermore show that the imposition of spherical symmetry and reflection invariance implies that any solution with these symmetries must have zero torsion. Here we observe that if a theory with a cosmological term is tolerated then neither of these statements is true.

The action form for such a theory generalising (10) may be written

$$\Lambda(\omega,e) = -\, \mathrm{Re}\, S\{\alpha\, \hat{R}\wedge{*}\hat{R} + i\beta\hat{R}\wedge e\wedge\bar{e} + i\gamma e\wedge\bar{e}\wedge e\wedge\bar{e}\} \tag{13}$$

where α, β, γ are (complex) coupling constants. The variational field equations become for real couplings:

$$2\alpha D{*}\hat{R} = -\, i\beta D(e\wedge\bar{e}) \tag{14}$$

$$2\beta\mathcal{H}(\hat{R}\wedge e) = -\, i\alpha\tau - 4\gamma e\wedge\bar{e}\wedge e \tag{15}$$

Normalising the action so that $\alpha = 2$ and choosing $\beta = \frac{2}{\lambda}$, $\gamma = \frac{1}{4\lambda^{2}}$ we can satisfy these (40 real) field equations with a curvature that satisfies the modified double duality condition

$$\left\{\hat{R} + \frac{1}{4\lambda}\, e\wedge\bar{e}\right\} = i{*}\left\{\hat{R} + \frac{1}{4\lambda}\, e\wedge\bar{e}\right\} \tag{16}$$

This follows since up to a local exact form the action (13) may be written $\Lambda_1 = -\operatorname{Re} S\{\hat{\sigma}\wedge *\hat{\sigma}\}$ where $\hat{\sigma} \equiv \hat{R} - i*\hat{R} + \frac{1}{2\lambda} e\wedge\bar{e}$. As the field equations are expressible in terms of $\hat{\sigma}$ and (22) is equivalent to $\hat{\sigma} = 0$, the assertion follows. The condition (16) admits as a special case the double self dual curvature $\hat{R} = -\frac{1}{4\lambda} e\wedge\bar{e}$ which generate the only vacuum Einstein solution to this ansatz. A spherically symmetric static solution with torsion is[1]

$$ds^2 = -\left\{1 + \frac{c}{r} + \frac{kr^2}{3}\right\}dt^2 + \frac{dr^2}{\left\{1 + \frac{c}{r} + \frac{kr^2}{3}\right\}} + r^2 d\Omega^2 \qquad (17)$$

$$\text{where } T = -i\,\frac{\left\{2\left(\frac{k}{3} - \frac{1}{\lambda}\right)r - \frac{c}{r^2}\right\}}{\left\{1 + \frac{c}{r} + \frac{kr^2}{3}\right\}^{\frac{1}{2}}}\, e^0\wedge e^1$$

and c,k are arbitrary real constants. It may be noted that solutions similar to (17) have subsequently been found in the framework of gauge theories which are interpreted in terms of the Poincaré group[7]. Although we do not assert that the particular cosmological term discussed here has physical relevance, our discussion illustrates that there can be no Birkhoff theorem for such a theory involving arbitrary couplings.

References

1 I M Benn, T Dereli, R W Tucker, Gen Rel Grav (to be published).
2 G Stephenson, Nuo Cim 9 (1958) 263. C N Yang, Phys Rev Lett (1974) 445.
3 R Pavelle, Phys Rev Lett 33 (1974) 1461; ibid 34 (1975) 1114; ibid, 37 (1976) 961; ibid, 40 (1978) 267.
A H Thompson, Phys Rev Lett 34 (1975) 507; ibid. 35 (1975) 320.
W-T. Ni, Phys Rev Lett 35 (1975) 314.
4 E E Fairchild, Jr, Phys Rev D14 (1976) 384; ibid, D16 (1977) 2438.
F Mansouri and L N Chang; Phys Rev D13 (1976) 3197.
K S Stelle; Gen. Rel. Grav. 9 (1978) 353.
5 G Debney, E E Fairchild, Jr, S T C Siklos, G.R.G. 9 (1978) 879.
6 S Ramaswamy and P B Yasskin, Phys Rev D19 (1979) 2264.
D E Neville, Phys Rev D21 (1980) 2770
7 F W Hehl, Fermions and Gravity, Cologne Preprint 1980).

DYNAMICAL SYSTEMS AND MICROPHYSICS: A WISH

S. Ciulli

Dublin Institute for Advanced Studies

Dedicated to the composer Aurel Stroe

1. INTRODUCTION

I am very grateful to Prof. A. Avez, Prof. A. Blaquière and to Prof. A. Marzollo for having invited me to give this talk. In fact, I would like to use this opportunity not for giving a lecture, but merely to express a wish: I am indeed convinced that a good deal of the mathematical prerequisites necessary to tackle the questions discussed below are already known to Dynamical Systems people, and I would be very happy if they would become active in this field.

Without any doubt, the Greatest Miracle in Nature is that there are no Miracles at all (or only very few). Laws inferred from experiments performed for some range of values of some parameters are usually valid also in regions where no experiments were performed at all (in future time, for instance), and hence, surprisingly enough, these laws will usually have positive predictive power.

The zero-order (naive) way of answering this question is to do as some philosophers are doing, to state that "the World is cognoscible" and to stop there. I might have myself stopped my

ISBN 0-12-068720-8

talk at this point if I had not had the enticing example of the XVIth century mathematician, François Viete, who was able to find some simple rules (= elementary algebra) which were a ready-to-use substitute for all the endeavour and personal talent necessary to solve awkward classroom arithmetic problems. Will Dynamical Systems and related theories turn into the algebra of Natural Science Laws?

2. RUELLE-TAKENS CRITICS OF LANDAU'S THEORY OF TURBULENCE

Ruelle-Takens critics of Landau's turbulence theory are a good example [1] of such a concern. In fact they have not dismissed Landau's theory, but have shown that his quasiperiodic-function approach is 'fragile' towards perturbations, and hence infinitely improbable. Indeed, since experiments are never able to determine individual points but rather work with diffuse open sets, physical concepts and laws (A) should be quite insensitive to small virtual perturbations, i.e. *structurally stable.*

A serious question arises at this point: how may one determine the class in which these virtual perturbations may lie? Since Natural Sciences are experimental Sciences we believe that this point has to be solved by experiment; in Section 5, while discussing Peixoto's symmetry breaking, we shall give a non-trivial example how that might be done.

3. CONTRADICTORY AIMS

Physical concepts have to cope not only with the imprecisions of the experiments but also with (B) the lack of knowledge of huge regions where no experiment has yet been done (e.g. at energies surpassing the present elementary particle accelerators). But a law which will be invariant to virtual changes both in regions A and B, will be an excellent description of the present knowledge, but will have no predictive value at all. A scientist would therefore like somehow in a contradictory way, to have laws stable versus the imprecisions of class (A), but extremely biased towards the region of interest (B).

Fortunately there exists some theoretical "levers" [2] by means of which one may control "the spread-off" of the predictions in (B). In this respect, Analytic Scattering Theory is an excellent playground, since analytic continuations off open curves have *both* the properties of being unique – if the data is absolutely exact – *and* to be *ill-posed, in the Hadamard sense,* i.e. giving extremely poor predictions if errors are present. In Section 6 we shall show, in a sketchy way, how these 'predictivity levers' might be used for practical purposes.

4. BROKEN SYMMETRIES AND BRANCHING POINTS

There has been much interest in spontaneous broken symmetries for the last twenty years (for a modern approach, see L. Michel [3]). Together with any other sort of branching, they represent the critical points of the theories, where some supplementary information is needed in order to restore predictivity. Sometimes these critical points look like the branching of the stream lines of a flow, and the supplementary information which restores determinism is of the form "go left" or "go right". In other cases, like in analytic continuation off open contours, the lines of flow inside the Banach Space are unique, but they diverge infinitely in the higher dimensions, which makes predictions impossible if errors are present (= "branchings of the second kind").

However the above examples do not exhaust the various forms of spontaneous branching: a non-trivial example will be discussed next.

5. PEIXOTO'S FLOW

A geodetic flow on a thorus is given by $d\theta/d\phi = c$, where c is a constant. Using a square chart ($0 \leq \phi < 2\pi$, $0 \leq \theta < 2\pi$) for the thorus (opposite points identified), a geodetic will be a straight line, disappearing, for instance, at the right, appearing again at the corresponding point at the left, and so on. This line is closed only if $c = p/q$ (p, q, integers), but as this happens with zero probability, in general a geodetic will tint the square in a uniform gray, in perfect agreement with the symmetries of the thorus. Until now, nothing unusual. But lets consider the generic (smooth) deformations of the thorus. The geodetics will now draw wavy lines, as their equation becomes $d\theta/d\phi = c + f(\phi,\theta)$. Again in complete agreement with the symmetries of the thorus *and* of the class of deformation, almost always they will not close but pass arbitrarely near to any point. But the unexpected and really strange result of Peixoto, is that this flow *has repulsors and attractors* whose equations are *almost always straight lines with* $d\theta^{a,r}/d\phi^{a,r} = p/q$ (!). In other words the symmetry of the problem will be spontaneously broken by the appearance of light-gray (the repulsors) and dark-gray (the attractors) strips!

This kind of symmetry breaking is entirely different (see R. Thom, [4]) of any symmetry breaking process used in Physics. It is interesting to note that these patterns *will not* appear for any class of deformations of the initial equation (for instance if $f(\phi,\theta)$ is a constant). Hence they work as a theoretical (infinite) magnifying glass, allowing us to decide experimentally of which class are the (unknown) perturbation which occur in Nature.

6. STABILITY IN ANALYTIC SCATTERING THEORY

Speaking about second class bifurcations (sect. 3) we said that there exists various kinds of informations which are redundant in the ideal zero-error case, but which have efficient stabilization properties for ill-posed problems with non-zero errors.

As an example let's take the continuation problem of the continuation problem of the Scattering Amplitude off space-like data points (off a segment lying in the holomorphy domain) to the time-like one, i.e. to the cuts. This is obviously an ill-posed problem. However if some special information[1] is available, in the form, say, of some function-bound for the derivative, the continuation problem is obviously stabilized.

But these stabilizing entities are useful even if the true value of the bound is unknown. I shall try to show that in a simple example [5].

Let $\delta_0[A]$ be the least value of the weighted L^2 - norm of the (tangent) derivative on the cuts of the Amplitude with respect to energy, still compatible with the data A and analyticity. The weight may be chosen so as to emphasise the region where the search is performed. If there will be much physical "structure" in that region, δ_0 will be large. One could check now the relative likelihood of different hypothesises (new particles, i.e. second Riemann-sheet poles, etc.) by substracting the latter and see how much δ_0 has decreased if computed for these modified functions, A'. As computer experiments have shown, if the errors are not too large, there are many orders of magnitude between the δ_0 corresponding to the correct hypothesis and that of the false ones.

These techniques have been used already with much success for instance to find values for vital parameters in Quantum Chromodynamics [6] and, in many other problems where Analyticity plays a leading role.

Although this kind of applications is less unprejudiced than those where the true value of the bound is known, these entities

1 *This information is certainly redundant in the zero-error case when the continuation is unique and the bound is identically satisfied.*

turn out to be strong investigation devices. Anyhow, since δ_0 is a characteristic *of the entire* set of 'Amplitudes' compatible with the given conditions, if this test fails no other criterion will work!

I wish to explain why I have decided to dedicate this paper to the great Rumanian composer Aurel Stroe. I have always been impressed by the profoundness of his thoughts and his exploring spirit, but now I have especially in mind his Canto Interrotto from his concert for Harpsichord and Orchestra. The rules of this composition were chosen so as to restrict the latter – in the absence of new themes coming from the exterior – to an Attractor of four notes, repeated to infinity. Aurel Stroe tried to suggest in this way that an Ideal, an Artistic Idea, or a whole national culture which is isolated from exterior inputs is foredoomed to shrink to a sterile limit cycle where no new structures are generated.

This applies also to Science.

REFERENCES

[1] D. Ruelle and F. Takens, Commun. math. Phys. 20, (1971), 167; also Commun. math. Phys. 23, (1971), 197.

[2] S. Ciulli and G. Nenciu, J. Math. Phys. 14, (1973), 1675; see also Physics Reports, 17C, (1975), 133 - 224.

[3] Louis Michel, Rev. Mod. Phys. 52, (1980) 617 *and* IHES/P/80/37.

[4] René Thom, Math. Phys. and Phys. Math., Polish Sci. Publ., Warszawa, (1976), 293 - 320.

[5] S. Ciulli and T. D. Spearman, J. Math. Phys., in press.

[6] I. Caprini and C. Verzegnassi, Nucl. Phys., in press.

NON-EQUILIBRIUM ENTROPY FOR KOLMOGOROV DYNAMICAL SYSTEMS

M. Courbage

Universitē Libre de Bruxelles
Campus Plaine C.P. 231
B - 1050 BRUXELLES, Belgique

§1. INTRODUCTION

The results we are to present here concern the problem of the obtention of time irreversible dynamics from time reversible ones and the physical interpretation of the procedure of this passage. This problem is closely linked to the Boltzmann problem : how to obtain a dynamic expression of the second law of thermodynamics. Recent results obtained jointly by Misra, Prigogine, Goldstein and the author /1/,/2/,/3/,/4/ present a new approach to this problem. In this note we recall briefly these results and we present a short physical interpretation of them.

The procedure of the passage to irreversible evolution here presented is physically relevent to a general coarse graining with this important feature : it takes into account the whole information contained in the fine grained initial statistical state. The procedure uses conditional expectation with respect to a sequance of partitions of the phase space becoming infi-

DYNAMICAL SYSTEMS
AND MICROPHYSICS

ISBN 0-12-068720-8

nitely fine. This is a generalized smoothing procedure which enables one to get a new physical representation. In this representation, the evolution is given by a semi-group of a Markov process with a Liapounov function. The systems admitting such a procedure are K-systems. Some final remarks may enlight how probability enters in this scheme which, in the same time, is a change of representation.

§2. GENERAL FORMALISM AND RESULTS

We consider a measurable dynamic system given by 1) a phase space Ω which is a measurable space with a σ-algebra $\mathcal{O}\!\!\mathcal{l}$ of measurable sets, 2) a one-to-one measurable dynamic transformation B acting at a regular time intervals 0, $\pm$ 1, $\pm$ 2,.. (here we take for simplicity a discrete time evolution, all results being true for flows) 3) a B-invariant normalized measure μ 4) A set of dynamic observables given by square integrable functions on Ω and a set of states $\mathcal{S}$ given by normalied density probabilities $\varrho(\omega)$, $\omega \in \Omega$, that belong to L^2_μ.

In the following, U will be the unitary operator associated to B by Koopman lemma :

$$(U\varrho)(\omega) = \varrho(B^{-1}\omega) \qquad \forall \varrho \in L^2_\mu \tag{1}$$

For a state ϱ, (1) expresses the Liouville equation with $U = e^{-iL}$, L being the Liouville operator.

Theorem 1 : If W is a contraction operator on L^2_μ (i.e. $\|Wf\| \leqslant \|f\|$) such that 1) W preserves positivity ($f \geqslant 0 \Longrightarrow Wf \geqslant 0$) 2) $W1 = W^*1 = 1$ a.e. then

$$(W^n \varphi_\Delta)(\omega) = P(n, \omega, \Delta) \; , \qquad \Delta \in \mathfrak{A} \tag{2}$$

is a transition probability for a Markov process with μ -invariant measure. If moreover $\|W^{*n}(\varrho - 1)\|$ decreases monotonically to zero then $\int (W^{*n}\varrho)(\omega) \log(W^{*n}\varrho)(\omega)\, d\mu$ decreases also monotonically. Here W^* is the evolution operator of states $\varrho \in \mathcal{S}$:

$$\varrho \longrightarrow \varrho_n = W^{*n}\varrho \Longleftrightarrow \int d\mu(\omega)\varrho(\omega)P(n,\omega,\Delta) = \int_\Delta \varrho_n(\omega)\, d\mu \tag{3}$$

Definition : A K-system is a dynamical system having a sub- σ -algebra $\mathfrak{F}_0 \subset \mathfrak{A}$ such that i) $B^n\mathfrak{F}_0 = \mathfrak{F}_n \subseteq \mathfrak{F}_{n'}$ if $n \leqslant n'$, ii) $\bigcup_{-\infty}^{+\infty} \mathfrak{F}_0 = \mathfrak{A}$ iii) $\bigcap_{-\infty}^{+\infty} \mathfrak{F}_n = \mathfrak{F}_{-\infty}$ is the trivial sub- σ algebra containing Ω and the sets of null measure. As well known it is possible to associate to any sub- σ -algebra $\mathfrak{F}_n$ a (measurable) partition ξ_n that generates $\mathfrak{F}_n$. Therefore $\xi_{n'}$ is finer than ξ_n for $n' \geqslant n$.

Let P_n be the orthogonal projection on $L^2_\mu(\mathfrak{F}_n) = \mathcal{H}_n$ the sub-space of all $\mathfrak{F}_n$ -measurable functions. P_n is the smoothing operation of $\mathfrak{A}$ -measurable functions over $\mathfrak{F}_n$ called conditionnal expectation $P_n f = E^{\mathfrak{F}_n}(f)$, $f \in L^2_\mu$. It follows that $P_n - P_{-\infty}$ is a spectral family over $\mathcal{H}^\perp_{-\infty}$ and $U P_n U^{-1} = P_{n+1}$.

We define Λ as a superposition of smoothings :

$$\Lambda \varrho = \sum_i \delta_i P_i \varrho \qquad \text{with} \quad \sum_i \delta_i = 1 \tag{4}$$

If we denote $\lambda_0 = \sum_{i=0}^{\infty} \delta_i$, $P_i - P_{i-1} = E_i$ and $\lambda_n - \lambda_{n+1} = \delta_n$ then we have the following theorem :

Theorem 2 : Let (Ω , $\mathfrak{A}$, B, μ) be a K-system. 1) The operator Λ defined as above is a bounded, states preserving, operator, defined by the spectral representation :

$$\Lambda = \sum_{i=-\infty}^{+\infty} \lambda_i E_i + P_{-\infty}$$

where λ_i is a decreasing sequence of positive numbers $\lambda_i \searrow 0$, $i \longrightarrow +\infty$ and $\lambda_i \nearrow 1$, $i \longrightarrow -\infty$.
2) If $\delta_i > 0$, $\forall i$ and $\nu_i = \lambda_{i+1}/\lambda_i$ is a decreasing sequence, then Λ^{-1} is a densely defined operator and $\Lambda U^{-1} \Lambda^{-1}$ extends to an operator W of a Markov process satisfying to the conditions of the theorem 1.

This theorem firstly proved for Bernoulli systems /1/,/2/ has been generalized to K-flows in /3/. Briefly speaking, the property 1 is a consequence of the positivity preserving of P_n. The idea of the proof of 2) results from the fact that $U^{-1}\Lambda U \Lambda^{-1} E_i = \nu_i E_i$, that is, $U^{-1}\Lambda U \Lambda^{-1} = \Sigma \nu_i E_i + P_{-\infty}$. Then, ν_i being decreasing, this operator takes the form (4) and then it preserves positivity. It has also the property that $\|\Lambda U^n \Lambda^{-1} \varrho\|^2 \searrow 1$, $n \longrightarrow \infty$ for any state ϱ. Now, on account of the theorems 1 and 2 one can define an entropy for this system by the functional :

$$\begin{aligned}\Omega(\tilde{\varrho}_n) &= -\int_\Omega (\Lambda U^n \varrho)(\omega) \log (\Lambda U^n \varrho)(\omega)\, d\mu \\ &= -\int_\Omega (W^{*n} \tilde{\varrho})(\omega) \log (W^{*n} \tilde{\varrho})(\omega)\, d\mu\end{aligned}$$

for the transformed density $\tilde{\varrho}_n = \Lambda U^n \varrho = \Lambda U^n \Lambda^{-1} \Lambda \varrho = W^{*n} \tilde{\varrho}$.

§3. REMARKS AND PHYSICAL INTERPRETATION

The above $\mathcal{H}$-theorem is given in terms of the transformed density $\tilde{\varrho}_n$ instead of ϱ_n and $\tilde{\varrho}_n$ evolves under a stochastic process. How probability enters here ?

For a deterministic system one can also define the evolution of ϱ by a transition probability $P(n, \omega, \Delta)$

$$P(n,\omega,\Delta) = 1 \qquad \text{if} \quad \omega \in B^{-n}\Delta$$

$$= 0 \qquad \text{if} \quad \omega \notin B^{-n}\Delta$$

$$P(n,\omega,\Delta) = \varphi_{B^{-n}\Delta}(\omega) = (U^{-n}\varphi_\Delta)(\omega)$$

for <u>any</u> Δ in $\mathfrak{A}$, where φ_Δ is the characteristic function of Δ. This transition probability can distinguish between points (the partition corresponding to $\mathfrak{A}_0$ is formed by points of Ω). If however we restrict the observable events to a sub-σ-algebra $\mathfrak{A}_0$, $\mathfrak{A}_0$ being, for simplicity, generated by a countable partition $\{A_i\}$, then one may not distinguish between different points of the same A_i. Denote by $A_i(\omega)$ the cell containing ω.

If now $B^n A_K(\omega)$ is included in some A_i then we may determine surely to which A_i belongs $B^n\omega$. But given a Δ of $\mathfrak{A}$ which is not in $\mathfrak{A}_0$, let us say strictly included in A_i the transition probability of ω to Δ is given by the conditional probability $\mu(A_i \cap \Delta)/\mu(A_i) = \mu(\Delta/A_i)$. Thus we obtain the transition probability given by :

i)
$$P(n,\omega,\Delta) = \sum_i \frac{\mu(A_i \cap \Delta)}{\mu(A_i)} \varphi_{B^{-n}A_i}(\omega)$$

ii)
$$B^{-n}\mathfrak{A}_0 \subseteq \mathfrak{A}_0 , \qquad n \geq 0$$

The second condition expresses the fact that $P(n,\omega,\Delta)$ takes constant values for all ω in A_i i.e. different points of A_i lead to the same A_j. The conditions i) and ii) can be written as follows :

i')
$$P(n,\omega,\Delta) = (U^{-n} P \varphi_\Delta)(\omega)$$

where P is the conditional expectation $E^{\mathfrak{A}_0}$ defined by :

$Pf = E^{\mathfrak{A}_0}f = \sum_i E_{A_i}(f)\varphi_{A_i}$ with $E_{A_i}(f) = \frac{1}{\mu(A_i)}\int_{A_i} f\,d\mu$
and ii')

$$U^{-n}L^2(\mathfrak{A}_0,\mu) \subseteq L^2(\mathfrak{A}_0,\mu) .$$

This implies that $U^{-n}P = PU^{-n}P$. Clearly $U^{-n}P$ is a semi group of a Markov process. In fact, $U^{-1}P = PU^{-1}P$ and therefore $U^{-n}P = (PU^{-1}P)^n$ where $PU^{-1}P$ is a contraction positivity preserving operator. Misra /4/ has shown that in order that some positivity preserving projection operator P generates an irreversible Markov process the system should be a K-flow. The irreversible Markov process means that $\|PU^nP(\varrho-1)\|^2$ decreases monotonically to zero and this implies that $\bigcap_{-\infty}^{+\infty} B^n\mathfrak{A}_0$ is the trivial σ-algebra generated by, $\{\hat{0},\Omega\}$ where $\hat{0}$ is the family of sets of null measure. Therefore the condition i)' and ii)' imply that (Ω , $\bigcup_{-\infty}^{+\infty} B^n\mathfrak{A}_0$, B , μ) is a K-system and necessarily the partition $\{A_i\}$ corresponding to $\mathfrak{A}_0$ should be uncountable. It is well known that the Radon-Nikodym theorem permits the extension of i) to any σ-algebra .

Now the signification of Λ constructed for K-systems in §2 is easy to see. Λ transforms the dynamic picture ϱ_n into the physical one $\tilde{\varrho}_n = \Lambda\varrho_n$ in the following way :

$$\tilde{\varrho}_n = \Lambda U^n\varrho = \sum_i \delta_i PU^n\varrho = \sum_i \delta_i P_i U^n P_i \varrho$$

$$= \sum_i \delta_i W_i^{*n}\varrho$$

where W_i^n is the Markov process corresponding to the coarse-graining with respect to F_i and δ_i is the weight associated to it in the total evolution. Thus $\tilde{\varrho}_n$ evolves as a weighted superposition of the different coarse-grained contributions of ϱ with respect to all $\mathfrak{F}_i$. In this superposition we take into account the whole information contained in $\varrho(\omega)$. It is important to note that $\tilde{\varrho}_n$ is not a reduced density probability (it is $\mathfrak{A}$-measurable) obtained via a projection operator. In fact, the range of Λ is a dense subset of L^2_μ and to any $\tilde{\varrho}_n \in \mathfrak{D}(\Lambda^{-1})$, Λ^{-1} associates a unique state ϱ in the original picture. Λ, thus, realises an equivalence between two representations /5/

$$\{f \in \mathfrak{D}(\Lambda^{-1}), U, \varrho \in \mathcal{S}\} \longleftrightarrow \{\tilde{f} \in L^2_\mu, W^*, \tilde{\varrho} \in \mathfrak{D}(\Lambda^{-1})\}$$

where $\tilde{f} = \Lambda^{-1} f$, $\tilde{\varrho} = \Lambda \varrho$ and

$$<f, U^n \varrho> = <\tilde{f}, W^* \varrho>$$

here $\mathfrak{D}(\Lambda^{-1})$ is the dense domain of Λ^{-1}.

Let us stress on the fact that the concept of instability plays an important role in the irreversible Markov process associated to sub-σ-algebras $\mathfrak{F}_n$. For Anosov systems $\mathfrak{F}_n$ corresponds to contracting fibers of B and dilating fibers of B^{-1}. This means that B maps any contracting fiber into a subset of some other contracting fiber in such a way that the future of a contracting fiber is known with probability one, on contrary the past of any such fiber is not determined uniquely and corresponds to a number of different fibers exponentially increasing with time, for $\mathfrak{F}_n$ is a bundle of dilating fibers of B^{-1}. In fact this property implies the irreversibility of the Markov processes here constructed. To see this, one may introduce the "velocity inversion" operation /5/ which axiomatizes the operation $I:(p,q) \longrightarrow (-p,q)$ in classical mechanics : that is, if $g_t(p,q) = (p_t, q_t)$ is the dynamic flow then $Ig_t\, Ig_t = 1$.

In the abstract formalism here used I is a unitary positivity preserving operator on L^2_μ such that $IUI^{-1} = U^{-1}$ and $I^2 = 1$. Now it is easy to check that W^{-1} does not preserve positivity /2/. This follows from a general theorem /6/ showing that if $\Lambda U \Lambda^{-1}$ and $\Lambda U^{-1} \Lambda^{-1}$ preserve together positivity then $\Lambda U \Lambda^{-1}$ should be unitary, which contradicts the monotonic decrease of $\|\Lambda U^n \varrho\|$. This entails that :

$$IWI^{-1} \neq W^{-1}$$

or equivalently $IW^* IW^* \varrho \neq \varrho$, that is, velocity inversion is not time inversion of W . On the other hand, it is clear that $\tilde{I} = \Lambda I \Lambda^{-1}$ is the time inversion of W^* :

$$\tilde{I} W^* \tilde{I} = W^{*-1}$$

but $\tilde{I}$ could not preserve positivity and it does not preserve probabilities, that is, it could not correspond to a physical operation.

It is also clear that $U^{-n}P$, which is a semi-group defined for $n \geq 0$ could not be extended for $n < 0$. This would imply that $B^{-1}\mathfrak{A}_0 \supset \mathfrak{A}_0$, which contradicts the fact that $B\mathfrak{A}_0 \supset \mathfrak{A}_0$. That is, the coarse-graining corresponding to the past is distinct from the coarse-graining corresponding to the future.

REFERENCES

/1/ B. Misra, I. Prigogine and M. Courbage, Physica <u>98A</u> (1979), p. 1.

/2/ M. Courbage and B. Misra, Physica <u>104A</u> (1980) p.359.

/3/ S. Goldstein, B. Misra and M. Courbage, J. Stat. Phys. 25 (1981) p.111.

/4/ B. Misra and I. Prigogine, Suppl. of the Progr. of Theor. Phys. 69, (1980) p. 101.

/5/ I. Prigogine, C. George, F. Henin and L. Rosenfeld, Chemica Scripta (1973) p.5.

/6/ K. Goodrich, K. Gustafson and B. Misra, Physica 102A (1980), p. 371.

The author thanks Professors B. Misra and I. Prigogine for many discussions concerning time symmetry breaking and coarse-graining.

BIMETRIC MACHIAN GRAVITATION : GENERAL THEORY AND COSMOLOGY

Riccardo Goldoni

University of Pisa, Istituto Matematico "L.Tonelli", Pisa, Italy
and G.N.F.M. of the C.N.R..

Einstein's Theory of Gravitation is based on three main guiding ideas : *a*) a *Physical Principle*, that is the local equality of fields of acceleration and of gravitation expressed by the Principle of Equivalence; *b*) a *Mathematical Principle*, that is the geometrization of the gravitational field equivalent to the statement that space-time (s.t.) and gravitation are described by a four-dimensional Riemannian manifold with Minkowskian signature and point particles travel along the geodesics of such a manifold; *c*) a *Philosophical Principle*, that is Mach's Principle.

General Relativity incorporates *a*) and *b*), but *not* *c*). Indeed we may assert with Fock (Cf. Ref.[1]) that *General Relativity is less relativistic than Special Relativity*.From General Relativity follows the existence of an absolute frame determined by the curvature of s.t., the gravitational equations (g.e.) have a solution also in the absence of matter at all and the theory does not incorporate the machian idea that the local properties of s.t. are determined not only by local matter, but also by the average distribution of cosmological matter.

Einstein himself tried to modify his theory in order to incorporate Mach's Principle, but he was not successeful in this respect. As far as we can see, the main motivations have been that *d*) any theory in explicit agreement with machian ideas on gravity and inertia cannot be a purely differential theory and that *e*) any such a theory should

DYNAMICAL SYSTEMS
AND MICROPHYSICS

ISBN 0-12-068720-8

be formulated in view of its philosophical validity in any conceivable Cosmos. Point *e)*, in particular, avoids in principle the existence of an absolute frame.

In a series of papers (Cf. [2-4]) we tried to formulate a new theory of gravitation based on the guiding ideas *a)*-*e)*. Based on the same general framework we have derived two different approaches, the second of which (Cf. Ref.s [3,4]) appears to be the most promising one and will be stated in the following.

We introduce two Riemannian manifolds (M^4, g) and (M^4, h), the first [second] of which describing s.t. and gravitation in atomic units (a.u.) [gravitational units (g.u.)] . M^4 denotes the differentiable manifold assumed to represent s.t., g and h two metric tensors with signature (1,-1,-1,-1). Light [gravitational interaction] is assumed to travel along null geodesics of (M^4, g) [(M^4, h)] . Latin (greek) indeces run from 0 to 3 (1 to 3) .

On the basis of Mach's Principle one cannot expect to detect inhomogeneities and unisotropies in Nature by using any system of units (s.u.) based on gravitational phenomena, so that h turns out to be equal to a De Sitter metric tensor.

To connect h and g we introduce a tensor H determined by the *imprint* of the Cosmos on the local structure (P, say) such that

$$g_{ab} = H^m{}_a \, h_{mn} \, H^n{}_b \tag{1}$$

From Machian requirements and from the consideration that the gravitational interaction travels on the null cone of h, follows that

$$\{H^a{}_b\} \equiv H = A \sinh P \quad , \quad P \equiv \{P^m{}_n\} \tag{2}$$

$$P^m{}_n(x) = -8\pi\alpha_N \int (-h(x'))^{\frac{1}{2}} \, d^4x' \, k(x,x') \, \delta^{(-)}(\sigma_h[x,x']) \;\; \times$$

$$h^m{}_{m'} \, h_{nn'} \left[T^{m'n'} - \tfrac{1}{2}(h_{a'b'} \, T^{a'b'}) \, h^{m'n'}\right] \tag{3}$$

As our Cosmos is concerned, A is assumed to depend on the whole content of matter in the Cosmos, but local matter : $A = A(x,\overline{P})$, where $\overline{P}$ denotes the contribution to P due to the large scale structure. We also assume that A has to be determined on the basis of the requirement that $H(x,\overline{P}) = I$ approximately holds in the solar system at the present Cosmos time t_0. Consequently the explicit determination of A depends on the cosmological model assumed to describe our Cosmos. On the contrary, under the assumption $|R_h| << 1$ the solutions of local problems is independent of its explicit determination and can be obtained by using the equation

$$H(x,P) = \exp(P_L) \tag{4}$$

holding with a very good approximation in the solar system at the present Cosmos time t_0. P_L denotes the contribution to P due to local matter . T denotes the energy-momentum-tensor (e.m.t.) in the system of atomic units we work, α_N the Newton coupling constant and $k(x,x') = (1/4\pi)\left[h(x)\ h(x')\right]^{-\frac{1}{2}} \det\{-\sigma_{h/mm'}\}$, where the symbol / denotes the covariant partial differentiation in (M^4,h). $h_{nn'}$ denotes the parallel displacement bi-vector in (M^4,h) and $h = \det h$. $\sigma_h(x,x') = \frac{1}{2}$(geodesic distance in (M^4,h) between x and x')^2.

The above definition of $k(x,x')$ allows P to be also well defined by the differential equation

$$(\Box_h - R_h/6)\ P^m{}_n = -8\pi\alpha_N\left[h_{na}T^{am} - \tfrac{1}{2}(h_{ab}T^{ab})\ \delta^m{}_n\right] \tag{5}$$

equipped with suitable boundary conditions. $\Box_h$ (R_h) denotes the covariant d'Alembertian (the scalar curvature) in (M^4,h).

We further assume a superuniverse model such that our Universe is represented by a local inhomogety in a homogeneous (in space and in time) and isotropic back-ground. With this assumption the inertial properties of local matter, as well the local gravitational coupling, are determined not only by the Universe structure, but also by the back-ground structure and by the relative dimensions of the back-ground and of the Universe, so that we have more freedom than in the standard approach for being in agreement with possible future experiments on the homogeneity and isotropy of inertia as well of the local gravitational coupling. Moreover such an assumption allows us to treat the cosmological problem as a, in some sense, local problem in a *a priori* fixed back-ground, that is in a fashion common to the most part of the other physical theories.

Under the simplest assumption that our Universe be filled of pure dust matter and $|R_h| \ll 1$ one further obtain two essentially distinct cosmological models. The first of them, characterized by a present deacceleration parameter $q(t_o) \geqslant 2.5$, is given by

$$\rho = |C| (\cos\beta)^{-6/5} (\cos\beta_o)^{-4/5} \tag{6}$$

$$g_o = (\cos\beta / \cos\beta_o)^{4/5} \eta \tag{7}$$

while the second of them, characterized by $0 \leqslant q(t_o) \leqslant 2.5$, is given by

$$\rho = |C| (\sinh\beta)^{-6/5} (\sinh\beta_o)^{-4/5} \tag{8}$$

$$g_o = (\sinh\beta / \sinh\beta_o)^{4/5} \eta \tag{9}$$

where C denotes a constant of integration, $\eta = \mathrm{diag}(1,-1,-1,-1)$, $\beta = \pm |10\pi\alpha_N C|^{\frac{1}{2}} t$ and $\beta_o = \beta(t_o)$.

For the first cosmological model one easily finds : $H(t_o) = \mp (2/5) |\beta/t|^{\frac{1}{2}} \mathrm{tg}\beta_o$; $q(t_o) = 5/2\sin^2\beta_o \geqslant 5/2$; $\sigma_o(t_o) = q(t_o)/3\alpha_N \geqslant 5/6\alpha_N \simeq 0.125\ 10^8\ g\ s^2/cm^3$, while for the second cosmological model one finds : $H(t_o) = \pm(2/5) |\beta/t|^{\frac{1}{2}} \coth\beta_o$; $q(t_o) = 5/2\cosh^2\beta_o \leqslant 5/2$; $\sigma_o(t_o) = q(t_o)/3\alpha_N \leqslant 5/6\alpha_N$. For both cosmological models one has $\rho(t_o) = q(t_o)\, H^2(t_o) (4\pi\alpha_N)^{-1}$. H denotes the Hubble parameter and $\sigma_o = 4\pi\rho/3H^2$.

Only the second cosmological model is compatible with the till now observed value $\rho(t_o) = 4\ 10^{-31}$ g/ cm^3. Assuming such a value and the mean observed value $H^{-1}(t_o) = 18\ 10^9$ years $\simeq 5.68\ 10^{17}$s, one easily finds : $q(t_o) \simeq 0.108$, $\sigma_o(t_o) \simeq 0.54\ 10^6$ g s^2/ cm^3, $C \simeq 88.6\ 10^{-31}$ g / cm^3 and $t_o \simeq 5.2\ 10^{17}$ s, which is less than the mean observed value of $H^{-1}(t_o)$, but of the same order . This last result appears to support the second cosmological model .

The theory now briefly outlined is assumed to hold in any conceivable Cosmos and is apparently compatible with Mach's principle . Bimetric theories have been introduced in the past by ROSEN (Cf . Ref . [5]) and PAPAPETROU (Cf . Ref .[6]) . More recently Rosen has proposed two well defined bimetric theories of gravitation (Cf . Refs [7,8]). The present theory, besides being observationally distinguishable from Rosen's ones, differs from Rosen's theories mainly for its machian behavior and for its simpler mathematical structure . At the orders of approximation which have been made, the first Rosen's approach (Cf . Ref.[7]) and ours supply us with the same metric tensor as the Schwarzschild-like problem is concerned , that is, using eq . (4), under the assumption $|R_h| \ll 1$, one has

$$g = \mathrm{diag}(\exp[-2\alpha_N\, m/ r],\ -\exp[2\alpha_N\, m/ r]\ ,\ -\exp[2\alpha_N\, m/ r]\ ,\ -\exp[2\alpha_N\, m/ r]\ ) \qquad (10)$$

As cosmology is concerned, we derived two cosmological models based on the simplest assumption that our Universe be filled of pure dust matter. One of the two cosmological models is compatible with the till now observed value of the density of dust mass and provides an age of the Universe of the same order of $H^{-1}(t_o)$, result which appears to support the model itself.

We are indebted to Prof. Nathan Rosen for his kind interest in this approach.

REFERENCES

[1] FOCK V., *The Theory of Space, Time and Gravitation*, Pergamon Press, New York, New York, 1976.

[2] GOLDONI R., *Gen. Rel. Grav., Vol. 7, p.731*, 1976; *ibid. Vol. 7, p. 743*, 1976; *ibid. Vol. 12, p. 9*, 1980; *Lett. Nuovo Cimento, Vol. 24, p. 297*, 1979; *ibid. Vol. 23, p. 103*, 1978,

[3] GOLDONI R., *Lett. Nuovo Cimento, Vol. 29, p. 403*, 1980.

[4] GOLDONI R., *Lett. Nuovo Cimento, Vol. 31,p.481*,1981; *ibid. ERRATUM (to appear); Università di Pisa, Istituto di Scienze dell'Informazione, Nota Scientifica S-13-81 (to be published in Lett. Nuovo Cimento).*

[5] ROSEN N., *Phys. Rev., Vol. 57, p.147*, 1940; *ibid. Vol. 57, p. 150*, 1940; *Ann. Phys. (N.Y.), Vol. 22, p.1*, 1963.

[6] PAPAPETROU A., *Proc. R. Ir. Acad. Set. A, Vol. 52, p. 11*, 1948.

[7] ROSEN N., *Gen. Rel. Grav., Vol. 4, p.435*, 1973; *ibid. Vol. 6, p. 259*, 1975; *ibid. Vol. 9, p. 339*, 1978; *Ann. Phys. (N.Y.), Vol. 34, p. 455*, 1974.

[8] ROSEN N., *Found. Phys., Vol. 10, p. 673*, 1980; *Gen. Rel. Grav., Vol. 12, p. 493*, 1980.

MECHANICS AND THE NOTION OF OBSERVABLES

André Heslot
Faculté des Sciences et Techniques de Monastir, Tunisie

INTRODUCTION.

An experimental point of view leads to define observables of a physical system as real valued regular functions of the state of the system.

Thus observables appear quite naturally in Classical Mechanics as the real valued smooth functions on the phase space. The set of observables is then a commutative and associative algebra.

The study of automorphisms of the phase space, i.e. canonical transformations, leads to another aspect of observables: that of generators of canonical transformations. More precisely, the set of observables is a Lie algebra under the Poisson bracket, closely related to the Lie algebra of the group of automorphisms of the phase space.

I suggest to consider this second aspect as more fundamental, that is, as a definition of the notion of observables. This point of view leads then to a natural generalization of Classical Mechanics, which turns out to apply successfully to Quantum Mechanics.

1. CLASSICAL MECHANICS AND ITS GENERALIZATION.

The space of states S of a physical system described by Classical Mechanics is its phase space, that is, an even dimensional smooth manifold provided with a symplectic structure, i.e.

DYNAMICAL SYSTEMS
AND MICROPHYSICS

ISBN 0-12-068720-8

with a 2-form ω, non-degenerate and closed.

The physical meaning of this symplectic structure is that the transformations of the states which do not modify the nature of the system, but merely correspond to a change of point of view, e.g. a rotation, a translation, or the time evolution, are the automorphisms of φ, that is, the ω-preserving diffeomorphisms. Such transformations will be called canonical .

The generator of a local one-parameter subgroup of canonical transformations is a vector X on S such that the Lie derivative $L_X\omega$ vanishes. The set A of all such vectors is a Lie algebra under the Lie bracket [,], which can be formally considered as the Lie algebra of the group of canonical transformations. Since ω is closed, $d(i_X\omega) = d(i_X\omega) + i_X d\omega = L_X\omega = 0$, and the inner product $i_X\omega$ is a closed 1-form. Let us now assume that the first Betti number $b_1(S)$ is zero. Then $i_X\omega$ is exact, that is, there exists some f belonging to the vector space $\mathcal{O}$ of real valued smooth functions on S such that $i_X\omega = -df$. Since ω is non-degenerate, the map $X \to i_X\omega$ is an isomorphism, and this conversely defines an $X \in A$ for every $f \in \mathcal{O}$: we shall write $X = X_f$.

We now define the Poisson bracket $\{f,g\}$ of $f, g \in \mathcal{O}$ by $\{f,g\} = \omega^{-1}(df,dg)$. It is easy to show (see for instance [1]) that $\mathcal{O}$ is a Lie algebra under the Poisson bracket and that $[X_f,X_g] = X_{\{f,g\}}$. Therefore, the linear map $f \in \mathcal{O} \to X_f \in A$ is an homomorphism of Lie algebras. This homomorphism is surjective, and its kernel is the set of the f such that $df = 0$. We shall assume that S is connected. Then this kernel reduces to the set of constant functions, and $f \in \mathcal{O}$ can be identified modulo an additive constant with $X_f \in A$.

Because they are real valued regular functions of the state of the system, the elements of $\mathcal{O}$ are considered as the observables of the system. But the meaning of $f \in \mathcal{O}$ as a function of the state is in fact determinated by the corresponding $X_f \in A$: for instance, the energy is the generator of time evolution. Thus the generator aspect of observables is more fundamental (see [3] for a more careful discussion), and we are led to the following natural generalization of Classical Mechanics :

a) We suppose that the space of states S of the system under consideration is provided with a structure whose physical meaning is that the transformations of the states which correspond to a change of point of view are the automorphisms of S .

b) We assume that this structure intrinsically induces on S a structure of symplectic manifold, such that S is connected and $b_1(S) = 0$: S is thus a classical phase space, provided in the general case with some complementary structure.

c) Then the observables of the system are those real valued smooth functions on S whose canonical transformations they generate are automorphisms of the whole structure of S .

From this definition, the set $\mathcal{O}$ of observables is clearly a Lie algebra under the Poisson bracket, which can be identified modulo additive constants with the Lie algebra of the group of automorphisms of S, as it is in conventional Classical Mechanics. But notice that the usual product of two observables, defined by product of their values, needs no longer to be an observable : thus the structure of commutative and associative algebra of $\mathcal{O}$ is lost in the general case.

This generalization of Classical Mechanics can be applied to a large class of systems (e.g., with some modifications, to classical systems with constraints - see [3]). In the following, we shall study the case of quantum systems. The first step will be of course the construction of an intrinsic symplectic manifold structure on the quantum-mechanical space of states.

2. SYMPLECTIC STRUCTURE IN QUANTUM MECHANICS.

The space of states S of a quantum system is the projective space of a complex separable Hilbert space $\mathcal{H}$, that is, a state M in S can be identified with a non-zero vector $|M\rangle$ in $\mathcal{H}$, defined up to a complex factor (We use Dirac's notations, so that the hermitian product $\langle \, | \, \rangle$ is linear on the right-hand side). We shall always choose $|M\rangle$ with norm equal to 1. $|M\rangle$ is then fixed up to a phase factor, and the well-defined map $(M_1,M_2) \in S\times S \to |\langle M_1 | M_2 \rangle|^2 \in R$ represents the structure of S, just as the symplectic form ω is the structure of the classical space of states. According to Wigner's theorem, the automorphisms of S, i.e. the transformations of S which preserve its structure, are the restrictions to S of the unitary and antiunitary operators in $\mathcal{H}$, two such operators being equivalent iff they differ from a phase factor (see for instance [4]). As we are interested with the generators of the automorphism group, we shall restrict ourselves to the consideration of unitary transformations. Then "intrinsic" will mean invariant under unitary transformations.

Following Cantoni [2], we shall now exhibit an intrinsic symplectic manifold structure on S.

Let M_0 be an arbitrary fixed state in S, whose representative $|M_0\rangle$ in $\mathcal{H}$ will be kept fixed, and $\mathcal{U}$ a neighborhood of M_0 in S such that $\langle M_0 | M \rangle \neq 0$ whenever $M \in \mathcal{U}$. For such an M, we can write $|M\rangle = \lambda^\circ |M_0\rangle + |P\rangle$, where $\lambda^\circ \neq 0$ and $\langle M_0 | P\rangle = 0$. Since $\lambda^\circ \neq 0$, it is always possible, multiplying $|M\rangle$ by a convenient phase factor, to suppose that λ° is real and strictly positive. The vector $|P\rangle$ is then uniquely determined, and completely characterizes $M \in \mathcal{U}$ (Since $|M_0\rangle$ and $|M\rangle$ are normalized, λ° is not an independant variable).

Now let us choose, in the subspace $\perp_{M_0}\mathcal{H}$ of $\mathcal{H}$ orthogonal to $|M_0\rangle$, an orthonormal basis $\{|M_k\rangle\}$, k = 1,2,... Then we can write $|P\rangle = \sum_k \lambda^k |M_k\rangle$, where the λ^k are complex numbers, and the real numbers $x^k = \mathcal{R}e\, \lambda^k$ and $x^{\bar{k}} = \mathcal{I}m\, \lambda^k$ may be considered as a local coordinate system in the domain $\mathcal{U}$ ($\mathcal{R}e$ and $\mathcal{I}m$ are respectively the real part and the imaginary part). It is clear, at least in the finite-dimensional case, that S is thus intrinsically provided with a smooth manifold structure, with S connected and $b_1(S)=0$. Notice also that if $\mathcal{H}$ has finite dimension n, then S has even dimension 2(n-1).

Moreover, the previous construction enables us to identify the space $T_{M_0}S$ of vectors tangent to S at the point M_0 with the space

$\perp_{M_0}\mathcal{H}$. More precisely, if $V \in T_{M_0}S$, let $M_0 \to M = M_0 + V\delta t$ be an infinitesimal displacement of the point M_0. Then there is an unique representative $|M>$ of M such that, to the first order in δt, $|M> = |M_0> + |V> \delta t$, where $|V>$ is some element of $\perp_{M_0}\mathcal{H}$. Conversely, this also uniquely determines V as a function of $|V>$, and it is clear that the correspondance is an isomorphism of vector spaces, provided $\perp_{M_0}\mathcal{H}$ is considered as a real vector space. Notice, however, that this isomorphism is not intrinsic : if $|M_0>' = \exp(i\theta)|M_0>$ is another representative of $M_0>$, the vectors $|M>$ and hence $|V>$ have to be multiplied by $\exp(i\theta)$. Thus, vectors $|V>$ are determined by vectors V up to a common phase factor.

Now let M be any point in S, V and W two vectors in T_MS, $|V>$ and $|W>$ the corresponding vectors in $\perp_M\mathcal{H}$. We define

$$\omega(V,W) = -\,2\hbar\,\mathcal{I}m < V|W > ,$$

where $\hbar$ is Planck's constant. Since $\omega(V,W)$ is invariant under the multiplication of $|V>$ and $|W>$ by a common phase factor, ω is for every $M \in S$ a well-defined intrinsic map $T_MS \times T_MS \to R$, and it can be shown that it is a symplectic form on S (sketch of the proof : modulo a $\sqrt{2\hbar}$ factor, the coordinates $(x^k, x^{\bar{k}})$ introduced previously are canonical in the sense of Darboux theorem - see for instance [1]).

3. OBSERVABLES IN QUANTUM MECHANICS.

Since the hilbertian structure of $\mathcal{H}$ induces on the projective space S a structure of symplectic manifold, with S connected and $b_1(S) = 0$, we can apply to Quantum Mechanics, at least in the finite dimensional case, the principles of Section 1. Therefore, the observables are those real valued smooth functions on S whose canonical transformations they generate are automorphisms of S, considered as the projective space of $\mathcal{H}$. Then we have the following result:

Theorem - Whenever $\mathcal{H}$ has finite dimension,

1°) The observables can be identified with the self-adjoint operators of $\mathcal{H}$, in such a way that the value at the point $M \in S$ of the observable f corresponding to the operator $\hat{f}$ is

$$f(M) = \langle M \mid \hat{f} M \rangle$$

2°) From this formula, it results that the correspondance $f \leftrightarrow \hat{f}$ is an isomorphism of vector spaces, and that $\hat{1}$ is the identity operator. But, moreover, this correspondance is an isomorphism of Lie algebras, i.e.

$$\widehat{\{f,g\}} = \frac{1}{i\hbar} [\hat{f},\hat{g}]$$

where [,] is the commutator (see also [2]).

3°) Let n be the dimension of $\mathcal{H}$, $\omega^{n-1} = \omega_\wedge\omega_\wedge\cdots_\wedge\omega$ the Liouville volume element of S, and Tr the trace. Then

$$\frac{\int_S f\omega^{n-1}}{\int_S \omega^{n-1}} = \frac{1}{n} \operatorname{Tr} \hat{f}$$

Sketch of the proof. The canonical transformation generated by $f(M) = \langle M \mid \hat{f}M \rangle$ is the restriction to S of the unitary transformation $\exp(\hat{f}t/i\hbar)$. The end of the proof is easy calculus.

Thus we have recovered the essential features of the quantum-mechanical formalism in such a way that Quantum Mechanics no longer appears as fundamentally distinct from Classical Mechanics. Moreover, through our interpretation of observables, the second part of the previous theorem is nothing but a justification of the canonical quantization rules.

REFERENCES.

[1] Abraham, R., and Marsden, J., *Foundations of Mechanics* (second edition), Benjamin, 1978

[2] Cantoni, V., Intrinsic geometry of the quantum-mechanical phase space, hamiltonian systems and correspondence Principle, Rend. Acad. Nat. Lincei, Vol. LXII, p.628, 1977.

[3] Heslot, A., Remarques sur la notion d'observables en Mécanique Classique et en Mécanique Quantique, Thèse de 3ème Cycle, Université Paris VI, 1979.

[4] Varadarajan, V.S., *Geometry of Quantum Theory*, Vol. I, Van Nostrand, 1968.

THE MEANING OF THE LAGRANGIAN

Z. Oziewicz
Institute of Theoretical Physics,
Wroclaw University, Poland.

Inverse problem. We attempt here to answer the question as to how to associate the Lagrangian function L with a given system R of ordinary differential equations. This is equivalent to the question of how to associate with a given congruence of rays, which are solutions of R, the wave fronts which are solutions of the Hamilton-Jacobi equation. This so-called inverse problem can be divided into two well-defined steps. Firstly, we must understand how rays $\{c\}$ can be grouped into families of rays $\{l\}$ which are Lagrangian sub-manifolds and which we shall call carpets for short. The second step is to associate with each carpet l its own wave fronts $l \to S_l$. Clearly, the existence of families of rays is the integrability condition for the local existence of wave fronts. It is exactly the (pre-)symplectic structure which relates the rays to the carpets. It is only in the second step that the Lagrangian function itself becomes involved in the relation of the wave fronts to the carpets. These two steps are not always properly recognized, because one might think that the Lagrangian function determines the system R directly. In fact, there is always a symplectic or a presymplectic structure Ω in an intermediate step. This Ω is closely related to the second partial derivatives of the Lagrangian function, i.e., to Jacobi's last multiplier (see the contribution by Della Riccia in this volume). In any case, the

ISBN 0-12-068720-8

role played by the symplectic structure in mechanics is that of a weaver (or spinner) producing carpets from rays.

A restatement of the first step, therefore, is how to associate the (pre-) symplectic structure Ω with the given system R, and then how to define and calculate the Lagrangian function for the given Ω. It would then appear to be natural to look for the equation for families of wave fronts, etc.

Poisson brackets. Let the system R be represented by the vector field X on the manifold M. X is the characteristic vector field of the closed ideal generated by Pfaff system P associated with X, $P = \{\omega \in \overset{1}{\Lambda}M,\ \omega X = 0\}$. Then, the equation for the rays, for example the Newton equation, takes the form $c^* P = 0$, where c^* is a pull-back mapping induced by the embedding $c : \mathbb{R} \to M$. With respect to the local chart x on M, $x : M \to \mathbb{R}^{dimM}$, the vector field X has the decomposition $X = X^i \partial_i$. P, therefore, is generated by the one-forms $\omega^{ij} \equiv X^i dx^j - X^j dx^i$.

The equation for carpets $\{l\}$ takes the form $l^* J = 0$, where J is closed ideal in the exterior algebra $J \subset \Lambda M$, which has the one-dimensional distribution generated by X as the unique characteristic distribution $i_X J \subset J$. Here i_X denotes the internal multiplication of the exterior algebra as defined in the papers by Thirring. As suggested by the standard calculus of variations (see below), the ideal J is generated by two elements, namely, by the closed 2-form Ω of maximal rank and by the closed 1-form $i_X \Omega \in P$. The bi-form Ω is called the pre-symplectic structure on M iff $dimM = odd$ and in the case of $dimM = even$, Ω is the symplectic structure on M. For brevity, we shall refer to Ω as the Poisson bracket on M, for all $dimM$. The 1-form $i_X \Omega$ is often referred to as the local Hamiltonian, and the vector field X is called the Hamiltonian vector field with respect to Ω. The defining equations for the Poisson bracket Ω for the given vector field X have the following forms.

$$di_X \Omega = 0 \qquad \text{if } \dim M = \text{even}$$
$$i_X \Omega = 0 \qquad \text{if } \dim M = \text{odd} \qquad (1)$$

where Ω is a closed bi-form of maximal rank. The first of equations (1) is invariant with respect to $X \to fX$ for $\forall f \in \Lambda M$, i.e., is invariant with respect to the reparametrizations of rays. Eq. (1) is the necessary and sufficient condition for the local existence of the Lagrangian function for the given vector field X. For the construction of the Poisson bracket Ω one can use either the canonical chart for Ω assured by the Darboux theorem, or the Pfaff system P which is the canonical non-holonomic "chart" for the vector field. In the first case $\Omega = dp_A \wedge dq^A$ and problem is reduced to determining the set of local functions $\{p_A, q^A\}$ in the given initial chart x. The popular identification $q^A = x^A$ is questionable in general.

Using P we see that $\overset{2}{\Lambda} M = \overset{2}{\Lambda} P + \tau \Lambda P$, where $\tau X \neq 0$, and it follows that Eq.(1) $\rightleftarrows \Omega \in \overset{2}{\Lambda} P$ in the odd case and, therefore, the problem is reduced (1) to the classification of all closed maximal rank elements in $\overset{2}{\Lambda} P$. In the even case, the problem is reduced to the classification of all closed elements in P i.e., to the classification of the local first integrals of X.

Lagrangian. Lagrangian function L for a <u>given</u> Ω is defined for all $\dim M$ by means of the following form of the Euler-Lagrange equation

$$L_X \alpha = dL \qquad (\rightleftarrows di_X \Omega = 0) \qquad (2)$$

(cf. lectures by Tulczyew). Here, locally $d\alpha = \Omega$ and Eq. (2) is not invariant with respect to the mapping $X \to fX$. From $d\alpha = \Omega$ it follows that the one-form α is defined up to closed one-form, and, in particular, up to the exact differential df of an <u>arbitrary</u> function f on M. We, therefore, have that $\alpha \to \alpha + df \rightleftarrows L \to L + Xf + \text{const.}$, where Xf is the "total" derivative of f along the vector field X.

In order to derive Eq.(2), consider the action functional

$$c \to \int_{c(A)} \alpha \in \mathbb{R} , \qquad \text{where} \quad A \subset \mathbb{R}$$

Then, the stationary condition with respect to the one-parameter variations generated by the vector field Y on M have the form

$$\int_A c^* \, L_Y \alpha = 0 \tag{3}$$

If Y is transversal [2] to c at ∂A for $\forall A$, then the Eq.(3) implies that

$$c^* \, i_Y d\alpha = 0 \qquad \text{for} \quad Y \in V \tag{4}$$

which is the Euler-Lagrange equation for the extremals $\{c\}$. In Eq.(4) the one-form α and the set V of the variational vector fields are considered as the given. Since we must identify (4) with the given Newton equation $c^* P = 0$ it follows that

$$i_V \, d\alpha = P$$

or

$$i_V (i_X d\alpha) = 0 \tag{5}$$

which, in the even case, is the only constraint on the set V of the variational vector fields. Eq.(5) implies Eq.(2) because of Cartan identity

$$i_X d\alpha = L_X \alpha - d \alpha X .$$

Therefore, $L = \alpha X + \text{const}$ for $\dim M = \text{odd}$

and $L = \alpha X - H + \text{const}$ for $\dim M = \text{even}$,

where $i_X \, d\alpha = - dH$.

Remark. It would be interesting to classify, without reference to the calculus of variations, all closed ideals $\{J\}$ in exterior algebra that have the given one-dimensional distribution as the unique characteristic distribution, and, moreover, to investigate the completeness of such ideals. The equations for J,

$$dJ \subset J \qquad \text{and} \qquad i_X J \subset J$$

can be considered as the generalization of the theory of Lagrangian sub-manifolds.

Acknowledgments.

I am greatly indebted to the many colleagues with whom I have had long conversations during the Seminar, among others Profs. D. Della Riccia, E. Tonti, W. M. Tulczyjew and G. Vitiello. I would also like to thank my friends in Wroclaw for many years of fruitful and stimulating discussions especially J. Kocik and G. Sobczyk.

REFERENCES

(1) Oziewicz, Z., Classical mechanics : inverse problems and symmetries, Université de Genève, preprint, UGVA-DPT 1981/08-307, March 1981.

(2) Hermann, R., *Differential Geometry and the Calculus of Variations*, Academic Press, New York and London 1968.

A CLASSICAL THEORY OF EXTENDED PARTICLES WITH THE PAULI EXCLUSION PRINCIPLE

Antonio F. Rañada

Departamento de Fisica Teorica
Universitad Complutense, Madrid, Spain

In this article, I shall discuss a model in which an extended particle with structure is represented as a solitary wave, bound state of a set of interacting nonlinear classical Dirac fields, which, therefore, are interpreted as its constituents. The interest of this approach lies in the fact that it offers a classical picture of the confinement, the triality and the Pauli exclusion principle. As the hadrons are extended objects with structure a theory of this kind might have, in addition to the more-or-less academic interest from the point of view of relativistic classical mechanics, the appeal of a new tool to deal with real existing particles and to study the relation between classical and quantum physics.

The model is based on a previous one proposed by Soler[1-2] which uses the Lagrangian density

$$L = \frac{i}{2} \bar{\psi} \gamma^{\mu} \partial_{\mu} \psi - m\bar{\psi}\psi + \lambda(\bar{\psi}\psi)^2 \ ; \ \lambda > 0 \qquad (1)$$

whose field equations have solitary wave particle-like solutions in which all the dynamical variables are concentrated in a space region with dimensions of the order of $1/m$. For instance, the S wave spin up and down solutions have the form:

ISBN 0-12-068720-8

$$\psi_\uparrow = e^{-i\Omega mt}\sqrt{\frac{m}{2\lambda}}\begin{pmatrix} G\begin{pmatrix}1\\0\end{pmatrix} \\ iF\begin{pmatrix}\cos\theta \\ \sin\theta e^{i\phi}\end{pmatrix}\end{pmatrix}, \quad \psi_\downarrow = e^{-i\Omega mt}\sqrt{\frac{m}{2\lambda}}\begin{pmatrix} G\begin{pmatrix}0\\1\end{pmatrix} \\ iF\begin{pmatrix}\sin\theta e^{-i\phi} \\ -\cos\theta\end{pmatrix}\end{pmatrix} \tag{2}$$

where F and G are dimensionless radial functions going exponentially to zero at infinity. They have been calculated by numerical analysis and it has been shown that, in sharp contrast with the scalar case, the solutions (2) are remarkably stable under deformations(3-5). This allows one to propose a model of the nucleon(6-8) which offers a fair account of its main static properties.

The fact that all the observed extended particles seem to be composite systems with spin 1/2 constituents suggests the convenience of developing a model based on the same ideas but in which the solutions are bound states of a set fundamental fields which would play the same role as the quarks. It turns out that this is feasible and that the important properties of confinement and triality as well as the Pauli behaviour can be given remarkably simple explanations. This model uses six Dirac fields (ψ_k, ϕ_k), $k = 1,2,3$ and a Lagrangian density which is the sum of 3 terms L_1, L_2, L_3, with the form

$$L_1 = \sum_1^3 (L_D(\psi_k) + L_{\bar{D}}(\phi_k)) \tag{3a}$$

$$L_2 = \frac{\lambda}{3}\{(\sum_k \bar{\psi}_k\psi_k)^2 + (\sum_k \bar{\phi}_k\phi_k)^2 + 4(\sum_k \bar{\psi}_k\psi_k)(\sum_j \bar{\phi}_j\phi_j)\} \quad \lambda > 0 \tag{3b}$$

$$L_3 = \lambda' \sum_{i<j} (\bar{\chi}_{ij}\,\chi_{ij})^2 \tag{3c}$$

where $L_D(\psi)$ is the usual linear Dirac Lagrangian density, as is $L_{\bar{D}}(\phi)$, but with a change in the sign of the derivative terms and

$$\bar{\chi}_{ij} = \chi_i - \chi_j \quad ; \quad \chi_i = \psi_i + \gamma^5\phi_i$$

Because of the gauge invariance, the current

$$j^\mu = \sum_1^3 (\bar{\psi}_k\gamma^\mu\psi_k - \bar{\phi}_k\gamma^\mu\phi_k) \tag{4}$$

is conserved. It must be stressed that the different terms in (4) are not separately conserved and that the minus sign comes from the above mentioned change of sign in $L_{\bar{D}}$. This leads to the interpretation of j^{μ} as the baryonic current and of ϕ_k as charge conjugate to ψ_k.

The field equations have the following important properties[9]:

1. they admit particle-like solitary wave solutions, which can be interpreted as composite systems, <u>all</u> of them being bound states of several fields. However there are no one field solitary waves or, in other words, the fields ψ_k, ϕ_k cannot manifest themselves as particles, although they appear as constituents of systems with structure. This curious property is due to the fact that each field acts as a source for the others. We, therefore, can say that <u>the fundamental fields ψ_k, ϕ_k are confined, just as the quarks are supposed to be</u>.
2. All the solutions have zero triality, i.e., in the particle-like solutions the difference between the number of non-vanishing ψ's and ϕ's is always a multiple of three. This is due to the term L_3 in (3).
3. If $m = 390$ Mev and $\lambda m^2 = 23$, the particle-like solutions are the following: i) a three quark baryon with spin 3/2 and a mass of 1,200 Mev together with its antiparticle; ii) three families of n(quark-antiquark) mesons, $n = 1,2,3$, with a mass of 800 Mev.

In order to evaluate the possibilities of this approach, we need to know if its predictions will agree with the consequences of the Pauli principle. This may seem strange since it is accepted that this principle is of a quantum nature and that its discussion is impossible in a classical frame. It turns out however that, contrary to this common belief, it can be embodied in a classical theory as is shown in the following. For this purpose it is necessary to use the fields in a way different from the usual one in QFT. We assume that the classical representation of a system of n spin 1/2 fermions is not given by a single Dirac field but by n spinors, each one representing one of the particles. Consequently we abandon for a while the system of the six fields ψ_k, ϕ_k and concentrate on a set of n Dirac fields with Lagrangian density

$$L = \sum^{n} L_D(\psi_k) + L_I + L_\eta \tag{5}$$

where L_I gives the interaction between ψ_k and other fields η with free Lagrangian density L_η. It is very easy to show[10] that if

$$\bar{\psi}_i \frac{\partial L_I}{\partial \bar{\psi}_j} - \frac{\partial L_I}{\partial \psi_i} \psi_j = 0 \tag{6}$$

a very slight restrictive condition which expresses the fact that ψ_i and ψ_j have the same couplings, the current $\bar{\psi}_i \gamma^\mu \psi_j$ is conserved, the scalar products

$$N_{ij} = (\psi_i, \psi_j) = \int \bar{\psi}_i \gamma^\mu \psi_j d\sigma_\mu \tag{7}$$

being constants of the motion. We now define the configuration space ε as the set of n component square integrable spinor functions $\{\psi(\bar{r})\}$ such that

$$N_{ij} = \delta_{ij} C \tag{8}$$

where C is a constant. This definition is a sensible one,because if the state is in ε at a certain time it will remain there for ever, i.e., ε is invariant under the time evolution, generated by (5) if (6) is verified.

If ψ_i and ψ_j are well separated packets in the remote past, they will always be orthogonal and will never go to the same state, implying that if the occupation numbers are 0 or 1 at a certain time, they will always be 0 or 1. As an example let us consider two charged fields ψ_1 and ψ_2 in the Coulomb potential of a nucleus. At $t=0$ let ψ_1 be bound in the state 1S spin up and let ψ_2 be a packet well separated from the nucleus so that $N_{12}=0$. The field ψ_2 can lose energy through radiation and become bound. But as it will always be orthogonal to ψ_1 if both go to the 1S wave, their spins will be opposite. A third field cannot be later bound in the 1S wave because of the same mechanism . Although the evolution is classical, the process of formation of atoms follows the well-known prescription of the Pauli principle.

Let us now show a very important property of the previous model of extended particles. If the Pauli principle is interpreted as before, the baryonic solutions behave as fermions while the mesonic ones as bosons. In order to represent n extended particles with structure we use n sets of six fields (ψ_{ik},ϕ_{ik}), $i = 1, \;\; , n$; $k = 1,2,3$ or, equivalently, n fields Ψ_i which are six component objects, the direct sum of ψ_{ik} and ϕ_{ik}. In addition to n terms as in (3), the Lagrangian density must contain an interaction part L_I. As a generalization of (6), let us assume that

$$\bar{\Psi}_i \frac{\partial L_I}{\partial \bar{\Psi}_j} - \frac{\partial L_I}{\partial \Psi_i} \Psi_j = \sum_k (\bar{\psi}_{ik} \frac{\partial L_I}{\partial \bar{\psi}_{jk}} + \bar{\phi}_{ik} \frac{\partial L_I}{\partial \bar{\phi}_{jk}} - \frac{\partial L_I}{\partial \psi_{ik}} \psi_{jk} - \frac{\partial L_I}{\partial \phi_{ik}} \phi_{jk}) \quad (9)$$

which implies that Ψ_i and Ψ_j have the same couplings. In this case the currents (10)

$$j^\mu_{(i,j)} = \bar{\Psi}_i \Gamma^\mu \Psi_j = \sum_k (\bar{\psi}_{ik}\gamma^\mu\psi_{jk} - \bar{\phi}_{ik}\gamma^\mu\phi_{jk}) \quad (10)$$

where $\Gamma^\mu = \mathrm{diag}\{\gamma^\mu,\gamma^\mu,\gamma^\mu,-\gamma^\mu,-\gamma^\mu,-\gamma^\mu\}$ is conserved, its constant integral being

$$N_{ij} = \int_{R^3} \bar{\Psi}_i \Gamma^\circ \Psi_j d^3\vec{r} = \sum_k^3 \{(\psi_{ik}, \psi_{jk}) - (\phi_{ik}, \phi_{jk})\} \quad (11)$$

In the baryonic solutions $\phi_k = 0$, while in the mesonic cases $(\psi_k,\psi_k) = (\phi_k,\phi_k)$. Consequently, in the cases of multibaryonic solutions the configuration space can be defined as before by the condition (8). However in the case of multimesonic solutions $N_{ij} = 0$. As a consequence <u>the multibaryonic solutions obey the exclusion principle while the multimesonic ones do not.</u>

The parallelism with the quark model is striking. Perhaps in order to achieve a succesful quantum description of the hadrons it may be necessary to have first a classical description which accounts for their internal structure. Here, the even more radical possibility that, contrary to leptons, the hadrons may be described classically suggests itself.

REFERENCES

1. Soler M (1970) Phys. Rev. D1, 2766.

2. Rañada A F (1980) To appear in "Essays in Honor of Wolfgang Yourgrau", edited by A. van der Merwe, Plenum Co.

3. Soler M (1975) Report GIFT 10-75, Madrid.

4. Alvarez A (1981) Preprint Univ. Complutense Madrid, UCFTM81/1

5. Alvarez A and Carreras B (1981) Preprint Univ. Complutense Madrid UCMFT 81/5. To appear in Physics Letters.

6. Rañada A F, Rañada M F, Soler M and Vázquez L (1974) Phys. Rev. D10, 517

7. Rañada A F and Vázquez L (1976) Prog. Theor. Phys. (Kyoto) 56, 311

8. García L and Rañada A F (1980) Prog. Theor. Phys. (Kyoto) 64, 671

9. Rañada A F and Rañada M F (1981) Preprints Univ. Complutense Madrid, UCMFT 80/1 and UCFTM 81/2, to be published.

10. Rañada A F (1981) Preprint Univ. Complutense UCFTM 81/9 to be published.

SOME REMARKS ON STATE REVERSIBILITY AND IRREVERSIBILITY IN SYSTEM THEORY

Paolo Serafini

University of Udine
Institute of Mathematics
Computer and System Science
Udine, Italy

1. INTRODUCTION

We are concerned here with linear systems defined by linear operators relating the input to the output, with both input and output being elements of a particular Hilbert space. Such linear systems may be thought of as mathematical models of some physical dynamical systems. We are particularly interested in the problem of the state space realization and in characterizing the free evolution of the state. More specifically we want to investigate whether the flow corresponding to the evolution of the state can be (at least locally) reversible, in the sense that two distinct trajectories do not collide into a single one.

We note that usual physical dynamical systems have no input structure since there should be no external influences acting on the system; their only effect has been to drive the system to some initial condition. The investigation carried out here reflects this view since we shall be mainly interested in the free evolution of the system.

From a system theoretical point of view one might dispense with the a priori distinction between inputs and outputs, as Willems [6] has shown. However we maintain the input-output framework, since it is still a fruitful conceptual setting in the investigation of system dynamics, as some research has shown (consider [1] as just one quotation).

ISBN 0-12-068720-8

2. MATHEMATICAL SETTING

We consider a linear system as a linear bounded operator K acting on the Hilbert space $\mathbb{H} = L^2(\mathbb{R}, \mathbb{G})$, i.e. the space of square summable G-valued functions of time, with G some Hilbert space. The domain of K is the input space, its range the output space. Physical dynamical systems can often be modeled in this way. Let us define the projections $P^t : \mathbb{H} \rightarrow \mathbb{H}$ as

$$(P^t u)(\tau) = u(\tau) \quad \text{if} \quad \tau \leqslant t$$

$$= 0 \quad \text{if} \quad \tau > t$$

Let $P_t = I - P^t$.

The translations $U_t : \mathbb{H} \rightarrow \mathbb{H}$ are defined as

$$(U_t u)(\tau) = u(\tau + t)$$

Both P^t and U_t are strongly continuous, i.e. $P^t x \rightarrow P^s x$ and $U_t x \rightarrow U_s x$ as $t \rightarrow s$ for any $x \in H$.

There is a deep relationship between U_t and P^t. In fact the infinitesimal generator of U_t is given by iH with H a selfadjoint unbounded operator (specifically d/dt). H is unitarily equivalent through the Fourier transform to the multiplication operator $(\widehat{H}x)(\lambda) = \lambda x(\lambda)$. The spectrum of $\widehat{H}$ is the entire real line and its spectral projections are exactly the operators P^t [7].

Let $V_t : S \rightarrow S$ be a semigroup. We say that V_t is reversible on a time interval T if $V_t x \neq V_t y \Leftrightarrow V_s x \neq V_s y$ for any $t, s \in T, x, y \in S$.

3. STATE EVOLUTION

The dependence of the output on the input u is (in most cases) dynamical, that is the output value at time t depends on the whole history of u and not merely on the input value at t. If the system is modeled casually, i.e. $P^t K P_t = 0$, then the output value at t depends on the values of u prior to t only. There are many reasons for the introduction of a new quantity called the state, defined so as to be entirely responsible for the system dynamics, with the interactions input-state and state-output being of an instantaneous type.

The following considerations are related to the research developed by several authors and whose state of art appeared in [3]. However the description given her of a state space differs considerably from the one given in [3].

In the sequel we shall focus our attention on the state dynamics only and shall not consider the interactions input state and state-output.

Let us split $P_t\, y$ as $P_t\, y = P_t\, KP_t\, u + P_t\, KP^t u$. The past input determines the future output through the term $P_t\, KP^t u$. In other words each element of the range of $P_t\, KP^t$ is the "memory" of the past inputs as far as the future output is concerned. It is therefore natural to call $S_t = \overline{\mathcal{R}(P_t KP^t)}$ (i.e. the closure of the range) the state space at time t, and to define the state space as the following topological space

$$\mathcal{S} = \bigcup_{t \in \mathbb{R}} (t, U_t S_t) \subset \mathrm{R} \times P_0\, \mathrm{H}$$

whose topology is inherited by $\mathbb{R} x P_0\, \mathbb{H}$.

Define the semigroup $T_t = P_0 U_t$ $(t \geqslant 0)$ and the following operators:

$$V_t : \mathcal{S} \to \mathcal{S},\ (\tau, z) \leftrightarrow (\tau + t,\ T_t z) \qquad t \geqslant 0$$

We want to show that the above definition is consistent (i.e. actually $V_t \mathcal{S} \subset \mathcal{S}$) and that V_t is a strongly continuous semigroup of operators. For this purpose let $z = U_t P_t KP^t u$ for some u; from $T_s z = P_0 U_s U_t P_t KP^t u = U_{s+t} P_{s+t} KP^{s+t} P^t u$, $(s \geqslant 0)$ it can be seen immediately that $T_s z \in U_{s+t}\, \mathcal{R}(P_{s+t} KP^{s+t})$; since $\|T_s z\| \leqslant \|z\|$, by a continuous extension to S_t, it follows $T_s U_t S_t \subset U_{s+t} S_{t+s}$ i.e. $V_s \mathcal{S} \subset \mathcal{S}$. The strong continuity of V_t and the properties $V_{s+t} = V_s V_t$, $V_0 = I$ follow easily.

Therefore the semigroup V_t gives rise on $\mathcal{S}$ to a continuous semi-flow. As apparent from the definition this semi-flow corresponds to the free evolution of the state.

An important subclass of systems is given by the time-invariant ones, for which, by definition

$$U_t K = KU_t$$

In this case $S_0 = U_t\, S_t$, whence $\mathcal{S} = \mathbb{R} \times S_0$ and the state space is actually a linear space.

Let us now investigate in more detail the semi-flow V_t. Since we are interested in the free evolution of the state, it makes sense to restrict V_t to the following set:

$$\mathcal{S}_t = \bigcup_{\tau > 0} V_\tau (t, S_t)$$

Let $V_\tau^t = V_\tau | \mathscr{S}_t$.

Note that $\mathscr{S}_t$ is the union of continuous trajectories. Now the map

$$\Phi_t : \mathbb{R}_+ \times U_t S_t \to \mathscr{S}_t$$

$$(\tau, z) \to (t + \tau, T_\tau z)$$

is clearly continuous and onto. Unfortunately it is not one-to-one in general; nor (in general) is its restriction to an arbitrary $(0, \delta) \times U_t S_t$ $(\delta > 0)$ (consider for instance a pure delay system). However note that if $z = U_t P_t K P^t u \neq 0$ and $P_\tau P_t K P^t u \neq 0$ for any $\tau \geqslant t$, then certainly $T_{\tau - t} z \neq 0$ $\tau \geqslant t$ as easily follows. Hence if the above condition is satisfied $\mathscr{S}_t$ is manifold parametrized by Φ_t; in other words if the system response to any input stopping at t is never identically zero from some $\tau > t$, then $\mathscr{S}_t$ is a manifold and, more important, the semi-flow V_τ^t is reversible for $\tau \geqslant 0$. This result has some physical interpretation: if the input space represents the totality of all possible inputs, then, in absence of these inputs, the free evolution obeys some kind of "inertial" law and does not go to zero. Irreversibility in this mathematical context is related to the intrinsic tendency of the system towards the zero state.

Alternatively, for finite dimensional systems (i.e. $\dim S_t < \infty$ $\forall$ t), it is possible to prove a rather partial answer to the above question, i.e. for any t there exists a $\delta_t > 0$ such that

$$\Phi_t : ((0, \delta_t), U_t S_t) \to \bigcup_{\delta_t > \tau > 0} V_\tau (t, S_t) \subset \mathscr{S}_t$$

is a homeomorphism.

Hence for any t there exists an interval $[t, t + \delta_t)$ on which the flow is reversible.

The question whether $\mathscr{S}_t$ is (at least locally) a differentiable manifold is more delicate. By formally computing the derivative one obtains:

$$D\Phi_t(\tau, z)(h_\tau, h_z) = (h_\tau, h_\tau \; T_\tau \; Az + T_\tau h_z)$$

with A the infinitesimal generator of T_t.

Hence $D\Phi_t(\tau,z)$ is defined only if $z \in \mathscr{D}(A)$ and $\mathscr{S}_t$ is a differentiable manifold if $U_t S_t \subset \mathscr{D}(A)$.

The wellknown formula $\dot{x} = A(t)x$ can be easily derived if the dimension of S_t is constant and finite and $\mathscr{S}$ is a differentiable manifold, i.e. there exist open time intervals Δ and diffeomorphisms $\Phi : (\Delta, \mathbb{R}^n) \to \mathscr{S}$. Then Φ can be defined as

$\tau \mapsto \tau$, $x \mapsto \varphi_\tau x$ with φ_τ a linear operator. Therefore the flow V_t induces a nonlinear flow $\hat{V}_t$ on $(\Delta, \mathbb{R}^n) : (\tau, x) \mapsto (\tau + t, \varphi_{\tau+t}^{-1} T_t \varphi_\tau x)$, from which one derives: $(\dot{\tau}, \dot{x}) = (1, A(\tau)x)$. It is not surprising that these concepts are closely related to the standard notions of reachability and observability of finite dimensional linear systems. In fact if $\dim S_t \leqslant n$ then $\forall t$ and $\dim S_\tau = n$, then S_τ can be written as the range of $\sum_i^n \varphi^\tau < u, \psi_i^\tau >$, for some linearly independent $\psi_i^\tau \in P^\tau H$ and $\varphi_i^\tau \in P_\tau \mathbb{H}$. This implies the positive definiteness of the matrices $< \varphi_i^\tau , \varphi_j^\tau >$ and $< \psi_i^\tau , \psi_j^\tau >$, which corresponds at a closer glance to the conditions of complete observability and complete reachability at τ in Kalman's terminology [4]. If however $\dim S_\tau < n$ then S_τ can again be written as the range of $\sum_i^n \varphi_i^\tau < u, \psi_i^\tau >$ with either φ_i^τ or ψ_i^τ (or both) no longer linearly independent.

4. TIME INVARIANT SYSTEMS

For time invariant systems the operators V_t constitute a semigroup of linear operators on some Hilbert space. The theory of semigroups of operators and its applications to system theory are wellknown. Therefore we shall limit ourselves to a few remarks. There are several ways of arriving at some conclusion about reversibility. The strongest result is perhaps the following: if V_t is a normal semigroup, then it admits an exponential representation $V_t = e^{At}$ with A a normal operator (possibly unbounded); it is known that e^{At} is an invertible operator and hence the flow V_t is reversible for $t \geqslant 0$. This is also trivially true if the state space is finite dimensional.

The most interesting flows are perhaps the unitary ones for which there exists a selfadjoint operator H such that $V_t = e^{iHt}$ (Stone's theorem [5]). Note that unitary semigroups V_t can be extended to a group of defining $V_{-t} = V_t^*$ $(t \geqslant 0)$ and hence reversibility is assured on the whole real axis.

It is generally assumed that a system isolated from the external world has an evolution governed by a unitary semigroup. So, if the system actually does not exhibit such an evolution, this could be ascribed to the system interaction with the external world. It is therefore natural to ask whether, given a semigroup V_t on a Hilbert space S, there exists a unitary group U_t on a Hilbert space $K \supset S$ such that V_t is obtained as the restriction of U_t to S, i.e.

$$\begin{array}{ccccc} & K & \xrightarrow{U_t} & K & \\ \text{proj.} & \downarrow & & \downarrow & \text{proj.} \\ & S & \xrightarrow{V_t} & S & \end{array}$$

The unitary dilation theorem [2] ensures that such an enlarged state space, in which the evolution is reversible, actually exists provided the system is dissipative ($\mathrm{Re} < Ax, x > \leqslant 0$ with A infinitesimal generator of V_t) or alternatively if the flow V_t is contractive, i.e. $\| V_t \| \leqslant 1$.

APPENDIX

The mathematical background of our investigations is some Hilbert space H together with a flow U_t which determines in turn P^t. The introduction of U_t and P^t is mainly caused by physical intuition. However it is natural to ask whether the choice of some other flow U^t (hence with some other P^t) could lead to physically significant conclusions. It seems that U_t constitutes a particular model of time for a system and hence some other U_t could be another, still physically significant model. We think that this interpretation deserves some investigation.

REFERENCES

[1] Brockett, R.W., Control Theory and Analytical Mechanics, in Geometric Control Theory, Math. Sci. Press, Brookline MA, 1977.

[2] Davies, E.B., Quantum Theory of Open Systems, Academic Press, 1976.

[3] Feintuch, A., State Space Theory for Resolution Space Operators, J. Math. Anal. Appl., 74, 164-191, 1980.

[4] Kalman, R.E., Falb, P.L., Arbib, M.A., Topics in Mathematical System Theory, McGraw-Hill, 1969.

[5] Rudin, W., Functional Analysis, McGraw-Hill, 1973.

[6] Willems, J.C., System Theoretic Foundations for Modelling Physical Systems, in Dynamical Systems and Microphysics, A. Blaquière, F. Fer, A. Marzollo (eds.), Springer, Wien 1980.

[7] Yosida, K., Functional Analysis, Sixth Edition, Springer 1980.

SYMPLECTIC STRUCTURES, ENERGY-MOMENTUM FUNCTIONS, AND HAMILTON EQUATIONS IN THEORIES OF GRAVITY

Wiktor Szczyrba

Institute of Mathematics
Polish Academy of Sciences
Warsaw, Poland

The geometry of space-time M is expressed by means of four linearly independent, Minkowski orthonormal one forms $\underline{e}^{(\alpha)} = e_{\mu}^{(\alpha)}dx^{\mu}$, and six connection one-forms $\underline{\Gamma}^{(\alpha)(\beta)} = \Gamma_{\mu}^{(\alpha)(\beta)}dx^{\mu}$. The skew-symmetry properties $\Gamma_{\mu}^{(\alpha)(\beta)} = -\Gamma_{\mu}^{(\alpha)(\beta)}$ are equivalent to the metric compatibility condition for connections. The local Lorentz group acts in the space of tetrad and connection one-forms. If $\underline{x} \to [L^{(\alpha)}_{(\beta)}(\underline{x})]$ is a field of special, orthochronous Lorentz matrices on M (an element of loc. Lor.), then

$$'\underline{e}^{(\alpha)} = (L^{-1})^{(\alpha)}_{(\beta)}\,\underline{e}^{(\beta)} \tag{1}$$

$$'\Gamma_{\mu}^{(\alpha)(\beta)} = (L^{-1})^{(\alpha)}_{(\varepsilon)}\,(L^{-1})^{(\beta)}_{(\tau)}\,\Gamma_{\mu}^{(\varepsilon)(\tau)} - \partial_{\mu}(L^{-1})^{(\alpha)}_{(\tau)}\,L^{(\tau)(\beta)}\ . \tag{2}$$

These transformations maintain the metric and affine structures of M. They also assure tensorial transformation properties for curvature $R_{\mu\nu}^{(\alpha)(\beta)}$, torsion $Q_{\mu\nu}^{(\alpha)}$ and for covariant derivatives $D_{\lambda}\varphi^{(A)}$

ISBN 0-12-068720-8

of tensor matter fields $\varphi^{(A)}$. The theories of gravity describing interactions between geometry and matter fields should be invariant with respect to the actions of the local Lorentz group and the group of diffeomorphisms of M. This implies that the corresponding Lagrangian densities $\mathcal{L}_{gr}$, $\mathcal{L}_{mat}$ are functions of $R_{\mu\nu}^{(\alpha)(\beta)}$, $Q_{\mu\nu}^{(\alpha)}$, $e_{\mu}^{(\alpha)}$ and $\varphi^{(A)}$, $D_{\lambda}\varphi^{(A)}$, $e_{\mu}^{(\alpha)}$, respectively (Hehl et al., 1976, 1980; Szczyrba, 1980). Such theories were investigated by several authors (Yang, 1974; Fairchild, 1976, 1977; Hehl et al., 1980; Trautman, 1980; Basombrió, 1980). The field equations for the theory in question are the Euler-Lagrange variational equations

$$(\xi q \cdot \mathrm{I})^{\mu}_{(\alpha)} = \delta(\mathcal{L}_{gr} + \mathcal{L}_{mat})/\delta e_{\mu}^{(\alpha)} = 0 \tag{3a}$$

$$(\xi q \cdot \mathrm{II})^{\mu}_{(\alpha)(\beta)} = \delta(\mathcal{L}_{gr} + \mathcal{L}_{mat})/\delta\Gamma_{\mu}^{(\alpha)(\beta)} = 0 \tag{3b}$$

$$(\xi q \cdot \mathrm{mat})_{(A)} = \delta\mathcal{L}_{mat}/\delta\varphi^{(A)} = 0 \tag{3c}$$

The Hamiltonian dynamics is discussed for the space-time diffeomorphic to the product $R \times \sigma$, where σ is a compact three-dimensional manifold without boundary. In the subsequent considerations, x^{o} is a coordinate in R and (x^{k}) are local coordinates in σ.

We would like to describe geometric configurations of the system by means of families of fields of geometric objects on σ (parametrized by x^{o}). Such an operation can be performed in two steps. First, we rotate fields of tetrads in such a way that the vector $\underline{e}_{(o)}$ coincides with the vector normal to σ. Then, we decompose the obtained 4-tensor fields $\tilde{e}_\mu^{(\alpha)}$, $\tilde{\Gamma}_\mu^{(\alpha)(\beta)}$, $\tilde{\varphi}^{(A)}$ into normal and tangential to σ parts (Szczyrba, 1981a). We get "new" potentials $\hat{e}_k^{(a)}$, $\hat{\Gamma}_k^{(a)(o)}$, $\hat{\Gamma}_k^{(a)(b)}$, $\hat{\varphi}^{(A)}$ and $\hat{\Gamma}_o^{(\alpha)(\beta)}$. The first three systems determine a field of triads of covectors on σ, the second fundamental form of σ, and a metric compatible connection on σ. We have $\hat{e}_o^{(a)} = 0$, $\hat{e}_k^{(o)} = 0$, $\hat{e}_o^{(o)} = 1$. We, therefore, must introduce 7 additional variables: the lapse function N, the shift vector N^k, and $(n^{(a)})_{a=1}^{3}$ - angle coefficients of the abovementioned Lorentz ("boost") transformation (Szczyrba, 1981 a, 1981 b).

If $U_{(\alpha)}^{\mu\nu} = \partial\mathcal{L}_{gr}/\partial(\partial_\mu e_\nu^{(\alpha)})$; $P_{(\alpha)(\beta)}^{\mu\nu} = \partial\mathcal{L}_{gr}/\partial(\partial_\mu\Gamma_\nu^{(\alpha)(\beta)})$; $P_{(A)}^{\lambda} = \partial\mathcal{L}_{mat}/\partial(\partial_\lambda\varphi^{(A)})$ are 4-momenta of the gravitational and matter fields respectively, $\hat{U}_{(a)}^{ok}$, $\hat{P}_{(\alpha)(\beta)}^{ok}$, $\hat{p}_{(A)}^{o}$ are the corresponding "$\wedge$" - variables, and if $\hat{X}_1$, $\hat{X}_2$ are vectors tangent to the space of geometric configurations, then the symplectic two-form is given by

$$\Omega(X_1,X_2) = \int_\sigma [\, \underset{1}{\delta}\hat{U}_{(a)}^{ok} \wedge \underset{2}{\delta}\hat{e}_k^{(a)} + \underset{1}{\delta}\hat{P}_{(\alpha)(\beta)}^{ok} \wedge \underset{2}{\delta}\hat{\Gamma}_k^{(\alpha)(\beta)} + \underset{1}{\delta}\hat{p}_{(A)}^{o} \wedge \underset{2}{\delta}\hat{\varphi}^{(A)} + \underset{1}{\delta}m_{(a)} \wedge \underset{2}{\delta}n^{(a)} \,]\, d^3x \tag{4}$$

Remark : The three scalar 3-densities $m_{(a)}$ are the conjugate symplectic momenta of $n^{(a)}$ and can be computed by means of $\mathcal{L}_{gr}$ (Szczyrba, 1981b).

In formula (4) variations $\delta \underset{i}{\hat{e}}{}^{(a)}_{k}$, $\delta \underset{i}{\hat{U}}{}^{ok}_{(a)}$,... $i = 1,2$ represent vectors $\hat{X}_i$.

The dynamics is given by the energy-momentum function

$$E_{\underline{Z}} = \int_{\sigma} \sum_{\lambda=o}^{3} (-1)^{\lambda} \overset{\lambda}{\xi}_{\underline{Z}} dx^o \wedge \cdots \hat{\lambda} \cdots \wedge dx^3 \tag{5}$$

where $\overset{\lambda}{\xi}_{\underline{Z}}$ is the energy momentum density of the system related to a given vector field $\underline{Z}$ on space-time. We, therefore, have the following formula (Szczyrba, 1980, 1981b)

$$\overset{\lambda}{\xi}_{\underline{Z}} = - [(\xi q \cdot I)^{\lambda}_{(\alpha)} e^{(\alpha)}_{\varepsilon} Z^{\varepsilon} + (\xi q \cdot II)^{\lambda}_{(\alpha)(\beta)} \Gamma^{(\alpha)(\beta)}_{\varepsilon} Z^{\varepsilon}] \tag{6}$$

For special cases, formula (6) was derived. (Komar, 1959 ; Kijowski, 1978). The dynamical evolution of the system is given by the Hamilton equation

$$dE_{\underline{Z}} \cdot \hat{V} = - \Omega(\hat{Y} \wedge \hat{V}) \tag{7}$$

where $\hat{V}$ is an arbitrary vector tangent to the space of geometric configurations. Let $\delta_Y \hat{e}^{(a)}_{k}$, $\delta_Y \hat{U}^{ok}_{(a)}$,... be variations of symplectic variables representing the Hamiltonian vector $\hat{Y}$, and let $\underline{Z} = \partial/\partial x^o$, then we get from (7) a system of dynamical equations

$$\begin{aligned} \partial_o \hat{U}^{ok}_{(a)} &= \delta_Y \hat{U}^{ok}_{(a)} \ ; \qquad \partial_o \hat{P}^{ok}_{(\alpha)(\beta)} = \delta_Y \hat{P}^{ok}_{(\alpha)(\beta)} \\ \partial_o \hat{P}^{o}_{(A)} &= \delta_Y \hat{P}^{o}_{(A)} \ ; \qquad \partial_o m_{(a)} = \delta_Y m_{(a)} \end{aligned} \tag{8}$$

and a system of ten constraint equations

$$(\xi q \cdot I)^{o}_{(\alpha)} = 0 \ ; \qquad (\xi q \cdot II)^{o}_{(\alpha)(\beta)} = 0 \tag{9}$$

Both systems are consistent by virtue of the contracted Bianchi identities (Szczyrba, 1980). There are no evolution equations for the variables N, N^k, $\Gamma_o^{(\alpha)(\beta)}$ and the constraint equations do not involve them. They are gauge variables of the theory and their values could be arbitrarily chosen on M. The existence of ten-gauge variables is closely related to the invariance properties of the theory with respect to the action of the semi-simple product of the local Lorentz group and the diffeomorphism group of space-time and with the degeneracy of the symplectic two-form Ω.

REFERENCES

Basombrió, F.G. (1980), GRG, 12, 109.

Fairchild, E.E., Jr. (1976) Phys. Rev. D, 14, 384
(1977) Phys. Rev. D, 16, 2438.

Hehl, F.W., von der Heyde, P., Kerlick, G., and Nester, J. (1976), Rev. Mod. Phys., 48, 393.

Hehl, F.W., Nitsch, J., and von der Heyde, P., (1980), in *General Relativity and Gravitation*, Held, A. (ed.) Plenum, New York.

Kijowski, J. (1978), GRG, 9, 857.

Kijowski, J., Szczyrba, W. (1976), Comm. Math. Phys. 46, 183.

Komar, A. (1959), Phys. Rev., 113, 934.

Szczyrba, W. (1980), in *Cosmology and Gravitation*, Bergmann, P. and De Sabbata, V. (eds.), Plenum, New York.

Szczyrba, W. (1981a), J. Math. Phys., 22.
(1981b), Phys. Rev. D, 24.

Trautman, A. (1980), in *General Relativity and Gravitation*, Held, A. (ed.), Plenum, New York.

Yang, C.N. (1974), Phys. Rev. Lett., 33, 445.

STOCHASTIC CALCULUS OF VARIATIONS, STOCHASTIC CONTROL, AND QUANTUM DYNAMICS

Kunio Yasue

Département de Physique Théorique
Université de Genève
Genève, Switzerland.

In the present paper I shall present some of the important aspects of the probabilistic representation of quantum dynamics discovered by E. Nelson. In particular, I shall emphasize the use of stochastic calculus of variations developed recently by J.C. Zambrini and myself [1,2]. Since the number of manuscript pages is limited, I apologize for omitting the proof, but the references in which the proof can be found are mentioned. Although several attempts have been made by physicists to translate Nelson's original formulation [3] into their familiar language, that of Nelson still remains the clearest and the most satisfactory one. The essence of his formulation, frequently called stochastic mechanics, is as follows: [4] let $\psi(x,t)$ be a solution to the Schrödinger equation

$$i\hbar \frac{\partial\psi}{\partial t} = \left(- \frac{\hbar^2}{2m} \Delta + V(x)\right)\psi \qquad (1)$$

with the Cauchy data $\psi(x,0) = \psi_0(x)$, where Δ is the Laplacian, $V(x)$ is the potential energy, $\hbar$ is the Planck constant divided by 2π and m is the mass. Suppose $\psi(x,t)$ admits a polar decomposition

DYNAMICAL SYSTEMS
AND MICROPHYSICS

ISBN 0-12-068720-8

$\psi(x,t) = \exp(\, R(x,t) + i\, S(x,t)\,)$ with R and S real valued. Let b $(x,t) = \text{grad}\,(\, R(x,t) + S(x,t)\,)$, where grad should be understood in the sense of distribution if needed. Consider a semigroup of operators $M = \{ M_t \mid 0 \leqq t < \infty \}$ with its infinitesimal generator G_t given by

$$G_t = (b(x,t),\text{grad}) + (\hbar/2m)\Delta \tag{2}$$

with domain $C_0(\mathbb{R}^n)$, where (,) is the Cartesian inner product. In terms of the product integral,[5] $M_t = \prod_0^t (\, 1 + G_t\, dt\,)$. If b(x,t) is sufficiently regular, the semigroup M generates a Markov family $\{X^x \mid x \in \mathbb{R}^n\}$, where $X^x = \{ X^x_t \mid 0 \leqq t < \infty \}$ is a Markov process with continuous sample paths such that $X^x_0 = x$ and

$$(\, M_t\, f\,)(x) = f(X^x_t), \quad 0 < t < \infty, \tag{3}$$

for any $f \in C_0(\mathbb{R}^n)$ [6]. If b(x,t) is continuous, the Markov family X^x can be generated directly by the Itô stochastic differential equation

$$dX_t = b(X_t,t)\, dt + (\hbar/2m)^{\frac{1}{2}}\, dW_t, \tag{4}$$

with initial condition $X_0 = x$, where $W = \{ W_t \mid 0 \leqq t < \infty \}$ is the standard Wiener process. The regularity condition on b in Eq. (4) can be actually relaxed and Eq. (4) makes sense even if b is divergent [7]. Nelson showed that the Markov family $\{X^x\}$ satisfies Newton's equation of motion in the mean

$$m\, \tfrac{1}{2}\, (\, D\, D_* X_t + D_* D\, X_t\,) = -\,\text{grad}\, V(X_t), \quad 0 < t < \infty \quad , \tag{5}$$

where D and D_* are the mean forward and backward derivatives of Nelson [4]. Besides the existence of this Markov family $\{X^x\}$

associated with the Schrödinger equation (1), he showed that $\Pr\{X_t^{\psi} \in d^n x\} = |\psi(x,t)|^2 d^n x$ if $X^{\psi} = \{X_t^{\psi} \mid 0 \leqq t < \infty\}$ is a Markov process generated by Eq. (4) with initial condition $\Pr\{X_0^{\psi} \in d^n x\} = |\psi_0(x)|^2 d^n x$, X^{ψ} satisfies Eq. (5) and represents quantum dynamics. The converse is also true. It is evident that X^x and X^{ψ} are Markov processes due to the Itô stochastic differential equation (4). The Markov family $\{X^x\}$ and the Markov process X^{ψ} well illustrate quantum dynamics and are compatible with the usual probabilistic interpretation of quantum mechanics. Although it differs from Heisenberg's matrix mechanics with physically unobservable quantities, the difference is indeed out of experimental justifications [3]. The equivalence of stochastic mechanics to matrix mechanics up to physically observable quantities does not make the probabilistic formulation of stochastic mechanics less interesting than the others from a physical point of view. I feel strongly that the only way to understand the non-locality inherent in quantum mechanics is provided by Nelson's stochastic mechanics. Furthermore, it is applicable to the problems of tunneling [8] and path-wise classical limit [9] in which other representatives of quantum mechanics are powerless. Other advantages of stochastic mechanics are shown in Refs. 10 and 11. Quantum mechanics is conceptually rich enough, and one might easily miss its total vision if one closes his mind and thinks only of a few similar mathematical representations of it.

As I mentioned before, in stochastic mechanics, quantum dynamics is represented by the Markov process subject to Eq. (5). I shall use the remaining pages for the exposition of a variational formulation of Nelson's probabilistic representation of quantum dynamics within the realm of stochastic calculus of variations. For the proof, see Ref. 1.

i) The Markov family $\{X^x\}$ and the Markov process X^ψ subject to Eq. (5) extremize the action functional

$$I(X) = E\left[\int_0^u L_{QM}(X_t, DX_t, D_* X_t)\, dt\right] \tag{6}$$

for any finite $u > 0$, where L_{QM} is the Lagrangian given by

$$L_{QM}(X, DX, D_* X) = \tfrac{1}{2}\left(\tfrac{1}{2}m|DX|^2 + \tfrac{1}{2}m|D_* X|^2 \right) - V(X). \tag{7}$$

This enables us to reformulate quantum mechanics as stochastic control problems [10].

ii) If the action functional I is invariant under a one-parameter group of diffeomorphisms $G = \{U_K\}_{K \in \mathbb{R}}$, $U_0 = 1$, then

$$\frac{d}{dt} E\left[\tfrac{1}{2}m\, (DX_t + D_* X_t) \cdot \frac{dU_K}{dK}\bigg|_{K=0} X_t \right] = 0. \tag{8}$$

From this I obtain, for example, the total energy conservation law and the total momentum conservation law in quantum mechanics [1].

iii) The wave function satisfying the Schrödinger equation (1) with the Cauchy data $\psi(x,0) = \psi_0(x) = u_0(x) \exp(\frac{i}{\hbar} S_0(x))$ admits the probabilistic representation in terms of the Markov process X^ψ;

$$\psi(x,t) = \left(Pr\{ X_t^\psi \in d^n x \} / d^n x \right)^{\frac{1}{2}}$$
$$\exp\left(\frac{i}{\hbar} E\left[S_0(X_0^\psi) + \int_0^t L_{QM}(X_s^\psi, DX_s^\psi, D_* X_s^\psi)\, ds \;\middle|\; X_t^\psi = x \right] \right). \tag{9}$$

iv) Quantum mechanics of constrained dynamical systems can be treated equally in terms of stochastic variational problem with constraint [2].

v) Path-wise semiclassical asymptotics of quantum mechanics can be investigated in terms of the Markov process X^{ψ} to obtain the same result as Maslov and Truman [9].

REFERENCES

1) K. Yasue, J. Funct. Anal. 41, 327 (1981).

2) J.-C. Zambrini, Lett. Math. Phys. 4, 457 (1980).

3) E. Nelson, Dynamical Theories of Brownian Motion (Princeton U. P., Princeton, 1969).

4) E. Nelson, Bull. Am. Math. Soc. 84, 121 (1978).

5) E. Nelson, Topics in Dynamics I: Flow (Princeton U. P., Princeton, 1969).

6) A. D. Wentzel, Theorie zufälliger Prozesse (Akademie, Berlin, 1979).

7) N. I. Portenko, Theor. Prob. Appl. 20, 27 (1975).

8) K. Yasue, Phys. Rev. Lett. 40, 665 (1978).

9) K. Yasue, to appear.

10) K. Yasue, J. Math. Phys. 22, 1010 (1981).

11) K. Yasue, Intern. J. Theor. Phys. 18, 861 (1979).